Heritage Matters

GEOHERITAGE AND GEOTOURISM

Heritage Matters

ISSN 1756-4832

Series Editors
Peter G. Stone
Peter Davis
Chris Whitehead

Heritage Matters is a series of edited and single-authored volumes which addresses the whole range of issues that confront the cultural heritage sector as we face the global challenges of the twenty-first century. The series follows the ethos of the International Centre for Cultural and Heritage Studies (ICCHS) at Newcastle University, where these issues are seen as part of an integrated whole, including both cultural and natural agendas, and thus encompasses challenges faced by all types of museums, art galleries, heritage sites and the organisations and individuals that work with, and are affected by them.

Previous volumes are listed at the back of this book.

Geoheritage and Geotourism
A European Perspective

Edited by

THOMAS A. HOSE

THE BOYDELL PRESS

First published 2016
The Boydell Press, Woodbridge

ISBN 978-1-78327-147-4

The Boydell Press is an imprint of Boydell & Brewer Ltd
PO Box 9, Woodbridge, Suffolk IP12 3DF, UK
and of Boydell & Brewer Inc.
668 Mt Hope Avenue, Rochester, NY 14620–2731, USA
website: www.boydellandbrewer.com

The publisher has no responsibility for the continued existence or accuracy
of URLs for external or third-party internet websites referred to in this book,
and does not guarantee that any content on such websites is,
or will remain, accurate or appropriate

A CIP record for this book is available
from the British Library

This publication is printed on acid-free paper

Typeset by
BBR (Sheffield)

Printed and bound in Great Britain by
TJ International Ltd, Padstow, Cornwall

Contents

Illustrations

TABLES

The editor, contributors and publisher are grateful to all the institutions and persons listed for permission to reproduce the materials in which they seemingly hold copyright. Every effort has been made to trace the copyright holders; apologies are offered for any omission(s), and the publishers will be pleased to add any necessary acknowledgment(s) in subsequent editions.

Preface

Europeans have engaged in and developed modern, scientific geological and geomorphological inquiry since the late 16th century. That engagement is today well represented by Europe's rich legacy of geoheritage material, held in its numerous museums, major private (and especially royal and aristocratic) collections, universities, archives and libraries. The names of some of its geological sites (or geosites) and geomorphological sites (or geomorphosites) and their host landscapes, together with their rocks, minerals and fossils (essentially the continent's geodiversity), populate the past and present-day geological literature and maps; together, they form Europe's cultural geoheritage. However, the cultural, historical and scientific significance of Europe's geodiversity is universally recognised by neither the scientific geological community nor its politicians and the general population. Hence, unlike archaeological and historical sites and their artefacts, it is not fully accorded the resources and statutory protection it justly deserves. It is, therefore, often threatened by the implementation of inappropriate land management strategies, by the construction of buildings and by other infrastructural developments. The perceived need for a broad societal understanding and constituency building for Europe's geology, identified initially by a mere handful of individuals within Europe's geological community, led to the development in the 1990s of the geodiversity, geoheritage and geotourism paradigms – the core elements examined in this volume.

Geoheritage is an applied scientific discipline that supports the place of geology and geomorphology within modern-day culture. It encompasses the localities and 'artefacts' that fulfil a key role in understanding the Earth's ancient history – its rocks, minerals, fossils, structures and physical landscapes. The latter bear witness to past and present geomorphological processes, underpinned and affected by a myriad of rock types and structures. A range of geoheritage sites serves both scientific and public interests. The sites have considerable potential for scientific study, as outdoor classrooms for the public understanding of science, for recreation and for economic development (especially in geoparks). Scientifically and educationally significant sites include those with textbook features and landscapes, distinctive rock and mineral types, unique or unusual fossils, or other geological and geomorphological phenomena important for education and research. Culturally significant sites are those where geological and geomorphological features or landscapes played a role in historical, artistic and scientific events. Aesthetically significant sites include landscapes that are visually appealing because of their geological and geomorphological features or processes; many have been superbly, and sometimes famously, recorded and interpreted by past and present painters, photographers, poets and travel writers.

Geotourism, initially developed as a means to promote and conserve geosites and geomorphosites, promotes the use and understanding of Europe's geoheritage and provides associated socio-economic and conservation benefits. This volume examines the major past and present underpinnings of geoheritage within the context of European understandings and practice. It provides an historical overview with a descriptive analysis of modern geoheritage developments.

Chapters outlining Europe's geoheritage are followed by several introducing and explaining its core elements with reference to global perspectives; a set of case studies, emphasising the interplay of various geoheritage and geotourism elements at a variety of scales, completes the volume. This timely volume is essential reading for anyone interested in the development and promotion of Europe's physical landscapes.

Thomas A Hose

Acknowledgments

In the preparation of this volume I have been fortunate in being able to call upon the expertise of several leading authorities in their respective country's geoheritage and geotourism provision; their patience, and their willingness to positively and swiftly respond to matters arising from the reviewing process over the extended genesis of this volume, have been much appreciated. I would also like to extend my thanks to them for their support and encouragement – especially that of Djordjije Vasiljević – when I occasionally despaired that the volume would ever be completed, due to the demands on their and my time from other professional, research and publication matters. I gratefully thank the Department of Earth Sciences, University of Bristol, for affording me the research facilities without which my contributions to this volume would have been much less well informed; likewise, the staff, especially Wendy Cawthorne and Michael McKimm, of the Library of the Geological Society of London. I owe a real debt of gratitude to Catherine Dauncey, former Publications Officer in the International Centre for Cultural and Heritage Studies, who has provided support above and beyond the call of her many duties in dealing with my demands as both volume editor and a contributing author. However, the most considerable and gratefully acknowledged debt for this volume is personally owed to Peter Davis, one of the editors of the Heritage Matters series. It was at his suggestion and invitation that the initial proposal for this volume was developed. He has been indefatigable in supporting and encouraging me to prepare and complete this volume to the high standards of the series. His advice and insightful reviewing, especially of my chapters, have been instrumental in ensuring the completeness of discussions, the logical development of the concepts, the adequacy of citations and the textual quality within and across the volume's various chapters.

Finally, the loving support and understanding (especially of my absences and frequent unavailability for family gatherings, and just occasional irritability due to IT issues, while completing this volume and other academic publications over the past three years in particular) of my wife, Sharon, is warmly acknowledged; similarly, my sons, Peter and Phillip, have added to that family support and understanding. I do hope that my family and the various contributors will consider that the volume has been well worth the time and effort.

Abbreviations

AIC	Associazione Italiana di Cartografia
ALSF	Aggregates Levy Sustainability Fund
AONB	Area of Outstanding Natural Beauty
ASSI	Area of Special Scientific Interest
AWRG	Association of Welsh RIGS Groups
BA	British Association for the Advancement of Science
BGS	British Geological Survey
bya	billion years ago
CCW	Countryside Council for Wales
CECA	Committee for Education and Cultural Action
CEE	Central and Eastern Europe
CEPRE	Centre for Economic Policy Research
CORINE	Coordination of Information on the Environment
DGG	Deutsche Geophysikalische Gesellschaft [German Geophysical Society]
EGMUS	European Group on Museum Statistics
EGN	European Geopark Network
ELC	European Landscape Convention
ESCC	Earth Science Conservation Classification
FAS	Fellow of the Antiquaries Society of London
FRS	Fellow of the Royal Society
GAM	Geosite Assessment Model
GCG	Geological Curators' Group
GCR	Geological Conservation Review
GGN	Global Geopark Network
GIS	Geographical Information System
GmbH	Gesellschaft mit beschränkter Haftung [Limited Liability Company]
G&M RIGS	Gwynedd & Môn Regionally Important Geological Site
HMSO	Her Majesty's Stationery Office
IAG	International Association of Geomorphologists
IBA	Important Bird Area
IBERS	Institute of Biological, Environmental and Rural Sciences
ICCHS	International Centre for Cultural and Heritage Studies
ICOM	International Council of Museums
ICONA	Instituto para la Conservación de la Naturaleza [Nature Conservation Institute]
IGC	International Geological Congress
IoACC	Isle of Anglesey County Council
IPA	Important Plant Area
ITGE	Instituto Tecnológico Geominero de España
IUCN	International Union for the Conservation of Nature

IUGS	International Union of Geological Sciences
JEMIRKO	Turkish Association for the Protection of Geological Heritage
ka	thousand years ago
LAPER	Laboratory for Palaeoenvironmental Reconstruction
LEADER	Liaison Entre Actions de Développement de l'Économie Rurale [Links Between Rural Economy Development Actions]
LGAP	Local Geodiversity Action Plan
LGS	Local Geological Site
LLD	Doctor of Law
MD	Medical Doctor
MNHN	Muséum National d'Histoire Naturelle
MTA	General Directorate of Mineral Research and Exploration [Turkey]
mya	million years ago
NEWRIGS	North East Wales Regionally Important Geological Site
NGO	Non-Governmental Organisation
NNR	National Nature Reserve
NORA	Nordic Atlantic Cooperation
NRW	Natural Resources Wales
OU	Open University
PGA	*Proceedings of the Geologists' Association*
ProGEO	European Association for the Conservation of Geological Heritage
PRS	President of the Royal Society
PTG	Polskie Towarzystwo Geologiczne [Polish Geological Society]
RCC	Regional Conservation Council
RIGS	Regionally Important Geological Site
RS	Royal Society
SDGG	Schriftenreihe der Deutschen Gesellschaft für Geowissenschaften [Series of the German Society for Geosciences]
SCC	Superior Council for the Conservation of Natural and Cultural Property
SIGEA	Società Italiana di Geologia Ambientale [Italian Society for Environmental Geology]
SNH	Scottish Natural Heritage
SSSI	Site of Special Scientific Interest
TECNA	Tecnología de la Naturaleza
UKGAP	United Kingdom Geodiversity Action Plan
UKRIGS	United Kingdom Regionally Important Geological Site
UNEP	United Nations Environment Programme
UNESCO	United Nations Educational, Scientific and Cultural Organisation
UNWTO	United Nations World Tourism Organisation
WCMC	World Conservation Monitoring Centre
WG-1	Working Group 1
ya	years ago

Introduction: Geoheritage and Geotourism

Thomas A Hose

Long engagement in geological inquiry explains Europe's rich legacy of geological material in its museums, universities, archives and libraries – its cultural geoheritage. Its geological and geomorphological sites (or geosites and geomorphosites), landforms, rocks, minerals and fossils (or geodiversity) – its natural geoheritage – populate the geological literature. The recognition that many of the localities mentioned in the literature (and from which the specimens in the collections had been gathered) were lost, degraded or no longer accessible led to the development of measures to protect them – that is, geoconservation. The recognition that geology in general, and geoconservation in particular (which demanded funding), were poorly regarded and understood by the public led to efforts by geologists and others to develop geotourism. The interrelationships between these various elements, examined within this volume, are summarised in Figure 1.1. In these various endeavours, Europe has led the field in the now globally accepted definitions and terminological development. However, the initial contribution of Australasia to geodiversity (Household and Sharples 2008, 269–70) and later to geotourism (Dowling and Newsome 2008a) should not be ignored. Likewise, whilst Europe led the way in developing geoparks, it is in Asia that they have particularly blossomed in size and popularity. For example, the People's Republic of China has more than a quarter (31 of 111) and the region as a whole over a third (39 of 111) of the United Nations Educational, Scientific and Cultural Organisation (UNESCO) Global Geoparks designated by September 2014; additionally, the People's Republic of China had at the same time 185 designated and 56 candidate national geoparks. It was the Ministry of Land and Resources of the People's Republic of China, along with UNESCO, that jointly held the First International Conference on Geoparks in Beijing in 2004. However, it is the practices and approaches originally developed in Europe that have guided the global development and recognition of geoparks.

Geoheritage: a modern geological paradigm

Geoheritage is a term in increasing usage in the European and global geological, geoconservation and nature conservation literature. Geoheritage is commonly interchanged with the similar terms 'geological heritage' (in Europe) and 'geologic heritage' (in North America). A major new summary of European geoheritage and geoconservation noted that: 'The natural heritage of any country includes its geological heritage, made up of many key geosites, as well as landscapes, profoundly shaped and defined by their geology. Fossils, rocks and minerals are just as much natural heritage as living plants and animals' (Wimbledon 2012, 7). It is also suggested that: 'Geoheritage is an applied scientific discipline which focuses on unique, special and representative geosites, supporting the science of geology and its place in modern culture' (ProGEO 2012,

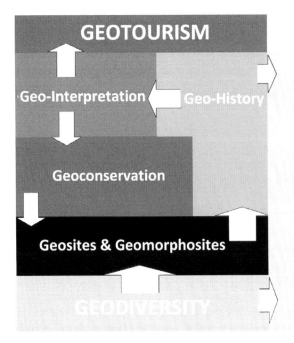

FIGURE 1.1 A TOPOLOGICAL REPRESENTATION OF GEOHERITAGE,
GEODIVERSITY AND GEOTOURISM AND THEIR CONTRIBUTORY ELEMENTS.

14). As the preamble to this definition makes clear, ProGEO's emphasis is on primary (that is, the most important) geosites and geomorphosites. For this volume, that definition is generally accepted but slightly expanded, to overtly recognise that it encompasses geomorphology, to: 'Geoheritage is an applied scientific discipline which focuses on unique, special and representative geosites and geomorphosites, supporting both the modern sciences of geology and geomorphology and their past development, and their place in modern culture.'

Some ProGEO member groups have employed the 'geotope' concept. This, like geodiversity, is an analogy with biodiversity's use of the term biotope (Krieg 1996). Wiedenbein (1994) explained that it had come into usage due to Stürm's (1992) work in Switzerland. Geotopes are 'those parts of the geosphere which are discernable on the earth's surface or are accessible from there; they are spatially limited and in a geoscientific sense clearly distinguishable from their surroundings' (Wiedenbein 1994, 117). Stürm considered that the geotope concept was aimed at 'the entire integration of geoconservation into town and country planning procedures' (Stürm 1994, 28) and that it 'is able to focus geological and geomorphological interest. It stimulates public awareness' (ibid). The geotope concept is almost exclusively employed in German-speaking countries and has been significant in advancing the recognition of their geoparks (see, for example, Reynard *et al* 2007; Röhling and Schmidt-Thomé 2004).

Within North America, and specifically the USA, the Geologic Resources Division of the National Park Service employs the working definition that:

geologic heritage encompasses the significant geologic features, landforms, and landscapes characteristic of our Nation which are preserved for the full range of values that society places on them, including scientific, aesthetic, cultural, ecosystem, educational, recreational, tourism, and other values. Geologic Heritage sites are conserved so that their lessons and beauty will remain as a legacy for future generations.

(National Park Service 2015)

Within North America, 'heritage geology' is sometimes employed when and where heritage geological features are recognised as special places of intertwined geology and landscape. Such features could include unique or exemplary outcrops, scenic views or other geologically significant features that together represent the geological diversity of a region.

It is largely due to European researchers and practitioners that geoheritage emerged at the 20th century's close as a field of study, publication and practice; the 2009 foundation of the journal *Geoheritage* was a landmark in the term's recognition. With the volume of published material and general usage (especially in policy documents) and practice, it could be argued that 'geoheritage' now has sufficient theoretical and conceptual status to qualify as a paradigm. The *Oxford English Dictionary* defines the basic meaning of a paradigm as 'a typical example or pattern of something; a pattern or model'. The *Merriam-Webster Online Dictionary* defines a paradigm as 'a philosophical and theoretical framework of a scientific school or discipline within which theories, laws, and generalizations and the experiments performed in support of them are formulated; broadly: a philosophical or theoretical framework of any kind'. In preparing this volume the author considers that, under the broad scope of such definitions, 'geoheritage' has unquestionably attained the status of a modern geological paradigm.

GEODIVERSITY

The major attention given to biodiversity and wildlife conservation within Europe up to the 1990s had reinforced the imbalance within nature conservation policy and practice between nature's abiotic and biotic elements. Then and now, national and international NGOs and governmental nature conservation bodies, although employing the general term 'nature conservation', used it synonymously with 'wildlife conservation' and focused their attention on the latter (Wimbledon 2012, 11). Indeed, two major reviews of the history of UK nature conservation draw attention to either the scant consideration given to geoconservation measures (Evans 1992, 78), or the lack of uptake of recommended measures (Sheail 1976, 118–19, 160–3) in the UK's emerging nature conservation legislation in the 20th century. Perhaps the best indication of the swing towards wildlife, rather than more general nature conservation, is provided by the eventual change in name of the UK's county-based voluntary Naturalists' Trusts to Wildlife Trusts. This followed the establishment in 1980 of the first urban Wildlife Trust (now the Wildlife Trust for Birmingham and the Black Country); with others formed soon after (in Bristol, London and Sheffield), this highlighted the overall strengthening of the Trusts' focus on wildlife and people. It was during this period that the Trusts changed their names from Naturalists' Societies to Trusts for Nature Conservation, and then to Wildlife Trusts. Exceptionally, some modern geoconservation groups (and also member groups of GeoConservationUK, the national coordinating body for the UK voluntary geoconservation movement), as in Cornwall and Cumbria, have opted to be an adjunct of their local Wildlife Trust.

Around the early 1990s, several geologists and geomorphologists recognised that their disciplines could be considered in much the same way as biology; it seemed obvious, since the Earth is geologically very diverse and its diversity is valuable but threatened and in need of protection. Hence, they began to use the term 'geodiversity' in the mid-1990s to describe the variety within abiotic nature that was conveyed by the use of the term 'biodiversity' for the organic or biotic world. Its first published usage was in Tasmania (Sharples 1993) and Germany (Wiedenbein 1994). Sharples (1993) used it to embrace 'the diversity of earth features and systems'. Dixon (1996), Eberhard (1997), Sharples (2002) and the Australian Heritage Commission (2002) have commonly defined it as 'the range or diversity of geological (bedrock), geomorphological (landform) and soil features, assemblages, systems and processes'. In the UK, Stanley (2001) suggested: 'It is the link between people, landscapes and culture; it is the variety of geological environments, phenomena and processes that make those landscapes, rocks, minerals, fossils and soils which provide the framework for life on Earth.' Within Europe, geodiversity has benefited from the widespread reportage and readership of a major text by Gray (2004), updated in 2014, with its widely published earlier definition as 'the natural range (diversity) of geological (rocks, minerals, fossils), geomorphological (landform, processes) and soil features. It includes their assemblages, relationships, properties, interpretations and systems' (Gray 2004, 8); following its rapid and widespread acceptance, Gray (2008a; 2008b) supported the opinion that geodiversity is a newly recognised geological paradigm.

Whilst the term geodiversity is less than 25 years old, the principles behind its application to European conservation are much older. The *Conservation of Nature in England and Wales: Report of the Wildlife Conservation Special Committee* (Huxley 1947) included a summary of geology's importance to the nation and noted that the UK had 'in a small area an extremely wide range of geological phenomena' (Huxley 1947, paragraph 64); terminologically, 'geodiversity' could substitute for a 'range of geological phenomena'. Indeed, European geoconservation has been practised, but not so labelled, case by case rather than as a general strategy since the mid-19th century (Doughty 2008; Erikstad 2008; Thomas and Warren 2008); it had not been recognised, perhaps because of the use of terms such as 'Earth heritage conservation' and 'natural features' (Gray 2001), across the broader nature conservation community. Consequently, some geologists and geomorphologists considered 'geodiversity' a means to promote geoconservation as well as a useful new way of conceptualising the abiotic world that would place it on a par with wildlife and nature conservation (Prosser 2002). Concomitant with the development of geodiversity, 'geotourism' (Hose 1995) emerged in Europe, and particularly in the UK, as another means to promote and support geoconservation.

GEOTOURISM: ANOTHER MODERN GEOLOGICAL PARADIGM

Largely due to European researchers and practitioners, geotourism emerged in the 1990s as a field of study, publication and practice. Given the volume of published material and general usage (especially in policy documents) and practice, it could be argued that 'geotourism' has sufficient theoretical and conceptual status to qualify as a paradigm. It was originally defined as: 'The provision of interpretive and service facilities to enable tourists to acquire knowledge and understanding of the geology and geomorphology of a site (including its contribution to the development of the Earth Sciences) beyond the level of mere aesthetic appreciation' (Hose

1995, 17). Today it has the benefit and support of significant literature and widely published, if sometimes divergent or vague, definitions.

Geotourism, as a geologically based contemporary approach to landscape promotion, was first recognised and defined in England. It developed following the late-1980s recognition by UK school, university and museum geologists of the increasing number of quarries and mines being lost to study due to reclamation schemes and unsympathetic after-uses. Natural geosites and geomorphosites were also being lost because of planning decisions permitting, for example, the obscuring of roadside exposures with soil and netting (Baird 1994) and the construction of hard coastal defences (Leafe 1998). Modern geotourism was initially seen as a way of promoting and potentially funding geoconservation by developing sustainable tourism provision, ranging from leaflets and geotrails to visitor centres.

In the 1990s several European geologists had made fleeting references to tourism and geology (De Bastion 1994; Jenkins 1992; Maini and Carlisle 1974; Martini 1994; Page 1998; Spiteri 1994), but they all failed to define their understanding of the term or to discuss its participants – that is, geotourists. Its first published definition (Hose 1995), with some of its associated concepts, was included within the *Geoparks Programme Feasibility Study* (UNESCO 2000; Patzak and Eder 1998); so too were the essential elements of a later redefinition to: 'The provision of interpretative facilities and services to promote the value and societal benefit of geologic and geomorphologic sites and their materials, and ensure their conservation, for the use of students, tourists and other recreationalists' (Hose 2000, 136). It thus encompassed geosite interpretative and promotional media, together with the artefacts, places and memorials of their associated Earth scientists. This approach was incorporated, and has the benefits of demonstrably building upon the author's previously widely accepted definitions (Hose 1995; 2000), including recent landscape studies (Hose 2008b; 2010a; 2010b). The most recent definition of geotourism is: 'The provision of interpretative and service facilities for geosites and geomorphosites and their encompassing topography, together with their associated *in situ* and *ex situ* artefacts, to constituency-build for their conservation by generating appreciation, learning and research by and for current and future generations' (Hose 2012, 11). It reinforces the initial geoconservation rationale and is a succinct summary, employing an easily translatable vocabulary, of the nature, focus and location of geology-based geotourism.

Other European authors have employed broader definitions. For Natural England (the UK government's nature conservation agency) geologists, it is 'travelling in order to experience, learn from and enjoy our Earth heritage' (Larwood and Prosser 1998, 98). In Germany, Frey (2008, 97–8) suggested: 'Geotourism means interdisciplinary cooperation within an economic, success-orientated and fast moving discipline that speaks its own language. Geotourism is a new occupational and business sector. The main tasks of geotourism are the transfer and communication of geoscientific knowledge and ideas to the general public.' This approach was used by Frey in her Vulkaneifel Geopark management role, where geoscientific, economic and political considerations combined to develop somewhat commercially orientated geotourism. In Poland, the introductory paper of the inaugural issue of the journal *Geoturystyka* suggested that geotourism is an 'offshoot of cognitive tourism and/or adventure tourism based upon visits to geological objects (geosites) and recognition of geological processes integrated with aesthetic experiences gained by the contact with a geosite' (Slomka and Kicinska-Swiderska 2004, 6). It was noted that there is, as the original approach (Hose 1995) stated, a tangible link between geotourism and geoconservation and its funding. For example: 'In today's economically stretched climate, tourists are a valuable source of local income. The encouragement of the tourist industry to include geodiversity within its remit is therefore high' (Burek 2012, 45).

In Australia, building upon their earlier collaborative text (Newsome *et al* 2002), Dowling and Newsome's (2008b) book was the first to be entitled *Geotourism* and clearly promoted geotourism's geological approach. In their later text, *Geotourism: The Tourism of Geology and Landscape* (Newsome and Dowling 2010), they stated:

> Geotourism is a form of natural area tourism that specifically focuses on geology and landscape. It promotes tourism to geosites and the conservation of geo-diversity and an understanding of earth sciences through appreciation and learning. This is achieved through independent visits to geological features, use of geo-trails and view points, guided tours, geoactivities and patronage of geosite visitor centres.
>
> (Newsome and Dowling 2010, 4)

As Dowling (2011) also recognised, 'the character of geotourism is such that it is geologically based and can occur in a range of environments from natural to built, it fosters geoheritage conservation through appropriate sustainability measures, it advances sound geological understanding through interpretation and education, and finally it generates tourist or visitor satisfaction' (Dowling 2011, 1).

Outside Europe, a much less geology-focused approach to geotourism has developed. In the USA, *National Geographic* ignored the high volume of already published European work on geotourism and erroneously claimed to have themselves singularly coined the term geotourism to describe a 'destination's geographic character – the entire combination of natural and human attributes that make one place distinct from another' (Stueve *et al* 2002, 1). Essentially, the *National Geographic* approach is a renaming of sustainable tourism with a holistic approach to landscape. In 2011 the European Geoparks Network proposed (without recourse to any other interested parties) to accept the *National Geographic* approach, with a geoheritage emphasis; the Organising Committee of the 11th European Geoparks Congress indicated in the published *Arouca Declaration* that:

> there is a need to clarify the concept of geotourism. We therefore believe that geotourism should be defined as tourism which sustains and enhances the identity of a territory, taking into consideration its geology, environment, culture, aesthetics, heritage and the well-being of its residents. Geological tourism is one of the multiple components of geotourism.
>
> (International Congress of Geotourism 2011)

This was a somewhat disputed break with the approach adopted by the majority of European governmental agencies, NGOs and authorities involved in geotourism. The same Organising Committee made various other suggestions (actually already well addressed by the published geotourism literature) and proposed a return in the nature of provision to the basic USA interpretation approach propounded in the 1950s by Tilden (1957), despite research (for example, Hose 1997; 1998; 2000; Taylor 1993; Uzzell 1998; Webb 1993; 1996) suggesting, albeit with some broad agreement on Freeman Tilden's principles, that a reappraisal of the basis and efficacy of that approach was long overdue. The Organising Committee had really embraced 'ecotourism', rather than true geotourism, which the United Nations World Tourism Organisation (UNWTO) defined as 'tourism which leads to management of all resources in such a way that economic, social and aesthetic needs can be fulfilled while maintaining cultural integrity, essential ecological processes, biological diversity and life support systems' (UNWTO 1997). However, the Organising Committee of the 11th European Geoparks Congress failed to consider that ecotourism itself might not be environmentally benign and could actually consume the very landscapes within which it is based – that is, 'geo-exploitation' (Hose 2008b; 2011). For example: 'People may be encouraged to visit spectacular remote mountain regions but be provided with western standards of comfort and accommodation resulting in associated environmental problems' (Acott and La Trobe 1998, 238). Any fragmentation of the European consensus on the geological approach to geotourism is at best unhelpful, and at worst divisive for its stakeholders and confusing to governments and funding bodies. Accepting its geological focus, knowledgable practitioners and researchers of true geotourism do not preclude the benefits of working closely with other nature conservation promotion interests; the incorporation of, especially aesthetic, landscape considerations (Hose 2010b) would seem to make this obvious.

However – and by whomsoever – defined (see Table 1.1), geotourism is a form of 'niche' (Hose 2005; Novelli 2005) or 'special interest' tourism in which the 'traveller's motivation and decision-making are primarily determined by a particular special interest … implies "active" or "experiential" travel' (Hall and Weiler 1992, 5). There is an obvious link with special interest travel for 'people who are going somewhere because they have a particular interest that can be pursued in a particular region or at a particular destination' (Read 1980, 195). As a form of niche tourism, with a geoconservation element, it should be 'more sustainable, less damaging and, importantly, more capable [than mass tourism] of delivering high spending tourists' (Novelli 2005, 1). For geotourists, it offers a 'meaningful set of experiences in the knowledge that their needs and wants are being met' (ibid). Modern geotourism provision meets geotourists' needs by attracting them to particular localities with spectacular or readily appreciated, and usually (on-site and/or off-site) interpreted, geological and/or geomorphological features. It could extend the tourism season in some coastal and upland areas.

Year	Author	Travel	Urban	Rural	Aesthetics	Appreciation	Enjoyment	Education	Learning	Understanding	Geo-history	Geo-conservation	Geo-interpretation
1995	Hose		●	●					●	●	●	●	●
1998	Larwood and Prosser	●					●	●				●	
1998	Frey							●	●				
2000	Hose		●	●		●		●	●		●	●	●
2004	Slomka and Kicinska-Swiderska	●			●			●					
2005	Ruchkys										●	●	●
2006	Joyce	●							●				
2007	Ruchys					●					●	●	●
2009	Nekouie-sadry			●					●			●	
2010	Amrikazemi	●				●			●				
2010	Dowling and Newsome			●		●			●	●		●	●
2011	Hose		●	●		●			●		●	●	●

TABLE 1.1 TABLE OF THE SUMMARISED CONTENT OF GEOTOURISM DEFINITIONS AND THEIR ASSOCIATED DISCUSSIONS. THIS SUMMARY INEVITABLY REQUIRED SOME QUALITATIVE JUDGEMENT IN ASSESSING THE MEANINGS AND EMPHASIS GIVEN TO PARTICULAR PHRASES WITHIN BOTH DEFINITIONS AND THEIR ASSOCIATED DISCUSSIONS. FOR EASE OF COMPARISON WITH OTHER DEFINITIONS, THE AUTHOR'S ARE SHOWN IN LIGHT GREY AND THE THREE KEY ELEMENTS ARE SHOWN IN DARK GREY.

Geotourism localities can be split into primary and secondary geosites and geomorphosites. The former have 'geological or geomorphological features, either naturally or artificially and generally permanently exposed, within a delimited outdoor area, that are at least locally significant for their scientific, educational or interpretative value' (Hose 2005, 29). The latter have 'some feature(s) and/or item(s), within or on a structure or delimited area, of at least local significance to the history, development, presentation or interpretation of geology and/or geomorphology' (Hose 2005, 29). 'Geotourists' – that is, those people visiting geosites and/or geomorphosites – can be split into 'educational' and 'recreational' (Hose 1997) and 'dedicated' and 'casual' geotourists (Hose 2000). The 'educational' and 'dedicated' geotourists are typically academic and amateur geologists, whilst the 'recreational' and 'casual' geotourists are particularly exemplified by holidaymakers, collecting fossils on beaches and visiting show caves. It has been suggested that tourism involving active components with some conservation focus, scholarship, science and environmental awareness is a small market (Heywood 1990, 46) – much as geotourism offers – predicated upon better-educated and wealthy tourists who broadly correspond to Plog's (1974) 'allocentrics' (Hall and Weiler 1992, 4). However, this is likely to apply only to 'educational' and 'dedicated' geotourists, who essentially visit geosites for professional and intellectual improvement. 'Recreational' and 'casual' geotourists are primarily pleasure-seekers, focused on social interaction at (preferably) interpreted geosites; significantly, they visit for informal educational experiences for themselves and for any accompanying (often grand)children (Hose 1997; 2000).

Successful geotourism needs to identify and promote its physical basis (geosites and geomorphosites), know and understand its user base, and develop and promote effective interpretative materials. Geotourism is a geoheritage promotional approach with antecedents in the landscape aesthetic movements that promoted travel into 'wild' areas, popularly followed by the social elite of Britain and Europe, particularly from the mid-18th century onwards (Hose 2008a). In defining and promoting the academic study of geotourism, the seminal work in England underpinned the first 'national' geotourism conference (indeed, the first anywhere), held at the Ulster Museum, Belfast, in 1998 (Robinson 1998). This had several sponsoring agencies, but that of the Geological Society's GeoConservation Commission in particular underlined one of the key elements of the initial geotourism approach (Hose 1995). Few of its wholly unpublished presentations made any attempt to define geotourism, and most – with the exception of Hose (1998) – focused on examples and case studies of its provision. This approach was reiterated by the presentations at two 'global' geotourism conferences in 2008 and 2010. There is now hardly a year that passes without a geotourism (global or otherwise) conference being held somewhere; many are associated with geoparks.

Conclusion

Whilst Europe's legacy of archaeological, historical and industrial sites and artefacts has long been widely recognised, valued, conserved and promoted as a tourism asset, its industrial and mining landscapes were almost obliterated before they began to relatively recently achieve comparable attention. Similarly, whilst Europe's art, music and literature – and the individuals who produced and patronised those achievements – have long been recognised and promoted (often as tourism products), its rich legacy of scientific and technological discoveries and innovations was comparatively neglected until recent times. This has particularly been the case for its geoscientific legacy – its geoheritage. Even at sites with a supposed geological or geomorphological basis, such as preserved coal and metalliferous mines, it is usually their social and industrial history that is explored and presented to tourists.

The paradox is that it was often the wealth generated by geodiversity's exploitation for energy and raw materials that funded Europe's rich cultural heritage. Further, to enhance and protect the physical cultural (particularly the architectural) heritage, it was at one time considered important to either mask or obliterate the sites that exploited Europe's geodiversity. It was the 1980s 'greening' of European economies that particularly accelerated the loss of mine and quarry sites to tourism, recreation-orientated landscape restoration, wildlife-focused forestry and water-based conservation schemes. They commonly became the barely accepted photogenic backdrops of aesthetically driven cultural and industrial heritage tourism that enjoyed considerable political support within postmodern de-industrialising Europe. They were often removed from tourists' potential gaze, on the pretext of safeguarding living and future generations of Europeans from toxins and from the sheer shock of past landscape despoliation (Barr 1969; Richards 1992); the nature conservation bodies were more than happy to acquire such sites as replacements for habitats lost to modern farming and coastal management practices. Of course, all of this was very forgetful of the fact that it was such exploitative industry landscapes that had inspired much of the great literature, social concerns and political debates of the 19th century. It was the latter two that led to the will to change Europe for the better, much of which was achieved from the mid-20th century onwards.

Despite, in particular, geology's contributions to the development and sustainment of a once industrial and now postmodern service-economy-focused Europe, there is little public and political awareness of the richness and cultural significance of its sites and collections, or of the threats to them. As one of the pioneers of modern European geoconservation noted: 'The public's opinion can be an effective support in the political arena, especially on the local level. However, the problem is that the general public, like the politicians, etc, knows almost nothing about earth sciences ... Popularization ... can be a great help to bring people in contact with earth-scientific aspects' (Gonggrijp 1997, 2946–7). Hence, the opening decades of the 21st century are a timely period in which to describe and critically examine the conservation, promotion and popularisation of Europe's geoheritage – a goal this volume seeks to achieve.

However, no single volume on Europe's geoheritage and geotourism could honestly claim to cover the current breadth and depth of theoretical underpinning, development and history of both topics, let alone provide a full set of case studies to reflect that breadth. Notwithstanding that caveat, the chapters in this volume offer a broad temporal, geographical (and generally not too technical for a broad readership) overview of Europe's geoheritage and the geotourism provision it inspires and supports. First and foremost, it should be noted that both concepts were developed and promoted by geologists and geomorphologists. However, they have drawn upon other, not always acknowledged, disciplines, especially from within the humanities. Geologists have not always been aware of the historical and cultural significance of geology.

Within several of the case study chapters, there is some understandable bemoaning that geoheritage is poorly understood and promoted outside the geological community; indeed, Reynard (Chapter 16) particularly makes the point that geotourism developments are generally limited to the geosciences community. This volume's contents have been directed towards remedying that situation, and also seek to act as an introduction to geoheritage and geotourism for readers with a broad interest in landscape history and the management of the countryside, natural history museums and rural tourism.

FURTHER READING AND REFERENCES

Acott, T G, and La Trobe, H L, 1998 An Evaluation of Deep and Shallow Ecotourism, *Journal of Sustainable Tourism* 6 (3), 238–53

Australian Heritage Commission, 2002 *Australian Natural Heritage Charter*, 2 edn, Australian Heritage Commission, Canberra

Baird, J C, 1994 Naked rock and the fear of exposure, in *Geological and Landscape Conservation* (eds D O'Halloran, C Green, M Harley, M Stanley and J Knill), The Geological Society, London, 335–6

Barr, J, 1969 *Derelict Britain*, Penguin, Harmondsworth

Burek, C V, 2012 The role of LGAPs (Local Geodiversity Action Plans) and Welsh RIGS as local drivers for geoconservation within geotourism in Wales, *Geoheritage* 4 (1–2), 45–63

De Bastion, R, 1994 The private sector – threat or opportunity?, in *Geological and Landscape Conservation* (eds D O'Halloran, C Green, M Harley, M Stanley and J Knill), The Geological Society, London, 391–5

Dixon, G, 1996 *Geoconservation: An International Review and Strategy for Tasmania: Occasional Paper 35*, Parks and Wildlife Service, Tasmania

Doughty, P, 2008 How things began: the origins of geological conservation, in *The History of Geoconservation: Special Publication No 300* (eds C V Burek and C D Prosser), The Geological Society, London, 7–16

Dowling, R K, 2011 Geotourism's Global Growth, *Geoheritage* 3 (1), 1–13

Dowling, R K, and Newsome, D (eds), 2008a *Inaugural Global Geotourism Conference (Australia 2008) Conference Proceedings*, Edith Cowan University, Perth

— 2008b *Geotourism*, Elsevier, London

Eberhard, R (ed), 1997 *Pattern & Process: Towards a Regional Approach to National Estate Assessment of Geodiversity*, Australian Heritage Commission, Canberra

Erikstad, L, 2008 History of geoconservation in Europe, in *The History of Geoconservation: Special Publication No 300* (eds C V Burek and C D Prosser), The Geological Society, London, 249–56

Evans, D, 1992 *A History of Nature Conservation in Britain*, Routledge, London

Frey, M-L, 2008 Geoparks – a regional European and global policy, in *Geotourism* (eds R K Dowling and D Newsome), Elsevier, London, 95–117

Gonggrijp, G P, 1997 Nature development: Biologist's experimental garden! Geologist's future sand box?, in *Engineering Geology and the Environment* (eds P G Marinos, G C Koukis, G C Tsiamaos and G C Stournass), A A Balkema, Rotterdam, 2939–48

Gray, J M, 2001 Geomorphological conservation and public policy in England: a geomorphological critique of English Nature's 'Natural Areas' approach, *Earth Surface Processes and Landforms* 26, 1009–23

— 2004 *Geodiversity: Valuing and Conserving Abiotic Nature*, Wiley, Chichester

— 2008a Geodiversity: a new paradigm for valuing and conserving geoheritage, *Journal of the Geological Association of Canada* 35, 2

— 2008b Geodiversity: the origin and evolution of a paradigm, in *The History of Geoconservation: Special Publication No 300* (eds C V Burek and C D Prosser), The Geological Society, London, 31–6

Hall, C M, and Weiler, B, 1992 What's special about special interest tourism?, in *Special Interest Tourism* (eds B Weiler and C M Hall), Belhaven, London, 1–14

Heywood, P, 1990 Truth and beauty in landscape – trends in landscape and leisure, *Landscape Australia* 12 (1), 43–7

Hose, T A, 1995 Selling the story of Britain's stone, *Environmental Interpretation* 10 (2), 16–17

— 1997 Geotourism – selling the Earth to Europe, in *Engineering Geology and the Environment* (eds P G Marinos, G C Koukis, G C Tsiamaos and G C Stournass), A A Balkema, Rotterdam, 2955–60

— 1998 Selling coastal geology to visitors, in *Coastal Defence and Earth Science Conservation* (ed J Hooke), The Geological Society, London, 179–93

— 2000 European geotourism – geological interpretation and geoconservation promotion for tourists, in *Geological Heritage: Its Conservation and Management* (eds D Barretino, W A P Wimbledon and E Gallego), Instituto Tecnológico Geominero de España, Madrid, 127–46

— 2005 Geotourism – appreciating the deep time of landscapes, in *Niche Tourism: Contemporary Issues, Trends and Cases* (ed M Novelli), Elsevier, Oxford, 27–37

— 2008a Towards a history of geotourism: definitions, antecedents and the future, in *The History of Geoconservation: Special Publication No 300* (eds C V Burek and C D Prosser), The Geological Society, London, 37–60

— 2008b The genesis of geotourism and its management implications, in *Abstracts Volume, 4th International Conference, Geotour 2008, Geotourism and Mining Heritage*, AGH University of Science and Technology, Krakow, 24–5

— 2010a Volcanic geotourism in West Coast Scotland, in *Volcano and Geothermal Tourism: Sustainable Geo-Resources for Leisure and Recreation* (eds P Erfurt-Cooper and M Cooper), Earthscan, London, 259–71

— 2010b The significance of aesthetic landscape appreciation to modern geotourism provision, in *Geotourism: The Tourism of Geology and Landscapes* (eds D Newsome and R K Dowling), Goodfellow, Oxford, 13–25

— 2011 The English origins of geotourism (as a vehicle for geoconservation) and their relevance to current studies, *Acta geographica Slovenica* 51 (2), 343–60

— 2012 3G's for Modern Geotourism, *Geoheritage* 4 (1–2), 7–24

Household, I, and Sharples, C, 2008 Geodiversity in the wilderness: a brief history of geoconservation in Tasmania, in *The History of Geoconservation: Special Publication No 300* (eds C V Burek and C D Prosser), The Geological Society, London, 257–72

Huxley, J, 1947 *Conservation of Nature in England and Wales: Report of the Wildlife Conservation Special Committee*, HMSO, London

International Congress of Geotourism, 2011 *Arouca Declaration*, available from: https://dl.dropboxusercontent.com/u/36358978/News/Declaration_Arouca_%5BEN%5D.pdf [2 September 2015]

Jenkins, J M, 1992 Fossickers and Rockhounds in Northern New South Wales, in *Special Interest Tourism* (eds B Weiler and C M Hall), Belhaven, London, 129–40

Krieg, W, 1996 Progress in the management for conservation of geotopes in Europe, *Geologica Balcania* 26 (1), 13–14

Larwood, J, and Prosser, C, 1998 Geotourism, conservation and tourism, *Geologica Balcania* 28, 97–100

Leafe, R, 1998 Conserving our coastal heritage – a conflict resolved?, in *Coastal Defence and Earth Science Conservation* (ed J Hooke), The Geological Society, London, 10–19

Maini, J S, and Carlisle, A, 1974 *Conservation in Canada: A Conspectus – Publication No 1340*, Department of the Environment/Canadian Forestry Service, Ottawa

Martini, G, 1994 The protection of geological heritage and economic development: the saga of the Digne ammonite slab in Japan, in *Geological and Landscape Conservation* (eds D O'Halloran, C Green, M Harley, M Stanley and J Knill), The Geological Society, London, 383–6

National Park Service, 2015 *America's Geologic Heritage* [online], available from: https://www.nature.nps.gov/geology/geoheritage [15 December 2015]

Newsome, D, and Dowling, R K, 2010 *Geotourism: The Tourism of Geology and Landscape*, Goodfellow, Oxford

Newsome, D, Moore, S A, and Dowling, R K, 2002 *Natural Area Tourism: Ecology, Impacts and Management*, Channel View Publications, Clevedon

Novelli, M (ed), 2005 *Niche Tourism: Contemporary Issues, Trends and Cases*, Elsevier, Oxford

Page, K N, 1998 England's earth heritage resource – an asset for everyone, in *Coastal Defence and Earth Science Conservation* (ed J Hooke), The Geological Society, London, 196–209

Patzak, M, and Eder, W, 1998 UNESCO Geopark: A new programme – a new UNESCO label, *Geologica Balcania* 28 (3–4), 33–5

Plog, S C, 1974 Why destination areas rise and fall in popularity, *The Cornell Hotel and Restaurant Administration Quarterly* 15, 55–8

ProGEO, 2012 Conserving our shared geoheritage, in *Geoheritage in Europe and its Conservation* (eds W A P Wimbledon and S Smith-Meyer), ProGEO, Oslo, 14–19

Prosser, C, 2002 Terms of endearment, *Earth Heritage* 17, 12–13

Read, S E, 1980 A prime force in the expansion of tourism in the next decade: special interest travel, in *Tourism Marketing and Management Issues* (eds D E Hawkins, E L Shafer and J M Rovelstad), George Washington University Press, Washington DC

Reynard, E, Baillifard, F, Berger, J-P, Felber, R M, Heitzmann, P, Hipp, R, Jeannin, P-Y, Vavrecka-Sidler, D, and Von Salis, K, 2007 *Geoparks in der Schweiz: Ein Strategie-Bericht*, Geosciences (Platform of the Swiss Academy of Sciences), Bern

Richards, L, 1992 Snailbeach Lead Mine – reclamation and conservation, *Earth Science Conservation* 31, 8–12

Robinson, E, 1998 Tourism in geological landscapes, *Geology Today* 14 (4), 151–3

Röhling, H-G, and Schmidt-Thomé, M, 2004 Geoscience for the public: Geotopes and National GeoParks in Germany, *Episodes* 27 (4), 279–83

Sharples, C, 1993 *A Methodology for the Identification of Significant Landforms and Geological Sites for Geoconservation Purposes*, Forestry Commission, Tasmania

— 2002 *Concepts and Principles of Geoconservation*, Tasmanian Parks and Wildlife Service, available from: http://dpipwe.tas.gov.au/Documents/geoconservation.pdf [16 December 2015]

Sheail, J, 1976 *Nature in Trust: The History of Nature Conservation in Britain*, Blackie, London

Slomka, T, and Kicinska-Swiderska, A, 2004 Geotourism – the basic concepts, *Geoturystyka* 1, 5–7

Spiteri, A, 1994 Malta: a model for the conservation of limestone regions, in *Geological and Landscape Conservation* (eds D O'Halloran, C Green, M Harley, M Stanley and J Knill), The Geological Society, London, 205–8

Stanley, M, 2001 Editorial, *Geodiversity Update* 1, 1

Stueve, A M, Cook, S D, and Drew, D, 2002 *The Geotourism Study: Phase 1 Executive Summary*, National Geographic, Washington DC

Stürm, B, 1992 Geotop – Grundzüge der Begriffsentwicklung und Definition, Materialien 1/1993 Ökologische Bildungsstätte, Oberfranken

— 1994 The geotope concept: geological nature conservation by town and country planning, in *Geological and Landscape Conservation* (eds D O'Halloran, C Green, M Harley, M Stanley and J Knill), The Geological Society, London, 27–31

Taylor, R, 1993 The influence of a visit on attitude and behaviour toward nature conservation, in *Visitor*

Studies: Theory Research and Practice, 6, Collected Papers from the 1993 Visitor Studies Conference, Albuquerque, New Mexico (eds D Thompson, S Bitgood, A Benefield, H Shettel and R Williams), The Center for Social Design, Jacksonville, 163–71

Thomas, B A, and Warren, L M, 2008 Geological conservation in the nineteenth and early twentieth centuries, in *The History of Geoconservation: Special Publication No 300* (eds C V Burek and C D Prosser), The Geological Society, London, 17–30

Tilden, F, 1957 *Interpreting our Heritage*, The University of North Carolina Press, Chapel Hill

UNESCO, 2000 *UNESCO Geoparks Programme Feasibility Study*, UNESCO, Paris

UNWTO, 1997 *What Tourism Managers Need to Know: A Practical Guide for the Development and Application of Indicators of Sustainable Tourism*, United Nations World Tourism Organisation, Madrid

Uzzell, D L, 1998 Interpreting our heritage: a theoretical interpretation, in *Contemporary Issues in Heritage and Environmental Interpretation* (eds D Uzzell and R Ballantyne), HMSO, London, 11–25

Webb, R C, 1993 The relevance of the consumer research literature to the visitor studies field: the case of involvement, in *Visitor Studies: Theory Research and Practice, 6, Collected Papers from the 1993 Visitor Studies Conference, Albuquerque, New Mexico* (eds D Thompson, S Bitgood, A Benefield, H Shettel and R Williams), The Center for Social Design, Jacksonville, 7–19

— 1996 Comparing high-involved and low-involved visitors: a review of the consumer behaviour literature, in *Visitor Studies: Theory Research and Practice, 9, Selected Papers from the 1996 Visitor Studies Conference* (eds M Wells and R Loomis), The Center for Social Design, Jacksonville, 276–87

Wiedenbein, F W, 1994 Origin and use of the term 'Geotope' in German-speaking countries, in *Geological and Landscape Conservation* (eds D O'Halloran, C Green, M Harley, M Stanley and J Knill), The Geological Society, London, 117–20

Wimbledon, W A P, 2012 Preface, in *Geoheritage in Europe and its Conservation* (eds W A P Wimbledon and S Smith-Meyer), ProGEO, Oslo, 6–13

2

Britain and Europe's Geoheritage

Thomas A Hose

Thomas A Hose

Dynamic Europe: geological forces and the creation of a continent

This chapter provides an overview of the geological/geomorphological history of Europe, and particularly Britain, in order to underpin the succeeding geological and geoheritage summary and case study chapters. It covers the nature of global geology (see Calder 1972; Cattermole 2000; Cocks 1981; Lamb and Sington 1998), the geological history of Europe (see Ager 1980), the geological history of Britain and its mineral wealth and rich fossil heritage (see Fortey 1993) as a precursor to noting its historical economic geology, and finally Britain's rich geoheritage.

Europe is bounded in the east by the Ural Mountains. Westwards and northwards it is delimited by the Arctic and Atlantic Oceans, and southwards by the Mediterranean Sea. It is the westernmost part of the Eurasian super-continent, which consists of several joined continental or 'crustal blocks'. Britain and Europe can be split into several broad areas, reflecting their degree of involvement in orogenic (mountain-building) episodes. A series of welded tectonic blocks forms the basement (the lowest geological layers) of Britain and Europe; it was and is the study of the rocks mainly atop these blocks that laid the foundations of modern scientific geology.

The Earth's surface is covered by a crust of relatively rigid rock sheets, or tectonic plates, moving over a viscous semi-molten rock layer. The plates are of two types – oceanic plates and continental plates – formed from rocks that, in terms of their mineral composition, are essentially basalts and granites respectively. When the edge of one plate runs over another, forcing the lower plate deep into the interior (in what is termed a 'subduction zone') of the Earth, a curving chain of volcanoes erupts along the edge of the upper plate. On land this forms a 'continental arc', with the erupted rocks created from an intermixing of melted old continental rocks and new volcanic magma. Good examples of volcanoes created in this way are Mount Etna and Mount Vesuvius in Italy. When the edges of tectonic plates meet under the ocean, the volcanoes form a curving string of islands – an 'island arc' – as seen in Japan.

When two continental plates collide, mountains are created along their leading edges and the continents are eventually welded together; this is termed a 'suture'. For example, the Alps and other mountains in southern Europe were formed by the collision between the plates of Europe and Africa. Sometimes continents are torn apart by sub-crustal forces pulling in opposite directions, initially creating a 'rift valley' racked by volcanic and earthquake activity. Eventually the rift widens and deepens to below sea-level and is consequently flooded, creating a new ocean basin. Another impact of this rifting process is that pieces of crust, torn away from the African plate and elsewhere, have journeyed across the Earth's surface and become welded along the edge of Europe; they underlie, for example, southern England, France, Germany and Greece. When a continental-sized slab of layered rocks is pushed from one side, the upper layers move much more readily than the lower ones. Here a horizontal crack, or 'thrust fault', forms between the near-surface (usually sedimentary, sometimes volcanic) rocks and the deep ancient crustal rocks. Where the thrust fault reaches the surface, 'fault-block' mountains are created, like the Jura and Carpathian Mountains.

These geological phenomena make us aware that neither Europe nor Britain was created in one piece and at one time. Plate tectonic activity brought the various parts of Europe together; although the continent began to form some 3bya, it is most obvious as an entity from around 600mya (Cocks and Torsvik 2006). Before and during the creation of the continent of Europe, the Earth was very unlike today. It was dominated by volcanoes and rifts with an atmosphere toxic to modern life. It had a solid crust (with oceans), with many small tectonic plates that now preserve its oldest rocks; such ancient rocks are found in Bulgaria, Finland, the Lofoten Islands, Russia and Scotland. This ancient crystalline basement lying deep beneath Scandinavia, the Baltic states, parts of Belarus, Russia and Ukraine is called 'Baltica'. It had independently drifted across the planet for hundreds of millions of years; it is the continental core (or 'craton') onto which other parts of Europe were welded. At the opening of the Mesozoic (245mya), a large area of western and southern Europe was squeezed up into the Central Pangean Mountains (Pangea being the ancient super-continent from which all of the younger continents later separated). Early Europe was almost completely landlocked, with a mountain chain in the south that stretched from Kazakhstan to North America's west coast. The birth (around 170mya) of the Atlantic Ocean, as Europe and America moved apart, signalled Pangea's end. By then the Central Pangean Mountains had been eroded to sea-level and below. A new ocean – the Ligurean Ocean – was created as Africa and Europe moved apart from each other, and thick sediments were deposited on the floor of a shallow sea where mountains had once stood tall.

Since the close of the Mesozoic (66mya), Europe has experienced the ongoing Alpine orogeny (mountain building). Prior to that, the Variscan orogeny, which lasted for some 100 million years, is especially significant in Europe's geological history. It made the folded mountain features known as the Hercynides and pushed several small tectonic blocks sideways into southern Europe. The Iberian Peninsula is one of the pieces of crust that was welded to Europe during the Variscan orogeny. Like Britain, Iberia is an unusual 'micro-continent' with a complex geological history. As the African plate shifted from the west to the south of the European plate, the two plates began an ongoing collision that destroyed the Tethys Ocean that originally separated them. The Alps are the result of the European plate's southern edge being pushed by the northern edge of the African plate. In Central Europe, early Mesozoic sedimentary rocks, along with their underlying older metamorphosed Hercynian rocks, were pushed into the continent and forced up and over each other into complex folds called 'nappes'. The Carpathians, Jura Mountains and Transylvanian Alps are made up of stacks of flat-lying sedimentary rock layers; they were thrust forward in giant sheets ahead of the rising Alps. The mountains were later carved into peaks and deep valleys by Pleistocene glaciers.

During the Palaeozoic (230–570mya), southern Europe acquired a tangled mass of continental crustal blocks from Africa; these form basement rocks which are found from the Carpathians south-westward to the Adriatic and Italy. Since the late Mesozoic (around 66–180mya), the widening Atlantic Ocean has been pushing Africa counter-clockwise. Throughout the Cainozoic (the present to 66mya), the islands of Corsica and Sardinia, with the Iberian Peninsula and two pieces of Africa (the 'Kabylies'), moved in various directions before assuming their present position. At this time the Apennines were formed as a volcanic island arc, when an oceanic plate to the east sank below the western Mediterranean floor; they later rotated counter-clockwise into their present position. The Tyrrhenian Sea formed as the crust was stretched behind this forward-moving island arc.

In the Balkans, crustal blocks have piled into each other over tens of millions of years. The compressed and deformed rocks of the mountain chains of the Dinarides and Hellenides, bordering the Adriatic Sea's eastern coast, have within them slices of a former ancient ocean floor; these mark the suture where two continental plates became welded into one when an ocean was closed some 65mya by their coming together. Just east of these mountains is a clearly recognisable plate boundary, where Europe and Africa meet. This extends from what is known as the Pannonian Basin (in Hungary, Romania and Yugoslavia) to Attica, near Athens. The Pannonian Basin resulted from crustal stretching as the Carpathians rotated in an easterly and northerly direction. The Aegean Sea formed as the continental crust was pulled apart in an east-west direction. The Pelagonian Massif underlying Attica, Euboea and Mount Olympus is a raised fragment of the ancient Aegean sea-floor. The Rhodopian Massif, located in Bulgaria, northern Greece and Macedonia, also extends beneath the Aegean Sea. The Balkans in Bulgaria probably marks the crumpled edge of the European craton, with the Proterozoic rocks (formed between 542mya and 2,500mya) extending northwards into Russia. Europe also has numerous isolated volcanoes that are related to structural troughs; these are rift valleys formed by two sets of parallel faults along which volcanoes can form. The River Rhine flows in such a trough – the Rhine Graben. The Rhine formerly flowed southward to join the River Rhône in France, but was diverted by upwarping of the crust around the Vogelsberg volcanic massif – Europe's largest basalt flows that began erupting around 19mya.

The most spectacular event in Europe's relatively recent geological history occurred around 6mya (in the late Neogene) on its western Atlantic seaboard. When sea-level fell below the level of the Straits of Gibraltar, the sea-water passage from the Atlantic Ocean to the Mediterranean Sea closed. Almost concomitantly, northward-moving Arabia closed the eastern ocean passage out of the Mediterranean Sea. The completely landlocked ocean basin began to dry out; this occurred on a cyclical basis up to 30 times. The ancestral Mediterranean, Black and Caspian Seas completely evaporated, producing successive thick layers of salts such as gypsum, sylvite and halite. It was a lifeless, arid place, similar to but on a much vaster scale than today's Death Valley. The major rivers of Europe, Asia and Africa cut deep valleys in their respective continental slopes as they dropped into the exposed ocean floor. Rises in global sea-level lifted water from the Atlantic Ocean over the barrier at Gibraltar; this water cascaded 4km down the exposed mountain flanks into the western Mediterranean basin. From Gibraltar to Central Asia, the bone-dry basin catastrophically filled with sea-water in a geological instant every few hundred years. The refilled ocean deposited deep-sea sediment directly atop the salt layers.

Around 2mya, in the Pleistocene, much of Europe's land and encircling seas were covered with ice sheets and glaciers. These thickened and spread, before melting and retreating, several times. Their ice was in constant but imperceptibly slow motion, scraping and scouring the land beneath them and carrying away huge amounts of boulders and ground rock. Alpine glaciers tore bedrock apart and moved the shattered pieces hundreds of miles away. Such forces shaped the sharp mountain peaks and deep U-shaped mountain valleys of modern Europe. They also created vast flat areas, or 'glacial outwash plains', from material carried by glacial rivers. Following the end of the Ice Age, around 12,000 years ago, only Greenland and Antarctica remain permanently ice-covered.

The geology of Britain

Geologically marginal lands

Europe's north-west margin has two major islands of ancient rocks – Britain and Ireland – together with their numerous smaller fringing islands. A brief account of Britain's geology provided here (but see Trueman 1971; Whittow 1992) illuminates in its small scale the richness of Europe's geoheritage. Northern Britain and Ireland are each made of three or four slices of continental crust that came together around 400mya. The scenery of Britain, albeit at a very small scale, embodies:

> almost all the rock types and landscape features found in countries of considerably greater extent. Its mountains are not of great height nor its rivers of great length, but it is impossible to travel many kilometres without crossing a geological boundary, and it is this irregular juxtaposition of contrasting rocks that gives British scenery its remarkable variety.
>
> (Whittow 1992, 1)

Europe has a similar geodiversity, but on a somewhat grander scale and over greater distances. Its mountains are much higher and its rivers are much broader and longer. That broad range of rock types, coupled with the impact of Ice Age erosion (especially marked in the uplands of northern Britain and Alpine Europe) and deposition (especially in eastern Britain and the Danube basin), has produced a variety of landscapes clothed with vegetation that is more influenced by the underlying geology than the prevailing climate.

Although Britain, like Europe, was not in any currently recognisable form for the first nine-tenths of Earth history, its rocks record ancient events in the evolution of the planet and its life; many of the periods of Earth history were first identified and named from British localities. This is because it has variously lain in two deep oceans, in several shallow seas and in the heart of a vast super-continent, as well as having two mountain chains. It has experienced searing tropical heat and the super-heated violence of catastrophic island-arc vulcanicity, the cold ocean depths of Chalk times and the deep-freezer conditions of the Ice Age. Its ancient landscapes have been traversed by a vast array of animals, from tiny shrew-like mammals to dinosaurs, and from dragonflies to pterosaurs and giant birds. Within the seas that bordered these lands, and in the oceans, there thrived a variety of animals, from the earliest jellyfish-like forms to giant sea scorpions, primitive jawless fishes, plesiosaurs and ichthyosaurs. The record of land plants is impressive, from the earliest known salt-marsh living rock samphire-like forms to early club-mosses and giant swamp and forest trees. These plants nourished a broad range of vertebrate and invertebrate life over the millennia. The world's oldest spider-like invertebrates and amphibian vertebrates have been found in Britain's rocks. Virtually all known fossil animal and plant lifeforms are represented within its fossil record. All of these events and the changing pattern of life-forms result from ancient environments reflecting Britain's gradual drift northwards, from an initial location in roughly the position of present-day South Africa, crossing the Equator in the process. Consequently, there is great potential for the observation of varied geological phenomena and specimen-collecting within Britain.

The recent complex geological development of Europe and Britain over the past 200 million years or so (see Figure 2.1), and individually for the latter over the past 500 million years (see Figure 2.2), can be summarised in a series of sketch maps. However, such maps cannot convey

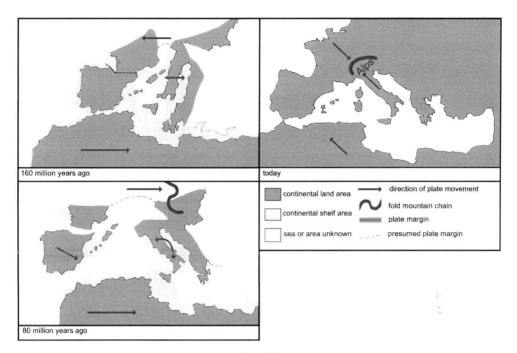

FIGURE 2.1 A CARTOON GEOLOGICAL HISTORY OF 'ALPINE' EUROPE. HOW THE MOVEMENTS OF TECTONIC PLATES CREATED THE ALPINE OROGENIC BELT AND SHAPED ITALY AND THE MEDITERRANEAN SEA.

the complexity of past, especially coastal and shallow marine, environments over 500 million years; hence, the following account describes how Britain's structure developed over geological time.

The geological evolution of Britain

Southern Scotland's rocks resulted from the collision (marked by the 'Iapetus Suture', a natural geological divide between Scotland in the north and England and Wales in the south) between a super-continent incorporating North America with the northern edge of Europe about 500mya. These events are described and illustrated in a major overview (Cope *et al* 1992) and a scholarly populist publication (Toghill 2000). Generally, the rocks of northern Britain are in north-east/ south-west trending belts; the oldest of the early Pre-Cambrian age rocks (1,700–3,000mya) are in north-west Scotland, well seen near Scourie and loosely known as the Lewisian Gneiss. The gneiss was produced by melting rock deep in the Earth's mantle, which then rose as magmas and crystallised at shallower levels, building up the planet's upper crust. These near-surface rocks were later folded and heated, and further magmas (usually as large dykes) were intruded during a phase of metamorphism. Hence, almost all the rocks now found are two types of gneiss, a rock that is distinctly banded due to alternating layers of different minerals. The minerals in the alternating bands in the two types of gneiss are a reflection of whether the original, or parent, rocks were granites or gabbros (including peridotites). Granites are generally pinkish or grey rocks, rich in glass-like quartz mineral, while gabbros are black or greenish-black rocks, rich

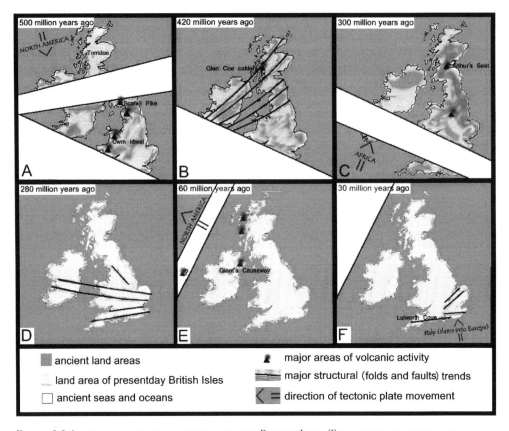

FIGURE 2.2 A CARTOON GEOLOGICAL HISTORY OF THE BRITISH ISLES. THE ACTION OF MAJOR PLATE MOVEMENTS IN FORMING THE STRATA OF THE BRITISH ISLES IS INDICATED OVER THE COURSE OF SIX SIGNIFICANT EVENTS THAT ARE ESSENTIALLY MOUNTAIN-BUILDING (OROGENIC) OR OCEAN-FORMING (RIFTING) EVENTS.

in iron and magnesium minerals, with peridotites particularly rich in them. Whatever their genesis, the gneisses were subjected to prolonged weathering and erosion which planed them to a roughly level surface, interrupted by ancient valleys cutting into them. These stable cratonic basements, such as that formed of the Lewisian Gneiss, are usually masked by thick sequences of younger (but older than 70 million years) mainly sedimentary rocks. For example, the Lewisian Gneiss cratonic platform is covered by a series of thick sandstones and shales, deposited by vast river deltas around 1,000mya in the Pre-Cambrian (540–4,000+mya), forming what are loosely termed the Torridon Sandstones, that occur near Gairloch and Torridon on Scotland's west coast.

The youngest of the basement rocks in Scotland date from the Cambrian (485–540mya). To the south of Scotland, in Northumbria and the Lake District, the rocks – even the cratonic basement rocks – tend to be much younger than their Scottish equivalents. The covering rocks are 40–500 million years old, except along the south-east coast of England where much younger rocks (about 2 million years old) are found. Substantially younger rocks, left behind by melting ice sheets and valley glaciers as little as 10,000 years ago, are found across most of lowland

Britain. Whereas the rocks from the Lake District northwards are much squeezed and broken, those of southern Britain were either gently squeezed or left much as when they were formed, in a volcanic island arc from the Lake District to Snowdonia (with great thicknesses of ash, shales and mudstones, later metamorphosed into slate). Cataclysmic volcanic eruptions generated welded volcanic tuffs (or 'ignimbrites') and fine pulverised ash (such as those presently formed in the Indonesian archipelago), that can be seen in Cwm Idwal (see Figure 2.3A) and Scafell Pike (see Figure 2.3B). Remnants of the earliest phases of this volcanic activity are found in Charnwood Forest, the Malvern Hills and northern Shropshire. In Anglesey (see Chapter 12) a major reheating and squeezing of ancient altered sediments, followed by the injection of granite, about 600mya, created what is called by geologists the Mona Complex.

Volcanic activity gradually built up a low-lying land-mass that, in the Silurian (420–440mya), was ideal for the establishment of the planet's first terrestrial ecosystem; it was frequently flooded by a clear tropical sea depositing great thicknesses of fossil-rich limestone. Eventually this sea was closed and a vast mountain chain (the Caledonides) formed, the granite roots of which lie today beneath the Lake District and Snowdonia. Britain in the Devonian (360–420mya) then lay at the centre of a vast super-continent. Lakes and shallow seas teemed with fish, and for the first time the land had woody plants. Finally, river deltas deposited great thicknesses of sandstone until the area became one vast hot desert. Evidence for these events is found in the rocks of the West Midlands, the Welsh Borderland and northern Scotland. In central Britain, great thicknesses of sedimentary rocks record the gradual wearing down and invasion by the sea of the old Variscanides mountain chain. Shallow tropical seas then covered most of northern and central Britain. Great thicknesses of Carboniferous Limestone and shale formed in environments ranging from reefs and lagoons to shallow banks of lime mud, similar to today's Bahamas; these can be seen in the Peak District and the Yorkshire Dales. A chain of volcanic islands stretched northwards from the Midlands; Arthur's Seat in Edinburgh and Glen Coe are evidence of these volcanoes.

Consequently, in the Upper Carboniferous (300–320mya), land slowly emerged from the sea with numerous interruptions, and great rivers poured out muds, sands and gravels into vast deltas upon which grew dense swamp forests – much like today's mangrove swamps, with animals ranging from dragonflies to amphibians. Many of these events are recorded in the rocks of the West Midlands. Unstable land was created by another mountain-building episode, about 300mya, as southern Britain was affected by the slamming of the African plate into the European plate, forming the Hercynides mountain chain; its granites form the backbone of Cornwall. During the Permian and Triassic (some 200–300mya), Britain lay at the centre of the vast super-continent of Pangea, an arid desert bordered by high hills off which torrents generated infrequent flash floods. Red-stained sandstones with gravel beds (river flood deposits) and thin shales (lake and flood silts), clearly seen in Cheshire and Staffordshire, formed. Vast salt lakes or inland seas also formed and contributed to the salt domes lying beneath Cheshire and the North Sea. Offshore shallow marine conditions also existed. These events are recorded in the rocks of the West Midlands and southern England.

In time, the mountains were worn down and the sea flooded the land, so that in the Jurassic (145–200mya) most of Britain was either low-lying deltas and river plains or stagnant shallow-water marine lagoons and seas. Great thicknesses of oolitic limestone formed in conditions much like today's Bermuda, and are now seen in the Cotswolds. Oolitic limestone (literally, 'egg stone') is so named because it is made up of tiny spheres of limestone (usually formed around a grain of sand or a minute piece of broken seashell) that look like fish eggs. The sea then deepened during

FIGURE 2.3 SOME ICONIC GEOLOGICAL LANDSCAPES OF THE BRITISH ISLES

A. CWM IDWAL IN SNOWDONIA SHOWS BEDDED VOLCANIC ASHES AND TUFFS, THE REMNANTS OF AN ANCIENT VOLCANIC ARC SOME 500MYA – ITS PRESENT FORM IS DUE TO THE RELATIVELY RECENT ACTION OF GLACIER ICE.

B. SCAFELL PIKE IN THE LAKE DISTRICT SHOWS BEDDED VOLCANIC ASHES AND TUFFS, THE REMNANTS OF AN ANCIENT VOLCANIC ARC SOME 500MYA – ITS PRESENT FORM IS DUE TO THE RELATIVELY RECENT ACTION OF GLACIER ICE.

C. THE GIANT'S CAUSEWAY IN COUNTY ANTRIM IS FAMOUS FOR ITS BASALT COLUMNS THAT MARK THE RIFTING PLATE MARGIN THAT OPENED UP THE ATLANTIC OCEAN SOME 60MYA.

D. STAIR HOLE IN DORSET, WHERE THE CRUMPLING OF THE JURASSIC LIMESTONES INTO A GREAT FOLD IS A DIRECT RESULT OF THE IMPACT OF A MICRO-PLATE CARRYING ITALY SLOWLY CRASHING INTO THE MAJOR EUROPEAN PLATE SOME 30MYA.

the Cretaceous (65–100mya), creating muddy sea-floors. In relatively deep, cool water, chalk formed – now exposed at Dover, the Weald, the Isle of Wight, the Chilterns and Flamborough Head. The seas of Chalk times swarmed with sharks and rays, together with crabs, shrimps, shellfish, sea urchins and corals. Sponges lived in the deeper water, their remains eventually forming the flint nodules commonly found in chalk rock. Dinosaurs roamed the land and small shrew-like mammals scurried about their feet. These events are recorded in the rocks of central southern and eastern England.

Eventually the sea retreated and Britain emerged as land for a very long period. Many of the earlier rocks, such as the chalk, that had covered most of Britain were eroded. The youngest major tectonic event was the departure of North America, opening up the Atlantic Ocean, about 60mya during the Palaeogene; this led to the eruption of great thicknesses of lavas, seen along the coasts of north-west Scotland and Antrim (see Figure 2.3C) and the Isle of Staffa, as basaltic rocks forced their way into older rocks and erupted from large volcanoes and fissure eruptions. Some of these rocks never made it to the surface, forming flat sheets ('sills') and vertical walls ('dykes') cutting older rocks, as at the Grinshill Quarries (Shropshire) and Great Ayton (North Yorkshire). At this time, Britain was generally a land-mass surrounded by shallow, rather muddy sub-tropical seas. Southern England was sometimes flooded by the sea; otherwise, it was covered by large river deltas depositing thick layers of clay and mudstone.

Sharks, turtles and crocodiles similar to modern forms swam in the sea. Palms were common on the land. During the Neogene (30mya), Italy slammed into southern Europe to help form the Alps; because this event occurred a long distance away, it only gently buckled the rocks of southern England into the Wealden dome and what is called the Stair Hole crumple in coastal Dorset (see Figure 2.3D).

However, marine conditions developed in the Lower Eocene (49–56mya), when the lower parts of southern Britain were flooded by another tropical shallow sea inhabited by whales, fishes and seashells similar to modern forms. Thick clays, such as the London Clay, were laid down in a sub-tropical swamp. This was then covered by the sands of a large river delta, with freshwater lagoons and lakes in which clays were deposited. Along the seashore, and across the delta, large and sometimes bizarre mammals roamed. Southern England has yielded one of the world's best-documented Tertiary (2.58–66mya) fossil bird faunas. It is also significant for the variety and quality of plant fossils.

The Ice Age began about 2mya, ending about 12,000 years ago. Some 8,000 years ago, southern England was separated from mainland Europe by the flooding of the English Channel as glacial melt-water filled the world's oceans. The eastern West Midlands and central south-west England bear witness to these events. The latter's coastal strip was never touched by ice sheets but was affected by periglaciation action; that is, the effects of geomorphological processes due to the seasonal snow thaw in permafrost areas, the run-off from which then refreezes in soft-ground ice wedges and hard-rock crevices. Much of southern England was smothered by glacial deposits. The West Midlands and Welsh Borderland were affected by much glacial deposition and some erosion.

Britain's mineral wealth

At various times, hot gases and superheated water, emanating from deep volcanic activity, led to minerals being deposited within rocks (commonly limestones and sandstones). Consequently, these great granitic igneous intrusions, such as those found in the south-west peninsula and the

Lake District, have rich deposits of the ores of copper, lead, tin, tungsten, zinc and precious metals such as gold and silver. Cornwall has the world's longest history of near-continuous metalliferous mining, including the world's richest and longest-mined copper and tin ores (Barton 1989). It has produced unique and beautiful mineral specimens (Embrey and Symes 1987). The Lake District has long provided rare, precious, unique and economically important minerals (Cooper and Stanley 1990). The Caldbeck Fells of the Lake District were once so important to the nation, particularly in the 16th and 17th centuries, for the production of copper and silver – with Europe's then largest silver smelter at Keswick – that they gave rise to the obscure proverb, 'Caldbeck and the Caldbeck Fells / Are worth all England else' (Hutchinson 1794). The iron ore mines of Cumberland, worked from the 18th to the 20th centuries, were some of the richest in Europe. The Peak District and the Welsh Borderland were rich sources of lead and zinc minerals. The former was mined for lead (Ford and Rieuwerts 1968) since at least Roman times (Harris 1971), as were the Pennines (Raistrick and Jennings 1965). Mines in Staffordshire (Robey and Porter 1972) and Shropshire have yielded copper and lead (Brook and Allbutt 1973) since at least the 17th century. The world's richest copper source, from the late 18th to the late 19th centuries, was Parys Mountain, Anglesey (Hooper 1994). Perhaps the best known mines are the gold mines of the Mawddach Estuary in North Wales; originally mined for copper, the gold was discovered in the 19th century.

Undoubtedly the least known quarried 'mineral', in the mid-19th century, was the phosphate found in eastern England, in an arc from Bedfordshire to Norfolk, that supported the world's first chemical fertiliser industry (Grove 1976; O'Connor 1998; 2001). Phosphate extraction has the dubious honour of being Britain's first large-scale opencast mining activity. Eastern England (especially the areas around Corby in Northamptonshire and Scunthorpe in Lincolnshire) was also home in the 20th century to vast iron ore mines working Jurassic ironstones.

Despite this richness and long history of mineral exploitation, there are comparatively few published texts on Britain's minerals (Firsoff 1971), other than those that are site-specific (Rogers 1968). Most are concerned purely with aspects of mining and mineral processing; that is, industrial archaeology. A number of older (but virtually no recent) texts examine both minerals and rocks at the populist level (Ellis 1954; Fletcher 1972; Rodgers 1975; 1979), indicating sources of specimens. Today, most of the academic and virtually all the populist interest in metalliferous mines is concentrated on their industrial archaeology, with several series of texts devoted to specific mining regions (such as Ford and Rieuwerts 1968) or mines (such as Robey and Porter 1972). A similar emphasis is evident for the iron ore and coal mining industries.

Britain's rich fossil heritage

Britain's rocks have yielded one of the most complete and nearly continuous records of life on Earth, for it is possible:

> to look back almost 3 billion years to the earliest beginnings of geological time. Because of the diversity of geological formations and phenomena this represents, many of the basic principles fundamental to the development of the geological sciences were evolved in Britain. As a result many British geological localities are regarded as 'type area' for rocks, fossils or minerals, or are taken as the standard reference section for time divisions of the geological column.
>
> (McKirdy 1987, 81)

This record includes some of the earliest-known primitive life forms, as well as some of the most advanced giant vertebrates. Just about every age, fossil type and method of preservation is recorded from Britain's rocks. Its fossils are the longest and most intensively researched in the world. New approaches to invertebrate taxonomy and phylogeny (Eager 1952; Hallam 1968; 1976; Trueman 1922), utilising innovative collection and extraction techniques, were pioneered in Britain. Much of the 19th century progress in stratigraphy and palaeontology was achieved utilising British material, most of which is still in its museums.

Internationally significant fossil discoveries are regularly unearthed in Britain. For example, in the 1950s the oldest (at around 560 million years) then-known European macro-fossils were recognised in rocks in Charnwood Forest. The fossils (Ford 1958; 1963) are similar to Australia's Ediacaran fauna; this is composed of the first multi-cellular animals (similar to modern jellyfish and sea pens) that needed oxygen for their growth. Geologists had noticed the Charnwood fossils in the 19th century (Hill and Bonny 1877; 1878; Bonny and Hill 1877a; 1877b) but dismissed them as inorganic concretions. In the 1970s similar material was discovered near Carmarthen in South Wales (Cope 1977; 1983). One of the most interesting, if inconspicuous, finds of ancient life has recently been described (Brasier and McIlroy 1998) from western Scotland; the faecal pellets of a worm pre-dating by at least 30 million years the Charnwood fossils, making them the world's oldest known multi-cellular animal life. An interesting point about this discovery – pertinent to the need to conserve geological collections – is that the specimens had been collected and kept in a university collection for 18 years before their significance was recognised.

Britain has the world's oldest known Silurian (about 420mya) land plants and animals, found at Rhynie in north-east Scotland and Ludlow in the Welsh Borderland. The great Upper Carboniferous (some 300mya) coal swamps led to the most comprehensive published accounts of fossil plants (Kidston 1923–25; Crookall 1955–76) and localities of this age from anywhere in the world. All of the major sites described by these authors have been reviewed (Cleal and Thomas 1995) for geoconservation purposes. The lush rainforests, with their cycads and ferns of early dinosaur times, are well preserved in the Jurassic (about 160mya) limestones of North Yorkshire (Harris 1961–79) and East Dorset. The forests of the Lower Eocene (49–56mya) London Clay (Collinson 1983) and of the Tertiary (2.58–66mya) of southern England have been intensively studied and published (Chandler 1961–64; 1978); many of the trees and smaller plants are similar to modern tropical forms. The geoconservation issues of Britain's plant fossil sites have been reviewed by Cleal (1988).

Small fossil animals, such as corals, crinoids and shellfish, are abundant in limestones and shales of various ages. However, it is the large vertebrates that most capture the public's imagination. Fossil reptile finds date from the 17th century, with the discovery of the first unrecognised dinosaur material. The Triassic rocks of Cheshire, Shropshire and Devon have yielded unique primitive reptile skeletons and spectacular finds of footprints (Tresise and Sarjeant 1997) for more than 100 years. Britain is home to some unique dinosaurs and is the birthplace of dinosaur studies. The Jurassic rocks of Yorkshire and Dorset have yielded spectacular marine reptiles. The fossil reptiles of Britain represent the full time range for these animals. Although it might be thought that intensive study would by now have revealed most of the important reptile material, relatively recent work shows that this is not so; for example, a limestone quarry near Edinburgh, disused since the 1840s, was where: 'Until very recently the only fossils known … were a few fossil plants … ostracods and the unusual eurypterid … it was not until 1984 that other fossils were found which were to make this site internationally significant' (Rolfe 1988, 22); it then yielded the oldest known fully terrestrial amphibian, along with smaller terrestrial invertebrates such as harvestman spiders, millipedes and the oldest known true scorpions. All of the major fossil reptile sites – famous for historical reasons and for their diversity and richness – have been reviewed (Benton 1988; Benton and Spencer 1995) for geoconservation purposes. For more than 150 years, Devonian fossil fishes from Thurso, Scotland (Andrews 1982) and the Welsh Borderland (Norton 1978) have been intensively studied. Lower Carboniferous fish, especially sharks, have been examined with spectacular success, and there have been significant finds of amphibians. Rocks in Oxfordshire have yielded the world's oldest, tiny, shrew-like mammals. British fossil mammals are significant from the Upper Jurassic to the Pleistocene, and include huge animals such as mammoths and even whales. A number of cave sites have yielded significant Pleistocene material. Britain's rocks are still the subject of palaeontological research and new significant discoveries can be expected.

BRITAIN'S AND EUROPE'S RICH CULTURAL GEOHERITAGE

Europe – and particularly Britain – are recognised for their global geoheritage significance, with type localities (that is, the places and specimens from which they were first described) for fossils, minerals and specific rock formations. Further, Britain's rocks and minerals underpinned the world's first full-scale industrial society in the late 18th and 19th centuries; it was fuelled by releasing the carbon captured in the fossilised remains of semi-tropical swamp forests (of Upper Carboniferous age) – that is, coal. Industrial and technological advances were forged by the smelting of iron minerals (using Lower Carboniferous and Jurassic tropical limestones as a flux), created from reducing conditions in stagnant lagoons and shallow seas (as in Lincolnshire and Northamptonshire). Iron minerals were also precipitated from hot fluids percolating through country rock during the end phase of volcanic activity (as in the western Lake District). Coal was also significant in the development of the heavy chemical industry; few people realise that coal was an essential component in manufacturing the world's first artificial drug (aspirin) and vital for the vivid pigments used in pre-Raphaelite paintings and Liberty's art-deco fabrics. Permo-Triassic salt-domes and sandstones acted as traps and reservoirs for the rotted remains of lower plant and animal life, migrating under heat and pressure from shales and clays, and fuelled the North Sea oil developments from the mid-1960s onwards.

What is particularly unappreciated is the association between Europe's geoheritage and tourism. The birth of modern mass tourism lies in the burgeoning inland and coastal spa resorts of the 18th century. The inland spas, such as Great Malvern, were often based around the remnants of ancient volcanic activity and their heated mineral waters. Volvic in the Auvergne derives its famous mineral waters, first tapped in 1922, from a nearby volcano that last erupted around 8,000 years ago. In Britain, the aptly named city of Bath has been a major spa centre since Roman times. The rise of seaside tourism in the 19th century was due to the perceived health benefits (in the 17th and 18th centuries, visitors actually drank the sea-water) of sea bathing and sea air, and to the expansion of the railways. The latter particularly expedited the development of geological fieldwork from the mid-19th century onwards, especially with the emergence of numerous local and regional natural history societies and regional and national geological societies. The better recognition of Britain's and Europe's past and recently exploited geoheritage will ensure that it continues to attract scholarly and casual interest into the future, and this geoheritage capital can only increase over time. It can perhaps best be appreciated by non-geologists through the exploration of its historical elements.

It is therefore worth noting that the landscapes of Britain and Europe have been visited and described by (geo)tourists for at least 400 years (Hose 2008). Aesthetic, especially mountainous, landscapes have inspired travellers, artists, poets and writers since the Renaissance, as they searched for new and different experiences and sights (Hose 2010). Some, such as Celia Fiennes, England's earliest published female traveller, even commented upon geological matters such as mining techniques (Fiennes 1949); others, such as J M W Turner and John Glover, painted and exhibited aesthetic landscapes that influenced public taste for, and interest in, areas such as the Lake District and the Scottish Highlands. Poets such as William Wordsworth and Samuel Taylor Coleridge eulogised those same landscapes. Wordsworth popularised the Lake District for elite travellers (Wordsworth 1835) rather than tourists, especially through his much reprinted guidebook (Wordsworth 1820) that also included notes on the area's geoheritage contributed by a leading geologist of the day, Adam Sedgwick. It was Sedgwick who helped unravel some of Britain's ancient history; he named the geological periods now known as the Devonian and the Cambrian. That and similar research work conducted by noteworthy individuals to unravel the disposition, geological history and history of ancient life is considered in Chapter 3. The depositaries of its material are covered in Chapter 4, and the location and sale of the material collected and published upon by these noteworthy individuals is explored in Chapter 5. Further, Chapters 7 and 6 respectively consider the (geo)conservation of localities at which much of the research was conducted and the means by which it was done (that is, fieldwork and its associated publications). Finally, Chapter 8 examines the history of European geotourism. Thus is established the groundwork to underpin the volume's case studies.

REFERENCES

Ager, D V, 1980 *The Geology of Europe*, McGraw Hill, London

Andrews, S M, 1982 *The Discovery of Fossil Fishes in Scotland up to 1845*, National Museums of Scotland, Edinburgh

Barton, D B, 1989 *A History of Tin Mining and Smelting in Cornwall* (revised edn 1969 – reprinted 1989), Cornwall Books, Marsh Barton

Benton, M J, 1988 British fossil reptile sites, in *Special Papers in Palaeontology No 40: The Use and Conservation of Palaeontological Sites* (eds P C Crowther and W A Wimbledon), Palaeontological Society, London, 73–84

Benton, M J, and Spencer, P S (eds), 1995 *Geological Conservation Review Series: Fossil Reptiles of Great Britain*, Chapman and Hall, London

Bonney, T G, and Hill, E, 1877a The rocks of Charnwood Forest, *Nature* 15, 470

— 1877b The rocks of Charnwood Forest, *Nature* 16, 8

Brasier, M D, and McIlroy, D, 1998 Neonereites uniserialis from c.600 Ma year old rocks in western Scotland and the emergence of animals, *The Journal of the Geological Society* 155 (1), 5–12

Brook, F, and Allbutt, M, 1973 *The Shropshire Lead Mines*, Moorland Publishing, Leek

Calder, N, 1972 *Restless Earth*, BBC Books, London

Cattermole, P, 2000 *Building Planet Earth: Five Billion Years of Earth History*, Cambridge University Press, Cambridge

Chandler, M E J, 1961–64 *The Lower Tertiary Floras of Southern England (Parts 1–4)*, British Museum (Natural History), London

— 1978 *Tertiary Research Special Paper Number 4: Supplement to the Lower Tertiary Floras of Southern England (Part 5)*, Tertiary Research Group, London

Cleal, C J, 1988 British palaeobotanical sites, in *Special Papers in Palaeontology No 40: The Use and Conservation of Palaeontological Sites* (eds P C Crowther and W A Wimbledon), Palaeontological Society, London, 57–71

Cleal, C J, and Thomas, B A, 1995 *Geological Conservation Review Series: Palaeozoic Palaeobotany of Great Britain*, Chapman and Hall, London

Cocks, L R M (ed), 1981 *The Evolving Earth*, British Museum (Natural History) and Cambridge University Press, London and Cambridge

Cocks, L R M, and Torsvik, T H, 2006 European geography in a global context from the Vendian to the end of the Palaeozoic, in *European Lithosphere Dynamics, Memoirs 32* (eds D G Gee and R A Stephenson), The Geological Society, London, 83–95

Collinson, M E, 1983 *Palaeontological Association Field Guide to Fossils No 1: Fossil Plants of the London Clay*, Palaeontological Association, London

Cooper, M P, and Stanley, C J, 1990 *Minerals of the English Lake District: Caldbeck Fells*, The Natural History Museum, London

Cope, J C W, 1977 An Ediacara-type fauna from South Wales, *Nature* 268, 624

— 1983 Precambrian faunas from the Carmarthen district, *Nature Wales*, NS1 (2), 11–16

Cope, J C W, Ingham, J K, and Rawson, P F, 1992 *Geological Society Memoir No 13: Atlas of Palaeogeography and Lithofacies*, The Geological Society, London

Crookall, R, 1955–76 *Memoirs of the Geological Survey of Great Britain (Palaeontology, Volume IV, Parts 1–7) Volume II: Fossil Plants of the Carboniferous Rocks of Great Britain [second section]*, HMSO, London

Eager, M, 1952 Growth and Variation in the Non-Marine Lamellibranch Fauna above the Sand Rock Mine of the Lancashire Millstone Grit, *Quarterly Journal of the Geological Society* 107, 339–73

Ellis, C, 1954 *The Pebbles on the Beach*, Faber and Faber, London

Embrey, P G, and Symes, R F, 1987 *Minerals of Cornwall and Devon*, The Natural History Museum, London

Fiennes, C, 1949 *The Journeys of Celia Fiennes* (ed C Morris), Cresset Press, London

Firsoff, V A, 1971 *Gemstones of the British Isles*, Oliver and Boyd, Edinburgh

Fletcher, E, 1972 *Pebble Polishing: A guide to collecting, tumble polishing and making baroque jewellery*, Blandford, London

Ford, T D, 1958 Pre-Cambrian Fossils from Charnwood Forest, *Proceedings of the Yorkshire Geological Society* 31, 211–17

— 1963 The Pre-Cambrian Fossils of Charnwood Forest, *Transactions of the Leicester Literary and Philosophical Society* 57, 62

Ford, T D, and Rieuwerts, J H (eds), 1968 *Lead Mining in the Peak District*, Peak Park Planning Board, Bakewell

Fortey, R, 1993 *The Hidden Landscape: A Journey into the Geological Past*, Pimlico, London

Grove, R, 1976 *The Cambridgeshire Coprolite Mining Rush*, Oleander Press, Cambridge

Hallam, A, 1968 Morphology, Palaeoecology and Evolution of the Genus Gryphaea in the British Lias, *Philosophical Transactions of the Royal Society of London, Series B* 254, 91–128

— 1976 Stratigraphic distribution and ecology of European Jurassic bivalves, *Lethaia* 9, 245–59

Harris, H, 1971 *Industrial Archaeology of the Peak District*, David and Charles, Newton Abbot

Harris, T M, 1961–79 *The Yorkshire Jurassic Flora (Volumes 1–5)*, British Museum (Natural History), London

Hill, E, and Bonny, T G, 1877 On the Precarboniferous rocks of Charnwood Forest, Part I, *Quarterly Journal of the Geological Society of London* 33, 754–89

— 1878 On the Precarboniferous rocks of Charnwood Forest, Part II, *Quarterly Journal of the Geological Society of London* 34, 199–239

Hooper, M, 1994 Life after death, *Geographical* 66 (8), 35–6

Hose, T A, 2008 Towards a history of geotourism: definitions, antecedents and the future, in *The History of Geoconservation: Special Publication No 300* (eds C V Burek and C Prosser), The Geological Society, London, 37–60

— 2010 The significance of aesthetic landscape appreciation to modern geotourism provision, in *Geotourism: The Tourism of Geology and Landscapes* (eds D Newsome and R K Dowling), Goodfellow, Oxford, 13–25

Hutchinson, W, 1794 *The History of the County of Cumberland*, F Jolly, Carlisle

Kidston, R, 1923–25 *Memoirs of the Geological Survey of Great Britain (Palaeontology, Volume II, Parts 1–6) Fossil Plants of the Carboniferous Rocks of Great Britain*, HMSO, London

Lamb, S, and Sington, D, 1998 *Earth Story: The Shaping of Our World*, BBC Books, London

McKirdy, A R, 1987 Protective works and geological conservation, in *Planning and Engineering Geology – Geological Society Engineering Geology Special Publication No 4 (Proceedings of the Twenty-Second Annual Conference of the Engineering Group of the Geological Society, Plymouth, September 1986)* (eds M G Culshaw, F G Bell, J C Cripps and M O'Hara), Geological Society, London, 81–5

Norton, J, 1978 *Old Red Sandstone Fishes of South Shropshire*, Shropshire County Museum, Shrewsbury

O'Connor, B, 1998 *The Dinosaurs of Sandy Heath: The Story of the Coprolite Industry around Potton*, Bernard O'Connor, Cambridge

— 2001 The Origins and Development of the British Coprolite Industry, *Mining History* 14 (5), 46–57

Raistrick, A, and Jennings, B, 1965 *Lead Mining in the Pennines*, Longmans, London

Robey, J A, and Porter, L, 1972 *The Copper & Lead Mines of Ecton Hill, Staffordshire*, Moorland Publishing/Peak District Mines Historical Society, Leek/Bakewell

Rodgers, P R, 1975 *Agate Collecting in Britain*, Batsford, London

— 1979 *Rock and Mineral Collecting in Britain*, Faber and Faber, London

Rogers, C, 1968 *A Collector's Guide to Minerals, Rocks and Gemstones in Cornwall and Devon*, Bradford Barton, Truro

Rolfe, I, 1988 Early life on land – the East Kirton discoveries, *Earth Science Conservation* 25, 22–8

Toghill, P, 2000 *The Geology of Britain: An Introduction*, Swan Hill, Shrewsbury

Tresise, G R, and Sarjeant, W A S, 1997 *The Tracks of Triassic Vertebrates: Fossil Evidence from North-west England*, National Museums and Galleries on Merseyside, Liverpool

Trueman, A E, 1922 The use of Gryphaea in the correlation of the Lower Lias, *Geological Magazine* 59 (06), 256–68

— 1971 *Geology and Scenery in England and Wales* (revised edn, eds J B Whittow and J A Hardy), Pelican, Harmondsworth

Whittow, J, 1992 *Geology and Scenery in Britain*, Chapman and Hall, London

Wordsworth, W, 1820 *A Guide through the District of the Lakes in the North of England, with a Description of the Scenery, &c. for the Use of Tourists and Residents*, 5 edn, Longman, Hurst, Rees, Orme and Brown, London

— 1835 *The River Duddon, a Series of Sonnets: Vaudracour and Julia: And Other Poems. To which is annexed, a Topographical Description of the Country of the Lakes, in the North of England*, Hudson and Nicholson, Kendal

3

Geological Inquiry in Britain and Europe: A Brief History

Thomas A Hose

Applied and scholarly approaches

This chapter summarises the development of scientific geology, and its consequent temporally and geographically diverse publications, as a major aspect of Europe's cultural geoheritage. It covers a period beginning in the late Renaissance (16th to 17th centuries), through to the 'heroic age of geology' (1775 to 1825) and the eventual demise of 'classic geology' at the end of the 19th century. Significant events, personalities and their work are considered for the UK, France, Germany, Italy, and parts of the old Austro-Hungarian Empire. South-eastern Europe in the Renaissance was where geological observations were first published. During the Enlightenment (1650s to 1780s), Italian influence was maintained and its pre-eminence in geological studies spread, as its scholars and travellers entered (especially abreast of the Austro-Hungarian Empire's expansion) central and northern Europe. By the late 18th century the focus of geological study had shifted to northern Europe. The majority of individuals who were influential in geology's development worked in the UK, France and Germany, and their study locations and the terms they employed are still used in today's geological literature; however, some Italian scholars made significant contributions at this time (Vai 2009).

The term 'geologia' was coined in England by Richard de Bury (1287–1345), who was Bishop of Durham, in *The Philobiblon* of 1345; this was concerned, amongst other subjects, with book collection and preservation (Durham County Council 1994, 4). The first widely published English reference to geology appears in the title, if not the content, of *Geologia; or, a Discourse concerning the Earth before the Deluge* (Warren 1690) by Erasmus Warren (d.1718), rector of Worlington, Suffolk. The term 'geologia' was employed by Ulisse Aldrovandi (1522–1605) in his published will of 1603 (Vai 2004) and in Fabrizio Sessa's *Geologia del Dottore* of 1687 (Sessa 1687). There are useful, if dated, overviews of the science of geology's development in Europe (Adams 1938; Bonney 1895; Edwards 1967; Gillispie 1996; Gohau 1991; Sarjeant 1980; Schneer 1969; Thompson 1988; Wendt 1968; Wheeler and Price 1985; Woodward 1911; Zittel 1901). Some deal with specific British geology movements (Desmond 1982; Fuller 1995; Morrell 1994; Oldroyd 1990; Porter 1977; Sarjeant and Harvey 1979; Secord 1986; Thackray 1976) and personalities (Bailey 1962, 1967; Clarke and Hughes 1895; Curwen 1940; Dean 1990; Eyles 1969, 1971a, 1971b; Geikie 1875; Lyell 1881; Morris 1989; Morton 2001; Phillips 1844; Secord 1982, 1985; Speakman 1982; Watts 1939; Wilding 1992; Winchester 2001). A few consider important British geological societies (Macnair and Mort 1908; Rudwick 1963, 1985; Vincent 1994; Sweeting 1958; Woodward 1907) and university departments (Vincent 1994; Wyatt 2000). There are four, now dated, bibliographies for Wales (Bassett 1961, 1963, 1967) and the United Kingdom (Challinor 1971). The first widely available account of 19th century British geology (Woodward 1911) was written more than a century ago and is, unsurprisingly for its time, descriptive and uncritical in nature. Modern accounts include a study of the political origins of museums and geology (Knell 2000) and a populist text that additionally covers the rock and fossil collecting craze, together with geology's portrayal in literature and art (Freeman 2004). There are several histories of the major agencies and bodies, such as the Geological Society of London (Davies 2007; Lewis and Knell 2009; Woodward 1907), the Geologists' Association (Sweeting 1958), the Geological Survey (Bailey 1952; Flett 1937; Wilson 1985) and regional bodies (Macnair and Mort 1908; Goddard 1929). Modern accounts generally critically review 19th century controversies of where, how and by whom rocks were mapped, named and stratigraphically assigned, when the 'demands of the early industrial revolution … forced scholarly philosophy to be translated into repeatable science' (Page 1998, 197). By the opening of the 19th century, two discrete strands of geological inquiry and publication had developed – namely, academic and applied geology. The protagonists of each strand had starkly different social standing and influence. The early writers on geology in the 18th century, including James Hutton (1726–97), and the 19th century Fellows of the Geological Society of London (the world's first such institution) were definitely representatives of the former strand. The applied geologists included the mineral dealer John Mawe (1764–1829) and the land surveyor William Smith (1769–1839). However, the contributions of practical geologists to geological publishing, especially with the initial shortage in the 19th century of appropriate journals, probably under-represents their contributions to geology. Some individuals, most notably Abraham Gottlob Werner (1749–1817), who was Professor at the Freiberg Mining Academy, managed to bridge the two strands professionally and personally; his practical discourses on geology were initially about mining.

Mapping geodiversity

Mineral exploitation, together with canal and then railway construction, led to many British geological publications and maps. The English antiquary John Aubrey's (1626–97) *Natural History of Wiltshire* (Aubrey 1847), written between 1656 and 1691 but never fully published, noted: 'I have often times wished for a mappe of England coloured according to the colours of the earth; with markes of the fossils and minerals.' The volume was not completed and published in his lifetime – probably due to a combination of other writing commitments (such as the biographical sketches which he worked on until 1693, eventually amounting to three folio volumes presented to the Ashmolean Museum and now in the Bodleian Library), and his constant diversions in the courts over lawsuits. These eventually rendered him homeless in 1670, with the loss of the ancestral home of Easton Piers, leaving him dependent on the hospitality of his friends, most notably Sir James Long (1617–92) of Drayton House in Wiltshire.

Interestingly, Aubrey (a distinguished conchologist) failed to recognise fossil shells as once-living animals. Martin Lister (1639–1712), a near contemporary of Aubrey, following his personal observation that different types of rocks were found in different parts of England, made to the Royal Society in 1684 'an ingenious proposal for a new sort of maps of countries, together with tables of sands and clays, such as are chiefly found in the north parts of England'. The break-through came with William Smith's correlation work in geological mapping, beginning with his 1799 map for the Bath area, which was a significant milestone in geology. He recognised that rocks included distinct fossil assemblages useful for correlating them across regions. In 1801 Smith (1769–1839) drew a rough sketch of what would become his famous *Geology Map of England and Wales* (Winchester 2001), published in 1815; the geological formations were indicated by different colours and were described in the accompanying memoir, *Delineation of the Strata of England*. It employed conventional symbols for major civil engineering works (canals, tunnels, tramways and roads) and exploited geo-resources (collieries, lead, copper and tin mines, together with salt and alum works). Not long afterwards, Smith published *Strata Identified by Organized Fossils* (1816 to 1819). In 1822 he designed and fitted out – according to his stratigraphy – the first modern geology museum, Scarborough's Rotunda Museum. Smith's life of mixed fortune (Eyles 1969; Morton 2001; Winchester 2001), including bankruptcy, exemplifies the contrast in lifestyles between the professional field geologists and the more gentlemanly geologists of the London scene (Desmond 1982), such as Roderick Impey Murchison (1792–1871) and George Bellas Greenough (1778–1855).

Greenough attended England's leading public schools, and Cambridge and Göttingen universities, but never actually graduated. Whilst in Germany, he met and holidayed with the Romantic poet Samuel Taylor Coleridge (1872–34) and undertook geological explorations. He was elected a Fellow of the Royal Society in 1812. He was associated with a group of mineralogists mentioned in Sir Humphrey Davy's (1778–1829) letter to William Hasledine Pepys (1775–1856) of 13 November 1807: 'We are forming a little talking Geological Dinner Club' (Davies 2007, 13). This 'Club' became the world's first national geology society, the Geological Society of London. Greenough was one of its founders in 1807, its first chairman, and in 1811 its first president – assuming the role twice again in his 50 years as a Fellow.

In 1819 Greenough published his *Geological Map of England and Wales* (Greenough 1819a); new editions followed in 1839 and 1865. These were the product of the Society's Committee of Maps, instigated and led by Greenough from 1809. Unlike Smith – whose approach was one of personal observation – Greenough relied upon the Society's Fellows to submit local rock information, which he then collated and plotted on a topographic map – probably the earliest collaborative geological project. He was considerably assisted by contemporary major researchers such as Henry Thomas De la Beche (1796–1855), the Reverend William Buckland (1784–1856), the Reverend William Daniel Conybeare (1787–1857), John Farey (1766–1826), Henry Warburton (1784–1858) and Thomas Webster (1773–1844). The latter had prepared the Isle of Wight's first major geological study (Englefield 1816). The first draft of Greenough's map, presented to the Society in 1812, suffered from a poor-quality topographic base-map, delaying plotting of the final version until 1814. When published in 1819 it had obviously drawn on Smith's work, although this was unacknowledged until the third edition. Greenough's map contained more geological detail and was cartographically superior to Smith's map. Fittingly, originals of both maps are now mounted side by side in the entrance foyer of the Society's apartments in Burlington House, London. In 1831 the Society conferred on Smith the first Wollaston Medal in recognition of his achievements; its then president, Adam Sedgwick (1785–1883), referred to Smith as the 'Father of English Geology'.

During the 18th century, attempts were made in northern Europe to elucidate the occurrence of mineral deposits and building and ornamental stones by preparing sections and maps (Oldroyd 1996). The first true geological sections were published in 1719 (Wilding 1992) by John Strachey (1671–1743). Some for the Peak District were prepared from 1806 by John Farey (1766–1826), a former estate agent for the Duke of Bedford, and latterly a surveyor and geologist

(Ford 1967). This was the same Duke who had employed William Smith as a land-drainage surveyor at Woburn Abbey, Bedfordshire. In the emerging German nation state, Christian Leopold von Buch produced a geological map in 1826. The first reliable economic geology maps were produced for areas of central and northern England in the 19th century. Likewise, mineral extraction and processing practices were developed in Britain and then exported by a skilled workforce, during the many declines in mining activity caused by fluctuations in metal prices; hence, 'Many of the sites have a value beyond their purely academic interest as they are an indispensable economic resource in terms of their usefulness as a training ground for future generations of geologists' (McKirdy 1987, 81). Their significance for geotourism has been only recently appreciated (Hose 1995).

EARLY SCIENTIFIC GEOLOGY IN MEDITERRANEAN EUROPE: A PUBLISHED LEGACY

From the 17th century, the Mediterranean was the initial centre of published geological observations, due to scholars who were born or domiciled in Italy. The best known domiciled scholar is the Dane Nicholas Steno (1638–86) from Copenhagen. After medical training at Leiden he moved to Italy in 1665, firstly as professor of anatomy at the University of Padua and then as in-house physician to the Grand Duke of Tuscany, Ferdinando II de' Medici (1610–70) in Florence. The Grand Duke invited Steno, who in return had to form a 'cabinet of curiosities', to live in his Palazzo Vecchio. After dissecting the head of a huge shark caught off Livorno in 1666, Steno published his findings, with probably one of the most widely known and published natural history images, in *Elementorum myologiae specimen* (1667). He recognised the similarity of the shark's teeth to the stony objects, *glossopetrae* or 'tongue stones', found in Malta's limestones. They were considered in medieval and Renaissance Europe to possess supernatural powers, especially against poison, due to their biblical association with Saint Paul. Shipwrecked on the island in AD 60, Saint Paul was bitten by a snake (Acts of the Apostles 28: 2–7); noting that he was unharmed, the islanders believed that he had either turned the island's snakes to stone or had at least removed their venom. Presumably, by some great stretch of imagination, the teeth were meant to look like either the forked tongue or fangs of snakes. The trade in fossil sharks' teeth was quite profitable, and it could be argued that Steno's revelations were probably the first by a geologist to undermine economically significant geotourism.

Ancient authorities, like Pliny the Elder (AD 23–79) in *Naturalis Historia* (AD 77–79) – which consists of 37 books collected into 10 volumes, and which was something of a model for later encyclopedias and scholarly works because of the breadth and depth of its coverage and attribution of its index – had suggested that the fossil teeth had fallen from the sky or the Moon. Steno's contemporary, Athanasius Kircher (1602–80), like some ancient authorities, suggested that they had grown within the limestones. The Italian naturalist Fabio Colonna (1567–1640) had stated that they were sharks' teeth in his *De glossopetris dissertatio* of 1616. Steno argued that they looked like sharks' teeth because that was what they were; differences in composition between *glossopetrae* and the teeth of living sharks had arisen because the fossils' chemical composition had been altered without changing their shape. He suggested that they were from once-living sharks, buried in mud or sand that had become dry land. His English contemporaries Robert Hooke (1635–1703) and John Ray (1627–1705) also argued that fossils were the remains of once-living organisms. Steno thought rocks formed when particles in fluids, such as water, settled out into horizontal layers; any deviations from this were due to later disturbances

– his 'principle of original horizontality'. He suggested that the youngest layers must be those at the top, with the oldest at the bottom – his 'law of superposition'. Steno noticed that in the Apennines near Florence the upper rocks were richly fossiliferous and the lower ones lacked fossils; he suggested that the former were created in the biblical flood (after the creation of life) and the latter before life existed – the first geological explanation distinguishing different periods of Earth history. He summarised his ideas in what was potentially his greatest geological work, *De solido intra solidum naturaliter contento dissertationis prodromus* (Figure 3.1A) of 1669. This was intended as an introduction to a much larger work that was never completed, due to his conversion to Roman Catholicism in 1667. Ordained as a priest in 1675, he became a bishop in 1677. However, his *prodromus* was widely circulated and translated into English. His published ideas earned him the title of 'Father of Stratigraphy'.

Italian mercenaries, scholars and travellers of considerable personal means or political and religious connections made their way northwards and eastwards across Europe during the 17th and 18th centuries; they frequently recorded what they saw, collecting specimens and often publishing their observations and interpretations. Their collections and texts form the core of some significant Italian museums and libraries. The most noteworthy of such travellers was the mercenary, naturalist and hydrographer Count Luigi Ferdinando Marsigli (1658–1730) (Stoye 1994). Today, in his native Bologna, he is celebrated as a general and military historian rather than as a scientist. Marsigli's work for the Austro-Hungarian Empire took him into what are now Albania, Austria, Bosnia, Bulgaria, Croatia, the Czech Republic, Hungary and Poland, where he recorded and published observations on topography, rivers, lakes and wildlife. He especially recorded and mapped the undersea topography and currents of the Mediterranean and the Bosphorus, the former noted in his *Histoire Physique de la Mer* of 1725. His seminal geomorphological work was on mapping and describing the Danube basin, finally published in his multi-volume *Danubius Pannonico-Mysicus* of 1726 (Figure 3.1C). He was elected Fellow of the Royal Society, as a foreign member, in 1691. Apart from his own publications, his significant geoheritage legacies are the art and science institutions he either founded (especially the Istituto delle Scienze in Bologna), and whose influence continues in some measure, or which have incorporated his natural history, art and book collections (especially the Museo di Palazzo Poggi in Bologna).

EARLY SCIENTIFIC GEOLOGY IN NORTHERN EUROPE: A PUBLISHED LEGACY

Published geological inquiries in 16th century Europe included those by the German physician Georgius Agricola (1494–1555) in *De natura fossilium* (1546), the first book specifically on minerals. The Swiss physician and botanist Conrad Gessner's (1516–65) *De rerum fossilium, lapidum et gemmarum maxime, figuris et similitudinibus liber* of 1565 was the first illustrated geology book. George Owen's (1552–1613) *Pamphlett conteiginge the description of Mylford Haven* (see Owen 1892) and *The course of the strata of coal and lime in Pembrokeshire* of 1595 (Owen 1595) were the first regional geological accounts for Wales. He also made one of the first accurate detailed recordings (Owen 1599) of the character and distribution of glacial till (John 1964). The major scientific journals began to publish geological papers from the middle of the 18th century; for example, the *Philosophical Transactions of the Royal Society of London* of 1758 carried 'An account of the impression of plants on the slates of coals' (Costa 1758) and, in 1791, 'Observations on the affinity between basalts and granite' (Beddoes and Banks 1791). There was a significant number of publications with geological content on England and Wales in the 17th

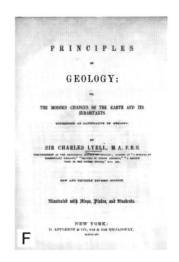

FIGURE 3.1 THE TITLE PAGES OF SOME 17TH TO 19TH CENTURY INFLUENTIAL GEOLOGICAL PUBLICATIONS

A. *DE SOLIDO INTRA SOLIDUM NATURALITER CONTENTO DISSERTATIONIS PRODROMUS* (STENO 1669).

B. 'BRIEF DIRECTIONS FOR MAKING OBSERVATIONS AND COLLECTIONS AND FOR COMPOSING A TRAVELLING REGISTER OF ALL SORTS OF FOSSILS' (WOODWARD 1728).

C. *DANUBIUS PANNONICO-MYSICUS* (MARSIGLI 1726).

D. *KURZE KLASSIFIKATION UND BESCHREIBUNG DER VERSCHIEDENEN GEBIRGSARTEN* (WERNER 1787).

E. *THEORY OF THE EARTH* (HUTTON 1795).

F. *PRINCIPLES OF GEOLOGY* (LYELL 1830–33).

NOTE THAT THE VOLUMES HAVE NOT BEEN REPRODUCED TO THE SAME SCALE.

and 18th centuries (Bassett 1963; Challinor 1971). *The Natural History of Lancashire, Cheshire, and the Peak in Derbyshire* by Charles Leigh (1662–1701) in 1700 is the first regional geology account for England. Meanwhile, observations similar to those of Nicholas Steno (1638–86) on structural geology were being published in England.

For example, William Hobbs, in *The Earth Generated and Anatomized* of 1715 (Porter 1981), suggested that the biblical 'days' of Genesis were much longer than those of the 18th century (see Roberts 2007) and noted that tilted layers of rock were originally deposited horizontally. Concomitantly, John Strachey (1671–1743) recognised the unconformable nature of some geological boundaries but left the full explanation to James Hutton (1726–97). Martin Lister (1639–1712), an English doctor, had suggested that if fossils were organic, which he had disputed in 1671, their originals were now extinct; he was one of the first to recognise that not every living creature had survived from the presumed biblical beginning. He was the first to systematically study and describe British spiders and their habits. However, it was his interest in shells that led him to look at fossil forms, especially ammonites.

Many topographic and travel accounts, contemporaneous with Lister's scientific publications, included minor geological references; for example, the Dutch artist William Schellinks visited England from 1661 to 1663 and wrote a journal (Exwood and Lehmann 1993) with some observations on Sussex's geology. Robert Hooke (1635–1703), Curator of Experiments to the Royal Society, presented from 1680 *Discourses on Earthquakes*, a series of lectures on the origins of earthquakes and mountains; these were eventually published in 1705 (Hooke 1705). Hooke recognised the organic origin of fossils and thought they might be useful in developing a chronology – an early suggestion for biostratigraphical studies. Hooke's exact contemporary was Robert Plot (1640–96), the first keeper of Oxford University's Ashmolean Museum. His 1677 *Natural History of Oxford-shire* (Plot 1677) illustrated some 300 'fossils', including the first dinosaur bone to be described in Europe, found in a quarry at Cornwell, Oxfordshire; it was sent to Plot, who discarded the idea that it was a thigh bone from a Roman elephant and concluded that it had come from a prehistoric giant. More than 100 years later, it was refigured and described as 'Scrotum humanum' in 1763 by Richard Brookes (fl. 1721–63) in *The Natural History of Waters, Earths, Stones, Fossils and Minerals, With Their Virtues, Properties and Medicinal Uses; To Which Is Added, the Method in Which Linnaeus Has Treated These Subjects* (Brookes 1763), which was the fifth volume of his six-volume *A System of Natural History*. The French philosopher Jean-Baptiste Robinet (1735–1820) described it in 1768 as a 'stony scrotum'! Fifty years later, the first-named dinosaur genus, 'Megalosaurus', was recognised (Buckland 1824; Eudes-Deslongchamps 1826; Meyer 1826; Parkinson 1822); the now lost bone was clearly assignable to that genus.

LATER SCIENTIFIC GEOLOGY: A PUBLISHED LEGACY

From the 18th century there was considerable European speculation on the origin and longevity of mountains; much was inspired by work in the extinct volcanic regions of France and Germany, especially the Auvergne and Eiffel districts. For example, the naturalist and mineralogist Jean-Etienne Guettard (1715–86) penned a short communication, *Mémoire sur quelques montagnes de la France qui ont été des volcans* (Guettard 1752), to the Académie des Sciences. At the close of the 19th century, studies on the extinct Tertiary volcanoes of Scotland led to an improved understanding of how molten rock makes its way to the surface and why volcanoes collapse. This work was somewhat popularised in the geological literature by Sir Archibald Geikie's two-volume

The Ancient Volcanoes of Great Britain of 1897. The field recognition and description that many volcanoes typically collapse in concentric rings around their cones – so-called caldera subsidence – was pioneered in Glen Coe (Clough, Maufe and Bailey 1909; Kokelaar and Moore 2006).

Mining was well researched after the installation of the world's first 'professor of mining', Abraham Gottlob von Werner (1749–1817), at Freiberg in 1775. His systematic and comprehensive study of rocks and minerals, *Kurze Klassifikation und Beschreibung der verschiedenen Gebirgsarten* (Werner 1787) (Figure 3.1D), was built upon his earlier work; as a student he had published a revolutionary treatment of mineral descriptions in *Von den äusserlichen Kennzeichen der Fossilien* (Werner 1774). Werner was a little travelled but most able teacher, and he incorporated many of the terms used by local miners and quarrymen into his few publications and many lectures. His students promoted his 'neptunism' theory (that all rocks were created in water, neatly fitting the biblical deluge), based on his field observations at the Scheibenberg. He introduced the first formal geology field excursions.

Werner's near contemporary was the Scot James Hutton (1726–97), whose *Theory of the Earth* (Figure 3.1E) of 1795 propounded (with reference to Arthur's Seat in Edinburgh, which has a similar rock succession to that at the Scheibenberg), the 'vulcanism' or 'plutonism' theory that all rocks were volcanic in origin. Hutton had previously published a paper on his theory in the *Transactions of the Royal Society of Edinburgh* in 1788, and had also privately printed a version of the paper's abstract, although this was actually read in 1785. Hutton's work has led some authors to argue that geology as a discrete discipline was a British creation and that Hutton's book 'dates the foundation of geology as a modern science' (Doughty 1981, 2). The conflicting theories of Werner and Hutton dominated much of the early development of scientific geology in Britain and Europe (Dott 1969; Greene 1982; Laudan 1987). Geology was the science 'in which almost anyone could hope to make exciting discoveries … geology appealed to the tastes of the middle classes, as much as it suited their resources, for unlike astronomy its active pursuit required little more than a reasonable degree of leisure, the means for at least a little travelling, and a taste for the open air and the countryside' (Rudwick 1976, 166). It opened up new markets for the dissemination of geological thought by populist and student publications.

The initial motivation for much 18th and 19th century natural history study (especially for the bulk of the material it generated that survives in today's museums and libraries), including geology, was to understand and show the handiwork of the essentially Christian creator (Allen 1976; Barber 1980; Gillispie 1954; Taylor 1997). Indeed, 'A fascination with natural history was particularly common amongst practising Christians' (Taylor 1997, 92). As well as furthering Anglican beliefs and moral rectitude, it had as much to do with self-improvement as with pure scientific inquiry. This interest arose just as lending and reference libraries, particularly those in mechanics' institutes and local learned societies, were founded. Further, it coincided with, and was promoted by, the emergence of local field clubs and natural history and philosophical societies, many of which published proceedings and transactions; for example, the Woolhope Naturalists' Field Club (Burek and Hose 2016) in 1852 and the Chester Society of Natural Science, Literature and Art (Burek and Hose 2016; Robinson 1971; Siddall 1911) in 1871. Local geological societies were also formed in the 19th century. The first, the Royal Geological Society of Cornwall, was established in 1814 and the second, at Newcastle-upon-Tyne, in 1829 (although it followed on from an earlier philosophical society established in 1793). A Dudley and Midland Geological Society was formed in 1842, but this seems to have survived only to 1843; however, its place was belatedly taken by the Dudley and Midland Geological and Scientific Society and

Field Club in 1862, which just survived into the 20th century. Three major, local geological societies – in terms of their publications – have survived from the 19th into the 21st century: the Edinburgh Geological Society, established in 1834; the Yorkshire Geological Society, established in 1837; and the Liverpool Geological Society, established in 1859. Papers by important geologists of the day were published in their journals.

Apart from through scientific publication, eminent geologists were also influential across society through their widely read textbooks; arguably the most influential was Charles Lyell's (1797–1875) multi-volume *Principles of Geology* (Lyell 1830–33) (Figure 3.1F), published between 1830 and 1833. Greenough's *A Critical Examination of the First Principles of Geology* (Greenough 1819b) was published somewhat earlier and was particularly useful at the time in refuting erroneous geological theories. Slightly earlier, a number of fossil guides had appeared as descriptive catalogues, such as the English surgeon James Parkinson's (1755–1824) *Outlines of Oryctology: An Introduction to the Study of Fossil Organic Remains; Especially of those found in the British Strata: Intended to aid the Student in his Inquiries respecting the Nature of Fossils and their Connection with the Formation of the Earth* of 1830; its introduction noted that it was 'dedicated to the service of those Admirers of Fossils who have not yet entered into a strict examination of the distinctive characters of these interesting substances' (Parkinson 1830, v). His earlier *Organic Remains of a Former World* (1804–11) was the first general treatise on fossils.

The first attempt at a taxonomic account of fossils was John Woodward's two-volume *An Attempt Towards a Natural Hiftory of the Fossils of England* of 1728 and 1729. Woodward (1665–1728) is considered the first major figure in English geology. Commonly for the time, his fossils broadly included objects other than those of purely organic origin. His 20-page pamphlet, *Brief Instructions for Making Observations in all Parts of the World and also for Collecting, Preserving and Sending over Natural Things* of 1696, is probably the first set of published instructions in English on the collection and preservation of natural history specimens, especially for travellers overseas. He also published 'Brief Directions for making Obfervations and Collections and for compofing a travelling Regifter of all Sorts of Fossils' (Figure 3.1B) in 1728, as pages 93 to 95 of *Fossils of all Kinds: Digested into a Method, Suitable to their mutual Relation and Affinity*. However, it is probable that these notes were issued in manuscript form around 1700 to 1705, to various other collectors employed by Woodward to add material to his collection (Eyles 1971b); they are the first published instructions on the collection and curation of geological specimens.

Woodward had earlier considered minerals in his *An Essay toward a Natural History of the Earth and Terrestrial Bodies, especially Minerals, &c* of 1695 (with further editions in 1702 and 1723). In this he explained that the Earth's stony surface was divided into strata and that their enclosed fossils were originally generated in the sea, or rather they were a result of the great biblical flood. He attempted a classification of minerals, under the heading of 'Fossils', in the Reverend John Harris' (1666–1719) *Lexicon chemicum* of 1704; this was reprinted with little change in Woodward's own *Naturalis historia telluris* of 1714. Woodward's *Fossils of all Kinds: Digested into a Method* of 1728 has a large fold-out systematic mineral classification table. The 56-page text describes some 200 minerals, some with their mode of occurrence. He had examined some of the minerals under the microscope and noted their characteristic features. The specimens upon which the two-volume work was based were housed within the 'Woodwardian Cabinets', as noted in the 'Publisher to the Reader' notes that precede the work's preface; the cabinets are preserved within the Sedgwick Museum, University of Cambridge.

William Martin's (1767–1810) *Outline of an Attempt to Establish a Knowledge of Extraneous*

Fossils on Scientific Principles (1809a) is the first modern introductory palaeontological textbook. It noted that the study of fossils is:

> useful to the geologist – it enables him to distinguish the relative ages of the various strata, which compose the surface of our globe; and to explain, in some degree, the processes of nature, in the formation of the mineral world … And the causes, that have operated to produce the distinction existing between plants and animals of the present day, and those of former unknown ages.
>
> (Martin 1809a, i)

It developed his *Petrificata Derbiensia; or figures and descriptions of petrifactions collected in Derbyshire* (also of 1809) into a more general work. Martin personally prepared the plates, almost certainly using specimens from his own collection. He promoted his text, suggesting that, 'when Natural History in general is cultivated with so much ardour, and introductory helps to its scientific attainment are daily increasing, I esteem it somewhat strange, that an elementary treatise on the subject of extraneous fossils, should hitherto be wanting' (Martin 1809b, i–ii).

The first British text specifically to illustrate fossil plants was Edward Lhuyd's (1660–1709) *Lithophylacci Britannici Iconographia* of 1699 (with a second edition in 1670). It was a catalogue of some English fossils in the collection of the Ashmolean Museum, prepared whilst he was its keeper from 1690 to 1709. John Lindley's (1799–1865) and William Hutton's (1797–1860) *The Fossil Flora of Great Britain; or Figures and Descriptions of the Vegetable Remains found in a Fossil State in this Country* (Lindley and Hutton 1831–37) followed Edmund Tyrell Artis's (1789–1847) *Antediluvian Phytology, Illustrated by a Collection of the Fossil Plants, peculiar to the Coal Formations of Great Britain* (Artis 1825); the latter was the first modern English book on palaeobotany. It appeared more than a century after the Swiss Johann Jakob Scheuchzer (1672–1733) had published *Herbarium Diluvianum*, the first palaeobotany text, in 1709. Plates from both of the British ground-breaking books were brought together into one volume by Gideon Algernon Mantell's (1790–1852) *A Pictorial Atlas of Fossil Remains, consisting of Coloured Illustrations selected from Parkinson's 'Organic Remains of a Former World' and Artis's 'Antediluvian Phytology', with Descriptions* of 1850. Around the same time as the major English palaeobotanical works were being published, the Frenchman Adolphe-Theodore Brongniart (1801–76), in *Histoire des végétaux fossiles, ou Recherches botaniques et géologiques sur les végétaux renfermés dans les divers couches du globe* (1828–38), and the German (although born in Prague) Count Kašpar Maria Sternberg (1761–1838), in *Versuch einer geogotisch-botanischen Darstellung der Flora der Vorwelt* (Sternberg 1820–38), were writing about and illustrating Europe's fossil flora. Georg Knorr (1705–73) and Johann Walch (1725–78) co-authored an early German general palaeontology text, *Die Naturgeschichte der Versteinerungen zur Erläuting der Knorrischen Sammlung von Merkwurdigkeiten der natur* (1755–73). Meanwhile, it was the Englishman Henry Thornton Mair Witham (1779–1844) who devised a method of examining fossil plants in thin section; his 48-page (with six plates) *Observations on Fossil Vegetables, accompanied by Representations of their Internal Structures as seen through the Microscope* (Witham 1831) revolutionised their future study.

THE EMERGENCE OF A BRITISH FOSSIL GEOLOGY

Whilst the last quarter of the 18th century saw geology recognised in Europe as a science, and the first widespread publication of geological information, British geology only came into prominence in the later years of King George IV's (b.1762, r.1820–30) and the early years of Queen Victoria's (b.1819, r.1837–1901) reigns. This was a consequence of the declining importance of the major continental centres of geological research, because of the political upheavals from rising nationalism and intra-national and civil wars. The decline of the Museum of Paris, so ably led and promoted by Baron Georges Cuvier (1769–1832), who had benefited from Napoleon's patronage, reflected the politics of the time, especially England's emergence as a major European nation following victory in the Napoleonic wars and overseas trade expansion. Cuvier was instrumental in establishing comparative anatomy, his publications comparing living animals with fossils. He recognised extinction in the fossil record, but rejected the early pre-Darwinian evolutionary theories of Jean-Baptiste de Lamarck (1744–1829) and Geoffroy Saint-Hilaire (1772–1844). Instead, he propounded 'catastrophism', in which major events killed off living things, to explain the sudden disappearance of fossils; it has found some present-day re-acceptance with regard to dinosaurs. He argued that he had found no evidence that one fossil form gradually changed into another. When Cuvier's paper on living and fossil elephants was published in 1796, it was the first scientific account to recognise extinction; until then, it had been assumed that such animals had simply migrated into more suitable areas as the climate changed.

Catastrophism was rejected in England by Charles Lyell (1797–1885), who propounded 'uniformitarianism' in the 1830s. His *Principles of Geology* was subtitled *An attempt to explain the former changes of the Earth's surface by reference to causes now in operation*. Uniformitarianism was built upon James Hutton's concept that the Earth was shaped entirely by slow-moving forces, still acting in the present day, but over considerable time; this suggested that the 'present is the key to the past'. England differed from the major European countries involved in geology in that the prime support came from private means, whereas in Europe study was supported by official government sources. Rudwick notes that:

> The contrast between the Geological Society in London, and the Paris Museum … reflects accurately the contrast in the position of the science in the two countries. Whereas the Museum and the Institute had been set up and liberally financed by the state … the Geological Society … was a purely private body almost entirely dependent on its Fellows' subscriptions.
>
> (Rudwick 1976, 165–6)

Consequently, geology and natural history were little evident in the curricula of schools and the English universities. It was the non-conformist educational institutions, as in Scotland, that initially most influenced the course of British scientific geology. There were limited attempts in England at public lectures (Siegfried and Dott 1980) for general, not necessarily gentlemanly, audiences. The significance of private and individual endeavour is exemplified by one of the ground-breaking geological discoveries made in the first quarter of the 19th century. Gideon Algernon Mantell (1790–1852), a Sussex-based country doctor with access to influential Fellows of the Geological Society, the Society of Antiquaries and the Royal Society, was the first to recognise the existence of large ancient reptiles. In 1821, whilst he was attending a patient near Cuckfield, his wife, Mary Ann, was walking in nearby Tilgate Forest and found some large teeth in a road-stone quarry's rock-pile. She noticed that they were different from anything in her husband's collection and had the wit to collect them; they are now displayed in The Natural History Museum, London. Mantell figured and described them within a year (Mantell 1822) in a general account of the local geology. Mantell arranged for Charles Lyell to convey one of the teeth to Cuvier, who suggested that it was from a rhinoceros – although one French author (Michard 1992) has suggested that he had recognised it as reptilian. Mantell was undeterred and, on a visit to the Royal College of Surgeons, he showed one of the teeth to Samuel Stutchbury (1798–1859), an eminent anatomist, who immediately identified it as from an iguana-like lizard. Stutchbury, who was curator of the Bristol Institution's museum from 1831 to 1850, later identified the dinosaur 'Thecodontosaurus' in 1836.

At Conybeare's suggestion, Mantell named the newly discovered reptile 'iguana tooth', or '*Iguanodon*', and presented a paper (Mantell 1825) to the Royal Society on the giant herbivore. He elaborated on this by creating the first populist geological term, the 'Age of Reptiles' (Mantell 1831). Within 20 years, great prehistoric terrestrial and marine beasts were established in the public's minds, due to Gideon Mantell, Mary Anning (1799–1847), who found and prepared ichthyosaurs and other reptiles at Lyme Regis, and Dean William Buckland (1784–1856), who had finally named '*Megalosaurus*' long after its first illustration by Plot. The names '*Iguanodon*', '*Megalosaurus*', '*Ichthyosaurus*' and '*Plesiosaurus*' entered the general vocabulary, and the animals were illustrated by Henry Thomas De la Beche (1796–1855) in the coloured *Duria antiquior* (1830), and by others (Hawkins 1854; Parley 1837; Rudwick 1992). Georg August Goldfuss (1772–1848), contemporaneously with De la Beche, included the line-drawn illustration 'Jura Formation' (in 1831) within his three-volume *Petrefacta Germaniae*, published between 1826 and 1844. An earlier attempt to accurately depict the geological past, despite the shortcoming of its biblical basis, was *Physica sacra* (1731–35) by the Swiss physician Johann Scheuchzer. The world's first life-sized three-dimensional reconstructions, by Benjamin Waterhouse Hawkins (1807–94), of the creatures in *Duria antiquior* and similar illustrations were exhibited at the Crystal Palace (Doyle 1993, 1995; Doyle and Robinson 1993; Hawkins 1854; McCarthy and Gilbert 1994; Owen 1854).

Due to Anning's collection and careful preparation of fossils from the Dorset coast, the first large marine and flying reptiles were scientifically described (Conybeare and De la Beche 1821). Most major European and American geology collections have at least one of her specimens; the finest can still be seen in the Natural History Museum, London. Some of her material and supporting documentation are in the Sedgwick Museum, Cambridge. From the 18th century there developed a tradition of fossil collecting, by which the 'landed gentry filled their cabinets with curiosities, and country people were continually on the lookout for objects they could sell to visitors' (Halstead 1982, 66), and this was particularly true of Lyme Regis. Mantell published extensively on its general geology (for example, Mantell 1838; 1844) and produced an illustrated field-guide to the area and the Isle of Wight (Mantell 1847).

NAMING THE AGES OF THE EARTH

More than half of the geological systems recognised today were named from British localities. The Carboniferous was established as the world's first formally recognised geological system in 1822 (Conybeare and Phillips 1822, 320). Its formal recognition was preceded by the naming by two Englishmen of its chief components, the 'Millstone Grit' in 1778 (Whitehurst 1778) and the

'Coal Measures' in 1811 (Farey 1811). William Daniel Conybeare and William Phillips coined the term 'Carboniferous Limestone' in 1822. Farey's work is a classic of British descriptive geology (Challinor 1947) because it covers the principles of structural geology, as well as probably the first account of Charnwood Forest's geology; its best feature is a fairly accurate 6.25 inches-to-the-mile (1:10,138) map of Derbyshire's geology. John Farey (1726–86) was also the first geologist to describe one of the examples of an anticline well known to British geology students (Himus and Sweeting 1972) at Ashover in Derbyshire.

The divisions established by Conybeare and Phillips, with some updating, have survived to the present day. In Wales, two researchers were outstanding in their fieldwork and publications; Roderick Impey Murchison (1792–1871) and Adam Sedgwick (1785–1873); the latter held Cambridge's Woodwardian Chair of Geology from 1818 until his death. Both had considerable financial means, affording them the wherewithal for extensive fieldwork. At that time, fieldwork was hampered by slow and unreliable transport, with the fastest travel by post-chaise averaging around seven miles per hour on turnpike roads, and by the limited availability of reliable maps. The first railways, which at least quadrupled the speed but not the safety of travel, were not built until the 1840s. Equally, the lack of accurate topographic maps was an issue. Although the origins of the Ordnance Survey (Hewitt 2010) can be traced to 1747, the first of its one-inch-to-one-mile (1:63,360) maps appeared only in 1801, and that was for Kent; its 'county' maps did not appear until the 1840s. A set of 46 county maps for England and Wales, at a scale of one-inch-to-three miles (1:190,000) was published in parts by Greenwood from 1829 to 1834. Thomas Moule also published a popular set of English county maps in 1830. A further fieldwork difficulty was that accommodation away from the taverns of towns and the inns on the main highways was uncommon.

The recognition of the geological periods known as the Cambrian and Silurian began when Murchison and Sedgwick set out in 1831 to record and classify the older rocks of north and central Wales and the Welsh Borders. They jointly presented their proposals for the Silurian and Cambrian at the British Association's August 1835 meeting. The Silurian and its divisions were first recognised in the Welsh Borders in the 1820s and 1830s by Murchison (Secord 1986). Torrens (1990) indicates that Murchison was reluctant, as was then common, to acknowledge the earlier research and assistance of others. The Silurian, named after an ancient tribe, the Silures, that inhabited the Welsh Borders before the Romans, was first published in 1835 (Murchison 1835). Its details were published in a full account four years later (Murchison 1839) and as a more popular, expanded text 15 years later still (Murchison 1854). The Cambrian system and its divisions were first recognised in central Wales in 1835 by Sedgwick (Sedgwick and Murchison 1836). Its name derives from the ancient name for Wales. Sedgwick was slow to publish the description, although a letter to the Geological Society of France describing his proposal was published in 1836 (Sedgwick 1836). Both geologists laid claim to the same rocks for their own system; the transition zone between them was acrimoniously disputed (Secord 1986) and was only resolved after their deaths by Charles Lapworth. Working in southern Scotland, he established the Ordovician system (Lapworth 1879), only formally accepted by the Stratigraphic Commission of the International Geological Congress in 1960. It was also named after an ancient Welsh tribe, the Ordovice. The global section for its base is Lapworth's Dob's Linn site in the Scottish Borders; perhaps it is fitting that the Scottish Borders provided the solution for a problem identified in the Welsh Borders by two Englishmen!

The younger rocks of Devon and Cornwall were also examined by Sedgwick and Murchison.

In 1839 they jointly founded the Devonian, named after the county (Sedgwick and Murchison 1839a; 1839b), to replace the older term, 'Old Red Sandstone'. The Devonian rocks are amongst the most diverse in Britain, ranging from clearly marine to desert and volcanic rocks. De la Beche (Hallam 1989) organised the first survey and memoir publication (De la Beche 1839) of Devon's rocks, crucial to Sedgwick and Murchison's final published classification. Murchison, who followed De la Beche as Director of the Geological Survey, published *The Geology of Russia and the Ural Mountains* in 1845, after an official invitation from the Tsar to visit and map the rocks of Russia. He established the Permian system, from the Perm region of Russia.

The Geological Survey was not the world's first such venture, but it is the oldest such national body. It originated in 1814 when John MacCullough was appointed Geologist to the Ordnance Survey, but was not officially constituted until 1835. Until his death in 1835 he was publishing papers (MacCullough 1817; 1819) on Scottish geology, and a geological map of Scotland (at four inches to one mile, 1:253,440) in 1836; a fine copy (sadly not an original) of the map hangs on the wall opposite and above the Smith and Greenough maps on the staircase of the Geological Society's apartments. MacCullough tragically died, following a carriage accident, whilst on his honeymoon in Cornwall. This tragedy was further compounded by the criticisms of his geological work by his contemporaries and later commentators (Bowden 2009). During the 1820s, attempts were made in Ireland to combine topographic and geographic mapping, but were abandoned in 1828 as too expensive; the latter restarted in 1830 and a Belfast geological office, with a museum, was opened in 1837.

Meanwhile, in England, Henry Thomas De la Beche (1796–1855) had been appointed in 1832 to map the geology of Devonshire. He was a prolific author of lengthy texts on stratigraphy and geological methods, such as *A Geological Manual* (1831), *How to Observe* (1835) and *The Geological Observer* (1851). He was well travelled, and this was reflected in his European-focused *Sections and Views, Illustrative of Geological Phenomena* (1830) with its many plates and descriptions of alpine and exotic locations (such as Jamaica and the Alps) – the geological equivalent of the 'grand tour' guidebook. In 1835 he became the Geological Survey's first officer. The Survey was granted parliamentary approval in 1845. It was established as the Ordnance Geological Survey and remained a branch of the Ordnance Survey until 1965. Its Museum of Practical Geology (subsumed in the late 20th century within the Natural History Museum in London) dates from 1841, although it had three earlier sites and various institutional titles. Many individuals involved in the Survey were also instrumental in developing the Geological Society of London. Dean William Buckland, Henry Thomas De la Beche, Charles Lapworth, Charles Lyell, Roderick Impey Murchison and Adam Sedgwick were all presidents of the Society. It established a museum in 1808 and a library in 1809; the museum (formerly in the Upper Library) was closed in 1911 and its collections were transferred to the Geological Museum and the British Museum (Natural History), but the library houses the UK's largest collection of geological reference works. It has one of the finest collections – approximately 3,500 volumes – of antiquarian books on geology and related subjects; most were published after 1800, but there are some 18th century volumes plus a few from the 16th century. The library houses impressive early volumes annotated by the 19th century geologists who originally owned them.

SOME BRITISH REGIONAL GEOLOGY PUBLICATIONS

Gideon Algernon Mantell (1790–1852) is credited (Hose 2008) with the first illustrated and readily pocketable regional geology guide (Mantell 1833) and the first true geology field-guide (Mantell 1847); the former was for south-east England and the latter for the Isle of Wight and coastal Dorset. The Geologists' Association's field excursions were particularly noteworthy vehicles for promoting geology (Green 1889), as were its populist lectures and regional groups. Many local natural history societies organised field excursions and occasionally published their itineraries; significant amongst these were the Woolhope Naturalists' Field Club and the Chester Society of Natural Science, Literature and Art (Burek 2008; Burek and Hose 2016). The Association's field excursions frequently made use of railways, coincidentally expanding their networks during its formative years, until after the First World War when motor vehicles took over. Significantly, it included reports of its field excursions in its *Proceedings*. The early excursions were twice published as bound volumes (Holmes and Sherborn 1891; Monckton and Herries 1909); they are a source of past localities and field customs. Such information and the original materials, especially museum collections, underpin cultural geoheritage.

EUROPE'S CULTURAL GEOHERITAGE SUMMARISED

The early Italian publications and collections underpinned and supported the scientific geology later developed in Britain and northern Europe. However, recognition of its key geoscientists is somewhat lacking. Britain has Europe's richest assemblage of significant geology collections and geosites; widespread recognition of its principal geoscientists lags far behind that of its artists and writers. Apart from specimens in museum collections, the other great physical legacy of five centuries of European geological inquiry is its publications. Many of these are scarce and lavishly illustrated volumes. They are a valuable source of images of lost, and surviving, geological specimens and geosites. However, it needs to be appreciated that these illustrations and accounts are supplemented by traditions of still-life and landscape painting, as well as Romantic poetry, which have themselves promoted nature study and landscape tourism for over 300 years (Hose 2008). Europe's published geoheritage is a temporally and geographically diverse and extensive cultural inheritance that demands better public recognition. As Chapter 4 shows, the physical basis (in terms of geological specimens illustrated within geological publications) is housed in today's museums. These museums are one of the prime vehicles, as they have been for several hundred years, for promoting cultural geoheritage to the public.

REFERENCES

Adams, F D, 1938 *The Birth and Development of the Geological Sciences*, Dover, Mineola

Agricola, G, 1546 *De natura fossilium* [*On the nature of fossils*], Froben, Basel

Allen, D E, 1978 *The Naturalist in Britain: A Social History*, Pelican Books, Harmondsworth

Artis, E T, 1825 *Antediluvian Phytology, Illustrated by a Collection of the Fossil Plants, peculiar to the Coal Formations of Great Britain*, printed for the author and published by J Cumberland *et al*, London [a source file for this is at: https://openlibrary.org/books/OL15195005M/Antediluvian_phytology]

Aubrey, J, 1847 *Memoires of Natural Remarques in the County of Wilts, To which are annexed Observables of the same kind in the County of Surrey and Flintshire. By Mr John Aubrey, FRS, 1685*, B Nicols, London

Bailey, E, 1952 *Geological Survey of Great Britain*, Thomas Murby and Co, London

Bailey, E B, 1962 *Charles Lyell*, Nelson, London

— 1967 *James Hutton: The Founder of Modern Geology*, Elsevier, Amsterdam

Barber, L, 1980 *The Heyday of Natural History: 1820–1870*, Jonathan Cape, London

Bassett, D A, 1961 *Bibliography and Index of Geology and Allied Sciences for Wales and the Welsh Borders 1897–1958*, National Museum of Wales, Cardiff

— 1963 *Bibliography and Index of Geology and Allied Sciences for Wales and the Welsh Borders 1536–1896*, National Museum of Wales, Cardiff

— 1967 *A Source-Book of Geological, Geomorphological and Soil Maps for Wales and the Welsh Borders (1800–1966)*, National Museum of Wales, Cardiff

Beddoes, T, and Banks, J, 1791 Observations on the Affinity between Basaltes and Granite. By Thomas Beddoes, MD; Communicated by Sir Joseph Banks, Bart., PRS, *Philosophical Transactions of the Royal Society of London*, 1 January 1791, 81, 48–70; doi:10.1098/rstl.1791.0005

Bonney, T G, 1895 *Charles Lyell and Modern Geology*, Cassell, London

Bowden, A J, 2009 Geology at the crossroads: aspects of the geological career of Dr John MacCulloch, in *The Making of the Geological Society* (eds C L E Lewis and S J Knell), The Geological Society, London, 255–78

Brongniart, A, 1828–38 *Histoire des végétaux fossiles, ou Recherches botaniques et géologiques sur les végétaux renfermés dans les divers couches du globe*, Chez G Dufour et E d'Ocagne, Paris and Amsterdam

Brookes, R, 1763 *The Natural History of Waters, Earths, Stones, Fossils, and Minerals, With Their Virtues, Properties and Medicinal Uses; To Which Is Added, the Method in Which Linnaeus Has Treated These Subjects*, Vol V [in 6 vols], Newberry, London

Burek, C V, 2008 The role of the voluntary sector in the evolving geoconservation movement, in *The History of Geoconservation: Special Publication No 300* (eds C V Burek and C Prosser), The Geological Society, London, 61–89

Burek, C V, and Hose, T A, 2016 The role of local societies in early modern geotourism: a case study of the Chester Society of Natural Science and the Woolhope Naturalists' Field Club, in *Appreciating Physical Landscapes: Three Hundred Years of Geotourism: Special Publication No 417* (ed T A Hose), The Geological Society, London, 95–116

Challinor, J, 1947 From Whitehurst's *Inquiry* to Farey's *Derbyshire*, *Transactions of the Staffordshire Field Club* 81, 52–88

— 1971 *The History of British Geology: A Bibliographical Study*, David and Charles, Newton Abbot

Clarke, J W, and Hughes, T McK, 1895 *The Life and Letters of Adam Sedgwick*, Cambridge University Press, Cambridge

Clough, C T, Maufe, H B, and Bailey, E B, 1909 The cauldron-subsidence of Glen Coe, and the associated igneous phenomena, *Quarterly Journal of the Geological Society of London* 65, 611–78

Colonna, F, 1616 *Fabii Columnae Lyncey De Glossopetris Dissertatio*

Conybeare, W D, and De la Beche, H T, 1821 Notice of the discovery of a new fossil animal forming a link between the Ichthyosaurus and crocodile, together with general remarks on the osteology of Ichthyosaurus, *Transactions of the Geological Society of London* 5 (2), 558–94

Conybeare, W D, and Phillips, W, 1822 *Outlines of the Geology of England and Wales*, Phillips, London

Costa, E M da, 1758 An Account of the Impressions of Plants on the Slates of Coals: In a Letter to the Right

Honourable George Earl of Macclesfield, President of the RS, from Mr Emanuel Mendes da Costa, FRS, *Philosophical Transactions of the Royal Society of London*, 1757–58, 50, 228–35; doi:10.1098/rstl.1757.0029

Curwen, E C (ed), 1940 *The Journal of Gideon Mantell*, Oxford University Press, Oxford

Davies, G L H, 2007 *Whatever is under the Earth: The Geological Society of London 1807–2007*, The Geological Society, London

De la Beche, H T, 1830 *Duria antiquior* [drawing converted to a lithograph by George Scharf]

— 1839 *Report on the Geology of Cornwall, Devon and West Somerset*, Longman, Orme, Brown, Green and Longmans, London

Dean, D R, 1990 A bicentenary retrospective on Gideon Algernon Mantell (1790–1852), *Journal of Geological Education* 38 (5), 434–43

Desmond, A, 1982 *Archetypes and Ancestors: Palaeontology in Victorian London 1850–1875*, University of Chicago Press, Chicago

Dott, R H, 1969 James Hutton and the concept of a dynamic earth, in *Toward a History of Geology: Proceedings of the New Hampshire Inter-Disciplinary Conference on the History of Geology, 7–12 September 1967* (ed C J Schneer), MIT Press, Cambridge, 122–41

Doughty, P S, 1981 *The State and Status of Geology in UK Museums: Miscellaneous Paper No 13*, The Geological Society, London

Doyle, P, 1993 The lessons of Crystal Palace, *Geology Today* 9 (3), 107–9

— 1995 *Crystal Palace Park Geological Time Trail: Paxton's Heritage Trail Number Two*, London Borough of Bromley, Croydon

Doyle, P, and Robinson, J E, 1993 The Victorian geological illustrations of Crystal Palace Park, *Proceedings of the Geologists' Association* 104 (4),181–94

Durham County Council, 1994 *County Durham Geological Conservation Strategy*, Durham County Council, Durham

Edwards, W N, 1967 *The Early History of Palaeontology*, British Museum (Natural History), London

Englefield, H C, 1816 *A Description of the Principal Picturesque Beauties, Antiquities and Geological Phenomena of the Isle of Wight. With Additional Observations on the Strata of the Island, and their Continuation in the Adjacent Parts of Dorsetshire by Thomas Webster*, T Webster, London

Exwood, M, and Lehmann, H L (eds), 1993 *The Journal of William Schellinks' Travels in England 1661–1663: Camden Fifth Series*, Vol 1, Royal Historical Society, London

Eyles, J M, 1969 William Smith: Some aspects of his life and work, in *Toward a History of Geology* (ed C J Schneer), MIT Press, Cambridge, 142–58

— 1971b Roderick Murchison, geologist and promoter of science, *Nature* 234, 387–9

— 1971c John Woodward 1665–1728: A biographical account of his life and work, *Journal of the Society for the Bibliography of Natural History* 5 (6), 399–427

Farey, J, 1811 *General View of the Agriculture and Minerals of Derbyshire*, Vol 1, G and W Nicol, London

Flett, J S, 1937 *The First Hundred Years of the Geological Survey of Great Britain*, HMSO, London

Ford, T D, 1967 The first detailed geological section across England, by John Farey, 1806–1808, *Mercian Geologist* 2, 41–9

Freeman, M, 2004 *Victorians and the Prehistoric: Tracks to a Lost World*, Yale University Press, New Haven

Fuller, J C G M, 1995 *'Strata Smith' and his stratigraphic cross sections, 1819: A review of facts worth knowing about the origin of stratigraphic geology in the mind of William Smith (1769–1839), an English country surveyor and civil engineer*, American Association of Petroleum Geologists/The Geological Society, London

Geikie, A, 1875 *Life of Sir Roderick Murchison*, John Murray, London

— 1897 *The Ancient Volcanoes of Great Britain* [in 2 vols], Macmillan, London

Gessner, C, 1565 *De rerum fossilium, lapidum et gemmarum maxime, figuris et similitudinibus liber*, Jacobus Gesnerus, Zurich

Gillispie, C C, 1996 *Genesis and Geology: A Study in the Relations of Scientific Thought, Natural Theology, and Social Opinion in Great Britain, 1790–1850*, Harvard University Press, Cambridge

Goddard, T R, 1929 *History of the Natural History Society of Northumberland, Durham and Newcastle-upon-Tyne 1829–1929*, Andrew Reid and Co, Newcastle-upon-Tyne

Gohau, G, 1991 *History of Geology*, Rutgers University Press, New Brunswick

Goldfuss, G A, 1826–44 *Petrefacta Germaniae – Abbildungen und Beschreibungen der Petrefacten Deutschlands und der angrenzenden Länder* [*German fossils – pictures and descriptions of fossils from Germany and neighbouring countries*] [in 3 vols], List and Franke, Leipzig

Goodrich, S G, 1837 *Peter Parley's Wonders of Earth, Sea and Sky*, Darton and Clark, London [a source file for this is: https://www.flickr.com/photos/thomasfisherlibrary/14598867766]

Green, C P, 1989 Excursions in the past: a review of the Field Meeting Reports in the first one hundred volumes of the *Proceedings*, *Proceedings of the Geologists' Association* 100 (1), 17–29

Greene, M T, 1982 *Geology in the 19th Century*, Cornell University Press, Ithaca

Greenough, G B, 1819a *Geological Map of England and Wales [together with a memoir]*, published under the direction of the Geological Society by Longman, Hurst, Rees, Orme and Brown, London

— 1819b *A Critical Examination of the First Principles of Geology: In a Series of Essays*, Longman, Hurst, Rees, Orme and Brown, London

Guettard, J-E, 1752 Mémoire sur quelques montagnes de la France qui ont été des volcans [Submission on some mountains of France which were once volcanoes], *Mémoires de l'Académie Royale des Sciences*, 1752 [1756], 27–59

Hallam, A, 1989 *Great Geological Controversies*, 2 edn, Oxford University Press, Oxford

Halstead, L B, 1982 *Hunting the Past: Fossils, Rocks, Tracks and Trails; The Search for the Origin of Life*, Hamish Hamilton, London

Harris, J, 1704 *Lexicon chemicum*, G D William Nealand, London

Hawkins, B W, 1854 On Visual Education as Applied to Geology, Illustrated by Diagrams and Models of the Geological Restorations at the Crystal Palace, *Journal of the Society of Arts* [and republished as a leaflet in London by James Tennant, 149 Strand]

Hewitt, R, 2010 *Map of a Nation: A Biography of the Ordnance Survey*, Granta, London

Himus, G W, and Sweeting, G S, 1972 *Elements of Field Geology*, 3 edn, University Tutorial Press, London

Homes, T V, and Sherborn, C D, 1891 *A Record of Excursions Made Between 1860 and 1890*, Geologists' Association, London

Hooke, R, 1705 *Discourses on Earthquakes and Subterraneous Eruptions* [1978 reprint with notes by A C Rossiter and R Walter, Arno Press, New York]

Hose, T A, 1995 Selling the story of Britain's stone, *Environmental Interpretation* 10, 16–17

— 2008 Towards a history of geotourism: definitions, antecedents and the future, in *The History of*

Geoconservation: Special Publication No 300 (eds C V Burek and C D Prosser), The Geological Society, London, 37–60

Hutton, J, 1788 Theory of the Earth, or an Investigation of the Laws Observable in the Composition, Dissolution, and Restoration of Land upon the Globe, *Transactions of the Royal Society of Edinburgh* 1 (ii), 209–304

— 1795 *Theory of the Earth, with Proofs and Illustrations* [in 2 vols], Creech, Edinburgh

John, B S, 1964 Early Discoveries XXI: A description and explanation of glacial till in 1603 George Owen (1552–1613), *Journal of Glaciology* 105 (39) 369–70

Knell, S J, 2000 *The Culture of English Geology, 1815–1851: A Science Revealed through its Collecting*, Ashgate, Aldershot

Knorr, G, and Walch, J, 1755–73 *Die Naturgeschichte der Versteinerungen zur Erläutering der Knorrischen Sammlung von Merkwurdigkeiten der natur*, Felßecker, Nuremberg

Kokelaar, B P, and Moore, I D, 2006 *Classical Areas of British Geology: Glencoe Caldera Volcano, Scotland*, British Geological Survey, Keyworth

Lapworth, C, 1879 On the tripartite classification of the Lower Palaeozoic rocks, *Geological Magazine* 16, 1–15

Laudan, R, 1987 *From Mineralogy to Geology: The Foundations of a Science, 1650–1830*, University of Chicago Press, Chicago

Leigh, C, 1700 *The Natural History of Lancashire, Cheshire, and the Peak in Derbyshire: with an Account of the British, Phoenician, Armenian, Gr. and Rom. Antiquities in Those Parts*, George Weft/Henry Clements, Oxford

Lewis, C L E, and Knell, S J (eds), 2009 *The Making of the Geological Society*, The Geological Society, London

Lhuyd, E, 1699 *Lithophylacci Britannici Iconographia*, London

Lindley, J, and Hutton, W, 1831–37 *The Fossil Flora of Great Britain; or Figures and Descriptions of the Vegetable Remains found in a Fossil State in this Country*, J Ridgeway, London

Lyell, C, 1830–33 *Principles of Geology, Being an Attempt to Explain the Former Changes of the Earth's Surface, by Reference to Causes now in Operation* [in 3 vols], John Murray, London

Lyell, K (ed), 1881 *Life, Letters and Journals of Sir Charles Lyell, Bart.*, John Murray, London

Macnair, P, and Mort, F (eds), 1908 *History of the Geological Society of Glasgow, 1858–1908*, Geological Society of Glasgow, Glasgow

Mantell, G A, 1822 *The Fossils of the South Downs; or Illustrations of the Geology of Sussex*, Relfe, London

— 1825 On the teeth of the Iguanodon, a newly-discovered fossil herbiverous reptile, *Philosophical Transactions of the Royal Society* 115, 179–86

— 1831 The geological age of reptiles, *Edinburgh New Philosophical Journal* 11, 181–5

— 1833 *The Geology of the South-East of England*, Longman, Rees, Orme, etc, London

— 1838 *The Wonders of Geology: Or, a Familiar Exposition of Geological Phenomena*, Henry G Bohn, London

— 1844 *Medals of Creation: Or, First Lessons in Geology and in the Study of Organic Remains*, Henry G Bohn, London

— 1847 *Geological Excursions Round the Isle of Wight and Along the Adjacent Coast of Dorsetshire, Illustrative of the Most Interesting Geological Phenomena and Organic Remains*, Henry G Bohn, London

— 1850 *A Pictorial Atlas of Fossil Remains, consisting of Coloured Illustrations selected from Parkinson's 'Organic Remains of a Former World' and Artis's 'Antediluvian Phytology'*, with Descriptions, H G Bohn, London

Marsigli, L F, 1725 *Histoire Physique de la Mer*, Amsterdam

— 1726 *Danubius Pannonico-Mysicus*, Uttwerf and Changuion, Amsterdam

Martin, W, 1809a *Outline of an Attempt to Establish a Knowledge of Extraneous Fossils on Scientific Principles*, J Wilson, Macclesfield [reprinted in facsimile in 1972, The Geological Society and Paul P B Minet, London and Chicheley]

— 1809b *Petrificata Derbiensia; or Figures and Descriptions of Petrifactions Collected in Derbyshire*, White and Co/Longman, Hurst, Rees and Orme, London

McCarthy, S, and Gilbert, M, 1994 *The Crystal Palace Dinosaurs: The Story of the World's First Prehistoric Sculptures*, Crystal Palace Foundation, London

McKirdy, A R, 1987 Protective works and geological conservation, in *Planning and Engineering Geology – Geological Society, Engineering Geology Special Publication No 4 (Proceedings of the Twenty-Second Annual Conference of the Engineering Group of the Geological Society, Plymouth, September 1986)* (eds M G Culshaw, F G Bell, J C Cripps and M O'Hara), The Geological Society, London, 81–5

Monckton, H W, and Herries, R S (eds), 1909 *Geology in the Field: The Jubilee Volume of the Geologists' Association, 1858–1908*, Stanford, London

Morrell, J, 1994 Perpetual excitement: the heroic age of British geology, *Geological Curator* 5 (8), 311–17

Morris, A D, 1989 *James Parkinson: His Life and Times*, Birkhauser, Boston

Morton, J K, 2001 *Strata: How William Smith Drew the First Map of the Earth in 1801 and Inspired the Science of Geology*, Tempus, Stroud

Murchison, R I, 1835 On the Silurian System of Rocks, *Philosophical Magazine*, 3 ser, 7, 46–53

— 1839 *The Silurian System, Founded on a Series of Geological Researches in the Counties of Salop, Hereford, Radnor, Montgomery, Caermarthen, Brecon, Pembroke, Monmouth, Gloucester, Worcester, and Stafford; With Descriptions of the Coal-fields and Overlying Formations*, John Murray, London

— 1845 *The Geology of Russia in Europe and the Ural Mountains*, John Murray, London

— 1854 *Siluria*, John Murray, London

Oldroyd, D R, 1990 *The Highlands Controversy: Constructing Geological Knowledge Through Fieldwork in 19th-Century Britain*, University of Chicago Press, Chicago

— 1996 *Thinking about the Earth: A History of Ideas in Geology*, Athlone, London

Owen, G, 1595 *The course of the strata of coal and lime in Pembrokeshire, from a MS of Geo. Owen, 1595*, in R Fenton, 1811 *A Historical Tour through Pembrokeshire*, Longman, Hurst, Rees, Orme & Co, London, 54–61 (Appendix 17) [a link for the Fenton book is: https://babel.hathitrust.org/cgi/pt?id=nyp.33433000 271043;view=1up;seq=718]

— 1599 Treatise on clay marl, unpublished manuscript written in 1599, in the Vairdre Book, National Library of Wales

— 1892 *The Description of Pembrokeshire, by George Owen of Henllys, Lord of Kemes* (ed H Dwell), Charles J Clark, London (Cymmrodorion Record Series, No 1)

Owen, R, 1854 *Crystal Palace Guidebooks: Geology and Inhabitants of the Ancient World*, Crystal Palace Library/Bradbury and Evans, London

Page, K N, 1998 England's Earth Heritage resource – an asset for everyone, in *Coastal Defence and Earth Science Conservation* (ed J Hooke), The Geological Society, London, 196–209

Parkinson, J, 1822 *Outlines of Oryctology: An Introduction to the Study of Fossil Organic Remains; Especially of those found in the British Strata*, M A Nattali, London

— 1830 *Outlines of Oryctology: An Introduction to the Study of Fossil Organic Remains; Especially of those found in the British Strata: Intended to aid the Student in his Inquiries respecting the Nature of Fossils and their Connection with the Formation of the Earth* (2 edn), T Comb and Son, Leicester [reprinted in facsimile in 1972, The Geological Society and Paul P B Minet, London and Chicheley]

Phillips, J, 1844 *Memoirs of William Smith*, Murray, London

Pliny the Elder, AD 77–79 *Naturalis Historia* [*The Natural History*] [37 books in 10 vols], English translation (1855) by J Bostock and H T Riley, Taylor and Francis, London

Plot, R, 1677 *The Natural History of Oxford-shire, Being an Essay toward the Natural History of England*, Printed at the Theater, Oxford

Porter, R, 1977 *The Making of Geology: Earth Science in Britain 1660–1815*, Cambridge University Press, Cambridge

— (ed), 1981 *The Earth Generated and Anatomized by William Hobbs: An Early 18th Century Theory of the Earth*, Cornell University Press and British Museum (Natural History), Ithaca and London

Roberts, M B, 2007 Genesis Chapter 1 and geological time from Hugo Grotius and Marin Mersenne to William Conybeare and Thomas Chalmers (1620–1825) in *Myth and Geology: Special Publication No 273* (eds L Piccardi and W B Masse), The Geological Society, London, 39–49, doi:10.1144/GSL.SP.2007.273.01.04

Robinson, H, 1971 *Chester Society of Natural Science, Literature and Art – The First Hundred Years 1871–1971*, Chester

Rudwick, M J S, 1963 The foundation of the Geological Society of London: its scheme for co-operative research in its struggle for independence, *British Journal of the History of Science* 1, 325–55

— 1976 *The Meaning of Fossils: Episodes in the History of Palaeontology* (2 edn), University of Chicago Press, Chicago

— 1985 *The Great Devonian Controversy: The Shaping of Scientific Knowledge Among Gentlemanly Specialists*, University of Chicago Press, Chicago

— 1992 *Scenes from Deep Time: Early Pictorial Representations of the Prehistoric World*, University of Chicago Press, Chicago

Sarjeant, W A S, 1980 *Geologists and the History of Geology: An International Bibliography from the Origins to 1978* [5 vols], Macmillan, London

Sarjeant, W A S, and Harvey, A P, 1979 Uriconian and Longmyndian: A history of the study of the Precambrian rocks of the Welsh Borderland, in *History of Concepts in Precambrian Geology: Geological Association of Canada Special Paper No 19* (eds W O Kupsch and W A S Sarjeant), 181–224

Scheuchzer, J J, 1709 *Herbarium Diluvianum* [*Herbarium of the Flood*], David Gessneri, Zurich

— 1731–35 *Physica Sacra* [in 4 vols], Christian Ulrich Wagner, Augsburg and Ulm

Schneer, C J (ed), 1969 *Toward a History of Geology: Proceedings of the New Hampshire Inter-Disciplinary Conference on the History of Geology*, 7–12 September 1967, MIT Press, Cambridge

Secord, J A, 1982 King of Siluria: Roderick Murchison and the imperial theme in 19th-century British geology, *Victorian Studies* 25, 413–42

— 1985 John W Salter: The rise and fall of a Victorian palaeontological career, in *From Linnaeus to Darwin* (eds A Wheeler and J Price), Society for the History of Natural History, London, 61–75

— 1986 *Controversy in Victorian Geology: The Cambro-Silurian Dispute*, Princeton University Press, Princeton

Sedgwick, A, and Murchison, R I, 1835 On the Silurian and Cambrian systems, exhibiting the order in which the older sedimentary strata succeed each other in England and Wales, in *Notices and Abstracts of Communications to the British Association for the Advancement of Science at the Dublin Meeting, August 1835, Report of the Fifth Meeting of the British Association for the Advancement of Science; held in Dublin in 1835*, 59–61

— 1839a Classification of the older stratified rocks of Devon and Cornwall, *Edinburgh Philosophical Magazine*, 3 ser, 14, 241–60

— 1839b On the classification of the older rocks of Devonshire and Cornwall, *Proceedings of the Geological Society of London* 3, 121–3

Sessa, F, 1687 *Geologia del Dottore*, Naples

Siddall, J D, 1911 *The Formation of the Chester Society of Natural Science, Literature and Art and an Epitome of its Subsequent History*, Chester Society of Natural Science, Literature and Art, Chester

Siegfried, R, and Dott, R H, 1980 *Humphrey Davy on Geology: The 1805 Lectures for the General Audience*, University of Wisconsin Press, Madison

Smith, W, 1815 *Delineation of the Strata of England with Part of Scotland; exhibiting the Collieries and Mines; the Marshes and Fen Lands originally overflowed by the Sea, and the Varieties of Soil according to the Variations in the Substrata, illustrated by the most descriptive Names* [memoir and map], John Cary and J Wyld, London

— 1816–19 *Strata Identified by Organized Fossils*, W Arding, London

Speakman, C, 1982 *Adam Sedgwick, Geologist and Dalesman, 1785–1783: A Biography in Twelve Themes*, Broad Oak Press, Heathfield

Steno, N, 1667 *Elementorum myologiae specimen, seu musculi descriptio geometrica. Cui accedunt canis carchariae dissectum caput, et dissectus piscis ex canum genere: ad serenissimum Ferdinandum II, magnum Etruriae ducem* [*An elementary account of muscle mechanics and a description of the dissected head of a dog-toothed shark fish: for the most serene Ferdinand II, a great leader of Etruria*], Ex typographia sub signo stellae, Florence

— 1669 *De solido intra solidum naturaliter contento dissertationis prodromus* [*A provisional treatise on solid bodies enclosed by the process of nature within a solid body*], Ex typographia sub signo stellae, Florence

Sternberg, K M, 1820–38 *Versuch einer geogotisch-botanischen Darstellung der Flora der Vorwelt* [in 8 parts], Fr Fleischer, Leipzig

Stoye, J, 1994 *Marsigli's Europe 1680–1730: The Life and Times of Luigi Ferdinando Marsigli, Soldier and Virtuoso*, Yale University Press, New Haven

Sweeting, G S, 1958 *The Geologists' Association 1858–1958*, Geologists' Association, London

Taylor, H, 1997 *A Claim on the Countryside: A History of the British Outdoor Movement*, Keele University Press, Edinburgh

Thackray, J C, 1976 The Murchison-Sedgwick controversy, *Journal of the Geological Society of London* 132 (4), 367–72

Thompson, S J, 1988 *A Chronology of Geological Thinking from Antiquity to 1899*, Scarecrow Press, Metuchen

Torrens, H S, 1990 The scientific ancestry and historiography of the Silurian System, *Journal of the Geological Society* 147 (4), 657–62

Vai, G B, 2004 Aldrovandi's Will introducing the term 'geology' in 1603, in *Four Centuries of the Word GEOLOGY: Ulisse Aldrovandi in 1603 in Bologna* (eds G B Vai and W Cavazza), Minerva Edizioni, Bologna, 65–110

— 2009 Light and shadow: the status of Italian geology around 1807, in *The Making of the Geological Society* (eds C L E Lewis and S J Knell), The Geological Society, London, 179–202

Vincent, E A, 1994 *Geology and Mineralogy at Oxford 1860–1986: History and Reminiscence*, Department of Earth Sciences, Oxford University, Oxford

Warren, E, 1690 *Geologia; or, a Discourse concerning the Earth before the Deluge*, R Chiswell, London

Watts, W W, 1939 The author of the Ordovician System: Charles Lapworth, *Proceedings of the Geologists' Society* 50 (2), 235–86

Wendt, H, 1963 *Before the Deluge*, Victor Gollancz, London

Werner, A G, 1774 *Von den äusserlichen Kennzeichen der Fossilien* [*On the external characters of fossils, including minerals*], Siegfried Lebrecht Crusius, Leipzig

— 1787 *Kurze Klassifikation und Beschreibung der verschiedenen Gebirgsarten* [*Short classification and description of the different kinds of rock*], Walther, Dresden

Wheeler, A, and Price, J (eds), 1985 *From Linnaeus to Darwin*, Society for the History of Natural History, London

Whitehurst, J, 1778 *An Inquiry into the Original State and Formation of the Earth*, J Cooper, London

Wilding, R, 1992 What about the (geological) workers?, *Geology Today* 8 (3), 99–101

Wilson, H E, 1985 *Down to Earth: One Hundred and Fifty Years of the British Geological Survey*, Scottish Academic Press, Edinburgh

Winchester, S, 2001 *The Map that Changed the World: The Tale of William Smith and the Birth of a Science*, Viking, London

Witham, H T M, 1831 *Observations on Fossil Vegetables, accompanied by Representations of their Internal Structure, as seen through the Microscope*, T Blackwood, Edinburgh

Woodward, J, 1695 *An Essay toward a Natural History of the Earth and Terreftrial Bodies, especially Minerals: and alfo of the Seas, Rivers, and Springs. With an Account of the Universal Deluge: and of the effects it had on the Earth*, A Bettesworth and W Taylor, London

— 1696 *Brief Instructions for Making Observations in all Parts of the World and also for Collecting, Preserving and Sending over Natural Things*, Richard Wilkin, London [1973 reprint with an introduction by V A Eyles, Sherborn Fund Facsimili Number 4, Society for the Bibliography of Natural History, London]

— 1714 *Naturalis historia telluris illustrata et aucta*, Richard Wilkin, London

— 1728 *Fossils of all Kinds, Digested into a Method, Suitable to their mutual Relation and Affinity; with the Names by which they were known to the Antients, and those by which they are at this Day known: and Notes conducing to the setting forth the Natural History, and the main Uses, of some of the most considerable of them. As also several Papers tending to the further Advancement of the Knowledge of Minerals, of the Ores of Metalls, and of all other Subterraneous Productions. By John Woodward, MD, late Professor of Physick at Gresham College, Fellow of the College of Physicians, and the Royal Society*, William Innys, London

— 1728–29 *An Attempt towards a Natural Hiftory of the Fossils of England; in a Catalogue of the Englifh Fossils in the Collection of J Woodward, MD. Containing a Description and Historical Account of each; with Obfervations and Experiments, made in order to difcover, as well the Origin and Nature of them, as their Medicinal, Mechanical, and other Ufes* [in 2 vols], J Osborn and T Longman, London

Woodward, H B, 1907 *The History of the Geological Society of London*, The Geological Society, London

— 1911 *History of Geology*, Watts and Co, London

Wyatt, A R, 2000 *A Turbid Tale: Geology at Aberystwyth*, privately published by the author

Zittel, K A von, 1901 *History of Geology and Palaeontology*, Scott, London

4

Museums and Geoheritage in Britain and Europe

Thomas A Hose

Europe's natural history museums have evolved over five centuries to house, research and present the material wealth of scientific endeavour. Museums with geological collections can explain the role of notable geologists; collections retain significant value, due to their association with major collectors and scientists. Geological collections also exemplify scientific and cultural progress and underpin taxonomic studies. However, these collections are not always valued, sometimes being relegated to storage and induced neglect, despite the fact that they are a very significant element of our geoheritage. This chapter explores the ways in which museums have evolved and responded to the needs of geological enquiry, and discusses some of the issues faced by geological collections in today's financially straitened times.

Early museums in Britain and Europe

Most European museums were founded as a result of post-18th century industrial wealth. However, their origins can be traced to the late Renaissance of the 16th century, when the aristocracy in middle and southern Europe created collections as an adjunct to their desire for self-improvement. It was the 16th century English philosopher Francis Bacon (1561–1626) who noted in *Gesta Grayorum* (1594), in advice on the study of philosophy, that the four essential requirements for gaining knowledge and enjoyment were a garden, a laboratory, a library and a 'cabinet'; within the last, 'whatsoever the hand of man … hath made rare in stuff, form or motion; whatsoever singularity, chance and the shuffle of things hath produced; whatsoever Nature hath wrought in things that want life and may be kept, shall be sorted and included' (Bacon 1594, 35). In Bacon's time, the nobility began to assemble collections to indicate their wealth, taste and position, and to define and order the world they saw around them. This approach offered an explanation of a complex world, based upon classifying as many objects and specimens as possible, concentrating the maximum number into the minimum space, and giving precedence to items valued for their curiosity, rarity and preciousness. The Church, nobility and universities embraced this collecting and display activity as a mark of their prestige; most of these 'proto-museums' were private, with access limited to those privileged by rank or scholastic acknowledgment. Most were located within the lavish interiors of English country houses, Italian villas and palaces, and French châteaux and palaces.

During the late Renaissance, especially in the Italian city states, scholars such as Pirro Ligorio (c.1510–83) and Bartolomeo Marliano (1488–1566) dedicated themselves to the study of antiquity in a broad sense, including geology. Ligorio was involved in seismology after the Ferrara earthquake of 1570, when he was placed in charge of a group summoned to the city to study seismological events and to conduct earthquake research – the first such European scientific effort. Ligorio, with numerous modern parallels, blamed the earthquake's extensive damage to buildings on inappropriate design and poor choice of construction materials; his treatise, *Rimedi contra terremoti per la sicurezza degli edifici* [*Remedies against earthquakes for the safety of buildings*] of 1571 included plans for an earthquake-proof building, the first such design anywhere. Both Ligorio and Marliano mapped and illustrated Rome's ancient buildings and topography. However, Renaissance scholars, rather than focusing on one narrow field, had numerous interests. The Dane, Ole Worm (1588–1655), was one such scholar; whilst preferring the study of antiquity, he assembled at his Copenhagen home a large collection of curiosities, ranging from New World native artefacts to stuffed animals and fossils. He compiled engravings of his collection, along with his speculations about their meanings and origins; these were posthumously published, edited by his son Willum as *Museum Wormianum* of 1655. Its frontispiece (Figure 4.1A) is one of the most published and best known illustrations of a 'cabinet of curiosities', the forerunner of the modern museum.

By the mid-17th century, antiquarians such as Worm embarked upon the study of antiquities as a discrete discipline to generate a theoretical framework, progressing from merely describing monuments to explaining their functions. They established the approach that scholars had only to examine the ground and excavate sites in order to flesh out the lifestyles of vanished peoples who had left no written history. Their work was commonly published in various county studies and later in the journal published by the Society of Antiquaries, established in London in 1751. They also developed the practice of touring sites. Published guides (although the 'guidebook' was only established as such in the early 19th century) began to appear in the late 18th century. In 17th-century England, Robert Plot (1640–96) and Edward Lhuyd (1660–1709), the Ashmolean Museum's first and second keepers respectively, were typical of a new breed of antiquarians who regarded the study of antiquities as part of natural history; fossils were amongst the antiquities illustrated in Plot's *Natural History of Oxford-shire* (1677) and *Natural History of Stafford-shire* (1686).

GREAT HOUSES AND CABINETS OF CURIOSITIES

The development of collections housed in the UK's great houses reflected and incorporated the gleanings of the Grand Tour. The artworks brought back as souvenirs were more than just ornaments; they were, like fine libraries, an essential symbol of status. Woburn Abbey in Bedfordshire has a fine example of such a collection and library, created by an aristocratic family. However, similar collections and libraries – if not on such a grand scale – were created by the burgeoning mercantile class. Few heritage tourists who visit these houses today are aware of their part in the development of scientific scholarly activity. Woburn Abbey in particular played a pivotal role in agricultural improvement and geological inquiry. William Smith (1769–1839), of geological map fame, undertook drainage improvement for the estate between 1801 and 1802 for John Farey (1766–1826), the Abbey's estate manager from 1792 to 1802. Farey's estate work acquainted him with Sir Joseph Banks (1743–1820), President of the Royal Society, to whom he reported enthusiastically in 1802 on Smith's geological insights. Farey established himself (using a term he coined) as a 'mineral surveyor' in 1808. He published his *General View of the Agriculture and Minerals of Derbyshire* in 1811, the first account of Derbyshire's geology. Like Smith, he hired himself out to landowners to appraise their estates for commercial mineral extraction, especially

FIGURE 4.1 SOME SIGNIFICANT GEOLOGICAL MUSEUMS
A. MUSEUM WORMIANUM, COPENHAGEN, FROM THE FRONTISPIECE OF *MUSEUM WORMIANUM* OF 1655;
NOTE THE ECLECTIC MIX OF NATURAL HISTORY AND ARCHAEOLOGICAL MATERIAL ON DISPLAY.
B. WOODWARDIAN MUSEUM, SEDGWICK MUSEUM, UNIVERSITY OF CAMBRIDGE; WOODWARD'S FOUR
ORIGINAL CABINETS ARE HOUSED WITHIN THE ENVIRONMENTALLY-CONTROLLED WOOD-PANELLED ROOM.
C. MUSEO DI PALEONTOLOGIA E GEOLOGIA 'G CAPELLINI', UNIVERSITY OF BOLOGNA; THE BUILDING
BELIES THE ANTIQUITY OF SOME OF THE SIGNIFICANT GEOLOGICAL HISTORIC COLLECTIONS IT HOUSES.
D. ROTUNDA MUSEUM, SCARBOROUGH; NOTE THE INNOVATIVE DRUM-SHAPED TOWER (OR ROTUNDA)
THAT HOUSES THE PRINCIPAL GEOLOGY GALLERY MUCH AS IT WAS WHEN FIRST INSTALLED, THAT WAS THE
ORIGINAL 1829 MUSEUM BUILDING BEFORE THE TWO ADDITIONAL GALLERY WINGS WERE ADDED IN 1860.

coal. Although a competitor, he maintained a close friendship with Smith and even rescued most of Smith's papers when his London home caught fire in 1804.

Chatsworth House in Derbyshire (the seat of the Cavendish family – the Dukes and Duchesses of Devonshire) also played a significant role in British geology by employing White Watson (1760–1835). Watson was a late 18th to early 19th century pioneer of Derbyshire and Peak District stratigraphy, best known for his published geological sections (1788 and 1811) and inlaid marble 'sections of strata' – amongst the first reasonably accurate geological sections to appear in England. He had a combined shop and 'museum' in Bakewell, from where he sold collections of minerals to wealthy travellers on Peak District tours. He assembled cabinets of minerals for collectors and was patronised by Georgiana, Duchess of Devonshire; a regular visitor to Chatsworth, he arranged and catalogued her collections and tutored her son in geology. Several of the Cavendishes of Chatsworth were keen mineral collectors. Three main mineral collections are still at Chatsworth: those of Georgiana, Duchess of Devonshire (1757–1806); her son the 6th Duke (1790–1858); and the 11th Duke (1920–2004). The most significant material is in the first two of these collections, which were restored by members of the Russell Society (founded for people interested in the study and description of the mineralogy of a geographic region) in the early 21st century. The main collection is in two mahogany traditional drawer units, purchased from the Hunterian Museum, and can be viewed on request to the curatorial staff. A small part is still displayed in Georgiana's original cases within the stable-block conference centre. A few spectacular mineral specimens are on display in various public rooms and some ornamental items, such as one of the largest Blue John vases ever made, are also on display. Blue John is a banded purple-blue or yellowish semi-precious mineral, mined only at Castleton in Derbyshire; in the 19th century it was mined for its ornamental value, and mining continues today on a small scale. Furniture with geoheritage interest is also on display, including two 19th-century tables, one with three stalactites suspended beneath its top and another with the top inlaid with various decorative marbles (see Thomas and Cooper 2008, Figures 7 and 10).

There are several collectors who provided specimens that underpinned taxonomic research and brought together interested parties who rarely published. One such 19th century collector was William Willoughby Cole (1807–86), the 3rd Earl of Enniskillen, who understood that: 'Science thrives on the interaction of the principal protagonists … In the 19th century the country house party and the post-meeting supper party with their attendant liaisons played a special and significant role' (Doughty 1996, 2). Cole invited leading geologists such as Louis Agassiz (1807–73), Sir Roderick Impey Murchison (1792–1871) and Adam Sedgwick (1785–1873) to his house parties. Cole was unusual for a gentleman of the time because, at Oxford University, he enrolled on Dean William Buckland's (1784–1856) geology classes. When the British Association met in Dublin in the summer of 1853, Cole hosted a meeting of some of the most important Irish and European geologists, including (apart from those already mentioned) Sir Richard Griffith (1784–1878), considered the 'Father of Irish Geology', John Phillips (1800–74), who eventually held professorships at three universities, and Philip de Malpas Grey Egerton (1806–81). The latter went up to Oxford in the 1820s, where he also attended lectures by Buckland and William Daniel Conybeare (1787–1857).

Whilst travelling in Switzerland, Egerton and Cole met Louis Agassiz; inspired by him, they decided to study fossil fish, and over the next 50 years they accumulated two of Europe's largest and finest private fossil fish collections. Egerton's, the basis of publications by the Geological Society (for which he was awarded its Wollaston Medal) and the Geological Survey, was housed at Oulton Park, Tarporley, Cheshire. Egerton's and Cole's collections were left to the Natural History Museum, London. Some 16,000 specimens in the two collections underpinned the still unsurpassed four-volume *Catalogue of Fossil Fishes in the British Museum (Natural History)* of 1889 to 1901 by Sir Arthur Smith Woodward (1864–1944).

UK MUSEUMS CONSIDERED

Museums, especially in the last quarter of the 20th century, were a European cultural growth phenomenon; by the 2000s mainland Europe boasted over 15,600 museums (Haan 2010). In the mid-1990s the UK already had some 2,300 museums, of which only around 80 had noteworthy geological collections and displays (Nudds 1994); most of these museums had their origins in the 19th century. From its 19th century publishing and field-collecting heyday (as noted in Chapters Three and Six respectively), geology within museums contracted in significance, exhibition space and staffing as archaeology became established (Morrell 1994). In the 20th century's last quarter, industrial heritage and social history dominated museums. Long-term shifts in museums' emphases and new communication technologies have led to the seeming redundancy, and actual loss, of earlier collections amassed for purposes different to modern-day needs, and with dissimilar intended audiences (see Arts Council England 2011; Burns Owen Partnership 2005). This was the fate of the Passmore Edwards Museum in London; its natural history (with major geological) collections were consigned to storage in an old fire station when its galleries and out-stations were closed in 1994. These changes in the fate of museums and their geology collections are discussed in detail below.

The first recorded British usage of the term 'museum' occurred in 1682, with reference to the collection bequeathed by Elias Ashmole (1617–92) to Oxford University – the basis of the Ashmolean Museum. Under Ashmole's bequest, it had to house schools of chemistry and natural history and was pedagogically arranged. Ashmole's bequest was the considerably augmented collection of the Tradescant family that he had inherited. John Tradescant the elder (c.1570–1638) and John Tradescant the younger (1608–62) were gardeners and plant collectors (Potter 2006) to Charles I (1600–49), and they amassed much natural history and anthropological material during their numerous foreign plant-collecting excursions. Eventually, the collection was split into natural and artificial sections and displayed to the public in their Lambeth house. By 1634 it was known as the 'Ark of Novelties', probably the UK's first truly public museum; amongst its exhibits were sea-shells, fossils, crystals, various stuffed animals, birds (perhaps most notably a dodo), fishes, snakes and insects. Its 1657 catalogue, *Musaeum Tradescantianum, or a Collection of Rarities preserved at South Lambeth near London by John Tradescant*, is the first such British publication.

Apart from natural history museums created by individuals, many were actually the product of the collections amassed by philosophical societies and naturalists' field clubs in the 19th century. The first of the former was founded by a group of like-minded, professional and mercantile gentlemen in Derby in 1784; its particular strength was its library, which acquired the leading scientific societies' publications and survived as a separate entity until the Society merged with Derby Museum in 1857. A major contribution to geological museum design and display was made by the Scarborough Philosophical Society, which funded the Rotunda Museum (see Figure 4.1D); the Society consulted with William Smith on its design and displays. It was the Society's president, Sir John Vanden-Bempde-Johnstone (1799–1869), who really promoted the Rotunda. He had employed Smith as his land steward, and was very much a patron of Smith and a staunch proponent of his ideas. He donated the Hackness Stone (an ideal building sandstone), from the estate where Smith was employed, used in the Rotunda's construction. Its original display of fossils illustrated Smith's ideas on stratigraphy, and the gallery featured a frieze, designed by John Phillips (Smith's nephew), of the local coastline's geology. The Rotunda, opened in 1829, is the world's first purpose-built modern geological museum.

Following the first enabling Museum Act of 1845, and the 1846 opening of the Sunderland Museum, local authorities became 19th century museum promoters. The real key to local authority museum involvement was the change in their powers consequent upon the 1835 Municipal

Corporations Act; municipal boroughs with a minimum population of 10,000 could then levy a half-penny rate for the development of a museum of art and science with a permitted admission charge, limited to one penny. The Act was amended by the 1850 Libraries and Museums Act; linkage of the two institutions was contentious in museum circles for the next 150 years, especially as libraries were a mandatory provision but museums were not.

Local authority involvement was often due to the inability of the founders of a society museum to afford its upkeep; civic pride, coupled with an awareness of the museum's cultural and educational potential, often resulted in a transfer to a local authority. Ludlow Museum (Lloyd 1983) is such an example. The town's first museum was established in 1833 by the Ludlow Natural History Society, but its collections were successively removed to new premises to accommodate their increasing size until the Assembly Rooms were completed in 1840. The museum and its parent Society flourished there until the end of the Great War, when the cost of maintaining the building proved too much for the Society's dwindling membership and finances. The collections were, unusually for the time, transferred to the (Shropshire) County Council as an adjunct of the library service, after which the Society was wound up in 1941. In 1952 some of the collections and exhibits were transferred to the Butter Cross building, whilst the rest were dispersed to other museums or disposed of (including by auction). Apart from a significant collection of fossils from Siwalik in India, virtually all the fossils (including specimens collected by Lewis and Murchison) were removed, and when the new Butter Cross Museum opened in 1955 it was something of a blank canvas from a geoheritage standpoint. In 1959 the first salaried curator, John Norton (b.1925), was appointed; he set about establishing a new natural history collection with a strong emphasis on fossils, underpinned by fieldwork and research. In 1995 the museum was returned to the Assembly Rooms, which were refurbished in 2008. Until its closure in 2015, one of its four public galleries was devoted to local geology; the John Norton gallery tells the story of Sir Roderick Impey Murchison's geological studies from 1831, which – with the especial help of the Reverend T T Lewis (1801–58) – led to the founding of the Silurian system. Noteworthy displayed items from the museum's early donations are those made by William Erskine Baker (1808–81) and John Colvin (1794–1871), Royal Engineers posted to India who collected fossils from the Siwalik area. Colvin was a key member of the original museum's founding body, the Ludlow Natural History Society. The geological collection has around 41,500 specimens, of which over 30,000 are fossils; of international and regional significance are the Siwalik vertebrates, Silurian cephalopods, Silurian and Devonian fish, Triassic Rhynchosaur (land reptile precursors of the dinosaurs), *Cheirotherium* (footprints of an ancient amphibian crocodile relative) and remarkably complete Quaternary mammoths. Most of this material is stored in a purpose-built Museum Resource Centre that replaced a former school building used as a museum store in 1999. Meanwhile, the museum is scheduled to return to one of its old homes, the Butter Cross building, in 2016.

Other examples of former society museums transferred to a local authority are the former Museum of Isle of Wight Geology (see Chapter 9) housed in Sandown Public Library, the Grosvenor Museum in Chester and the Passmore Edwards Museum in East Ham, London. The latter is named after its benefactor, John Passmore Edwards (1823–1911), and was established in 1900 to house the Essex Field Club collections. The Club had been formed in 1880 to promote the study of the natural history, geology and prehistoric archaeology in and around Essex. Having campaigned for the adoption of the Free Libraries Acts, in just 14 years Passmore Edwards was responsible for the construction of more than 70 art galleries, libraries, museum, hospitals and convalescent homes. The Grosvenor Museum (see Henry and Hose 2015) is slightly unusual, in that the 1885 building was originally supported by the donation of land and monies by the local established landowner after whom it is named, rather than the society which formed its collections; it transferred to the local authority in 1915.

This section examines museums whose histories exemplify a plurality of patronage, funding and development. They show the role of key individuals in establishing, building the reputation of, and maintaining major geological museums. The UK examples described are the Sedgwick Museum (and the included Woodwardian Museum), Cambridge; the Lapworth Museum, Birmingham; the Geological Museum, London; and the Yorkshire Museum, York. The museums in Europe featured here are the Museo di Paleontologia e Geologia 'G Capellini', Bologna; the Natural History Museum, Florence; the National Museum of Natural History, Paris; the Bergakademie, Freiberg; and Terra Mineralia, Freiberg.

The development of the museums in Cambridge and Birmingham reflects differences between the two universities and the growth of their collections over different time-frames. The Yorkshire Museum reflects the changing engagement and role of the voluntary and local authority sectors in museum development and management. The Geological Museum exemplifies the changing fortunes of 'official geology' at the national level. The development of the Bologna and Florence museums is entwined with the complex political history of Italy. The development of the Freiberg museums reflects two distinct strands of university teaching needs and public engagement through a major bequest.

The Woodwardian and Sedgwick Museums, University of Cambridge, UK

The Sedgwick Museum originated in the first quarter of the 18th century through the work and bequest of Dr John Woodward (1665–1728), professor of physic (medicine) and a collector of natural history material from around the world; he paid others to collect on his behalf and amassed 10,000 specimens over 35 years. His fully catalogued collection was housed in five especially constructed cabinets. Woodward developed a classification system for his collection which was posthumously published, as the first such museum curatorial text, as the two-volume, *An Attempt towards a Natural History of the Fossils of England* (Woodward 1728–29); it was a detailed 600-page catalogue of his collection of some 4,000 to 5,000 minerals and fossils; it provided specimen localities, often with the names of correspondents who supplied them; it included an enlarged version of his classification and was used by geologists for almost a century after its publication.

His first classification attempt appeared in 1704, entitled 'Fossils', in *Lexicon Technicum* by John Harris (1666–1719); it was reprinted with little change in his *Naturalis historia telluris* [*Natural history of the Earth*] (Woodward 1714). Later, Woodward greatly enlarged this classification in *Fossils of all Kinds Digested into a Method* (1728), with a large folding table setting out his systematic mineral classification and a text with 56 pages describing 200 minerals; some were examined under the microscope, and Woodward mentions the characteristics used in making his determinations. He also published *An Essay toward a Natural History of the Earth and Terrestrial Bodies, especially Minerals, etc.* in 1695 (with further editions in 1702 and 1723), followed by *Brief Instructions for Making Observations in all Parts of the World* (1696). He later wrote *Naturalis historia telluris illustrata et aucta* (1714). When he died, he bequeathed to Cambridge University half of his collection (in two cabinets); it later bought another two, with the last added in the 1840s. His will stipulated that his collection should be available 'to all such curious and intelligent persons as shall desire a view of for their information and instruction'; this Woodwardian Museum (Figure 4.1B), probably the world's oldest surviving intact collection of its type, is housed within the galleries of the Sedgwick Museum. Woodward's bequest left funds to establish a geology lectureship, stipulating that the lecturer should 'attend daily in the room where [the collections] are reposited [deposited] from the hour of nine of the clock in the morning until eleven, and again from the hour of two in the afternoon until four three days a week to show the said Fossils, gratis, to all such curious and intelligent persons as shall desire a view of them for their information and instruction'. The tradition of the 'Woodwardian' lecturer continues today, as the Woodwardian Professor – the oldest such post in the world.

Until Adam Sedgwick (1785–1873) was appointed Woodwardian Professor in 1818, little had been added to the geological collections. Sedgwick himself is significant in the development of scientific geology, for he proposed (based on personal fieldwork) the Devonian and Cambrian systems. He was well acquainted with some of the major collectors of his time, including Mary Anning (1799–1847), from whom he purchased several ichthyosaurs. He persuaded the university to set aside space in the Cockerill Building for a museum, opened in 1840, for his collections. By the time Sedgwick died, they had outgrown the Cockerill and a new building was suggested as a posthumous memorial. Its construction was supervised by Sedgwick's successor, Thomas McKenny Hughes (1832–1917), who negotiated a site with the university and then collected more than £95,000 by public subscription towards its cost. King Edward VII attended its 1904 ceremonial opening. Eventually it was recognised that a full-time curator was needed. In 1931, Albert George Brighton (1901–88) was appointed, and he only retired in 1968. Brighton is recognised for his major contribution to the development of the Sedgwick Museum in the 20th century, especially documentation and displays of the collection (Price 1989). The museum today houses some well documented 1.5 million rocks, minerals and fossils.

The Sedgwick retains its active role in teaching and research in the university's Department of Earth Sciences. The galleries are housed, albeit with modern supporting graphics, in traditional wooden museum display cases with under-drawers. There is a separate William Whewell Mineral Gallery, holding some of the collection of 40,000 minerals and gemstones; information about their chemical compositions and uses is provided. It is named after William Whewell (1794–1866), a former professor of mineralogy (1828–32) and of philosophy (1838–55). He was a historian of science, writing *History of the Inductive Sciences* (1837) and *The Philosophy of the Inductive Sciences, Founded upon their History* (1840). Admission to the museum is free, and it is a regular venue for Earth Science public outreach events. The Sedgwick's key geoheritage interest is its association with Woodward and Sedgwick (respectively, one of the UK's earliest 17th century and one of the last major 19th century geologists), coupled with its remarkable collections and exhibits.

The Lapworth Museum of Geology, University of Birmingham, UK

The Lapworth Museum is housed within the University of Birmingham's Aston Webb building; it is named after Charles Lapworth (1842–1920), first professor of geology at Mason College, the University's forerunner. Established in 1880, and still retaining its Edwardian interior, it is one of the UK's oldest specialist geological museums. Lapworth came to geology without any formal training, quite late in life. As a schoolmaster he took an interest in a small area of Scotland's Southern Uplands, Assynt. His approach was different from that of the professional Geological Survey geologists, who made rapid traverses of large areas, only superficially examining the exposed rocks. He focused on the detail in a small area. Lapworth's ten years of meticulous fieldwork established the Ordovician system and secured him the professorship of geology in 1881. He received the greatest scientific accolade in 1891 when presented with the Royal Society's Gold Medal. It took the Geological Society until 1899 to award him its Wollaston Medal; he was its president from 1902 to 1904. Although he originally defined the Ordovician in 1879 (in the January issue of the *Geological Magazine*), it was not formally recognised until 1960. His system separated and included rocks from within Murchison's Silurian and Sedgwick's Cambrian. The Lapworth Archive contains one of the most complete records of the work of a geologist of his time. The museum houses the Midland region's finest and most extensive (over 250,000 specimens) collections of fossils, minerals and rocks; additionally, it has large collections of early geological maps, equipment, models and photographic material, collected by many influential geologists, scientists and collectors of the 19th and early 20th centuries, encompassing Lapworth's collection and the Birmingham Museum and Art Gallery's geological collection – which includes the mineral collection of the Industrial Revolution pioneer Matthew Boulton (1728–1809).

The fossil collections are scientifically and historically important, with fine material from Dudley's Wenlock Limestone and the Midlands coalfields (fossil plants, fish, insects, arachnids, fossil footprints and animal tracks), and beautifully preserved dragonflies, crabs, lobsters, fish and pterosaurs from the Solnholfen Limestone. There are also outstanding fish collections from Brazil, Italy, Lebanon and the USA, together with extraordinary invertebrates from the Burgess Shale. The mineral collection houses around 15,000 specimens, many of them rare, from around the world; the old mining areas of the UK are especially well represented. Of historical interest is the mineral collection of William Murdoch (1754–1839), who worked with James Watt (1736–1819) and Matthew Boulton at Soho House in Birmingham. Soho House was Matthew Boulton's from 1766 until his death, and a regular meeting-place of an informal scholarly group, the Lunar Society of Birmingham, so named because it met on the nights of the full moon so its members could get home by its light in the days before street lighting (Uglow 2002).

The museum supports geology and natural history research and teaching within the university, together with schools and colleges in the West Midlands region. The Lapworth Museum is currently (2015) undergoing a major redevelopment; a £130,000 Heritage Lottery Fund development grant was made, to develop plans for a significantly improved visitor experience and widened access. This is to be achieved, retaining original features and display cases within the Edwardian interior, by using interactive technology and more tactile displays, whilst new learning resources will include a dedicated education room for schools and visiting groups to engage with a diverse range of topical geological and environmental issues. Access will be improved to current university research that focuses on environmental change, the history of life and the Earth's resources.

The Lapworth Museum's key geoheritage interest is its association with probably the UK's last major 19th century classic stratigrapher geologists. Unlike the Sedgwick Museum, with its significant collections dating back to the 17th century, the Lapworth Museum's collections principally reflect late 19th and early 20th century interests. The strength of 19th century material lies in its association with men of practical science and geology, whereas the Sedgwick Museum's major historical collections reflect a more scholarly and gentlemanly purpose, as befits its location in an ancient English university.

The Geological Museum, London, UK

The UK's only stand-alone national geology museum was subsumed by the Natural History Museum in 1988 as its 'Earth Galleries'. It originally opened in 1837, at the instigation of Henry Thomas De la Beche (1796–1855) – the Geological Survey's first director – as the Museum of Practical Geology, in premises at Craig's Court in Whitehall. Purpose-built premises were soon needed, and were provided on a site fronting Piccadilly and Jermyn Street. By 1843 a substantial library, with De la Beche's material as its core collection, had been established. Opened in 1851 by Prince Albert, the new building housed exhibition galleries, a library, lecture theatre, and the Survey's offices and laboratories; its descriptive guide gave its purpose as 'to exhibit the rocks, minerals and organic remains, illustrating the maps and sections of the Geological Survey of the United Kingdom: also to exemplify the applications of the Mineral productions of these Islands to purposes of use and ornament' (Hunt and Rudler 1867). Consequently, the exhibits were in two main sections, covering natural materials and the products manufactured from them in the UK; three minor sections detailed machinery used to process raw materials, specimens of historical products, and overseas materials as imported in their raw state. In 1935, having outgrown its premises and with the establishment of the 'South Kensington Museums Quarter', it moved to its current Exhibition Road premises as the Geological Museum. The new exhibitions were especially noted for their large dioramas and the specimen-rich displays illustrating the Survey's regional geology guides. In 1965 it merged with the British Geological Survey and the various Overseas Geological Surveys into the 'Institute of Geological Sciences' (an adjunct of the Natural Environment Research Council). In the mid-1970s, following yet another renaming, the British Geological Survey removed itself to Keyworth in Nottinghamshire, although it retained control of the museum. The museum was eventually transferred from the Geological Survey, which retained an office and shop in the building, to the Natural History Museum in 1986.

The Natural History Museum was originally formed from the natural history departments of the British Museum, and only became a separate entity as the British Museum (Natural History) through the British Museum Act 1963. It became the Natural History Museum following the passing of the Museums and Galleries Act 1992. An administrative merger of the Geological Museum with the Natural History Museum had been effected by 1986. The former Geological Museum's galleries, original and refurbished, are now known as the 'Red Zone' in the Natural History Museum. The geological exhibitions have frequently been at the cutting edge of both scientific geology and museum interpretative practice. In 1971, Queen Elizabeth II opened its ground-breaking 'The Story of the Earth' exhibition on plate tectonics – the first anywhere on the concept. In the early 1970s, its own newly-formed design team, working with its scientists and technicians, produced some outstanding exhibitions. Starting with 'Gemstones', it then produced 'Early Days of Geology in Britain', 'Black Gold' (topically on North Sea oil), 'Britain Before Man', 'Journey to the Planets', 'British Fossils', 'Pebbles', 'Treasures of the Earth' and 'British Offshore Oil and Gas', which opened in 1988. 'Treasures of the Earth' (Fifield 1985), which opened in 1985, was probably the world's first major geology exhibition to employ computers to provide additional visitor information by presenting images and text next to specimens and artefacts.

Most of the exhibitions had gallery guides that were also stand-alone publications, summarising their content and current scientific thinking. The 'Earth Lab', opened in the early 1990s, provided facilities for enthusiastic amateurs to explore British geology, but was only open to pre-booked parties for a fee. It had 2,000 geological specimens within cased displays and also offered access to microscopes, a small library and an in-house (and online) computerised database. 'Restless Surface', with some interactive elements, is the only dedicated geomorphology exhibition in the UK, and possibly in Europe. The other current exhibitions are 'Earth's Treasury' (rocks, minerals and gemstones), 'Earth Today and Tomorrow' (apart from exhibiting the museum's largest meteorite, it also explores humankind's uses and abuses of the planet), 'From the Beginning' (which covers the entire span of geological time and the evolution of life), 'Lasting Impressions' (a small gallery with many rocks, minerals and plants that the visitor can touch), and 'Visions of Earth'. 'The Power Within', covering seismic activity and plate tectonics, included a reconstruction of a supermarket caught up in the 1995 Kobe earthquake, where visitors could experience the drama of being caught in a major geological event (Robinson 1996); it was included in the 'Volcanoes and Earthquakes' exhibition (opened in 2014) that clearly marks a shift away from the spectacular specimen-rich and grand reconstruction approach of many of the late 20th century exhibitions. Following the Museums and Galleries Act 1992, the museum charged a not inconsiderable entrance fee, waived except for major temporary exhibitions after 2001.

The Yorkshire Museum, York, UK

The Yorkshire Museum was founded, and originally housed in Low Ousegate, in 1823 by the Yorkshire Philosophical Society to accommodate its geological and archaeological collections. The Society was formed in York in December 1822 with the aim of gaining and disseminating knowledge related to science and history. In 1828 the Society received, by royal grant, ten acres (0.040km²) of land for a new building; this Grade I listed building was designed and built in the Greek Revival style. A condition of the royal grant was that the land surrounding the building should be a botanic garden; this was opened in the 1830s and is now the Museum Gardens. Officially opened in February 1830, the museum is one of England's longest-established regional museums. The 300-seat Tempest Anderson Hall (one of the world's first modern concrete public buildings) was added in 1912 as a bequest of Tempest Anderson (1846–1913) to provide adequate facilities for scientific lectures; he was a local ophthalmic surgeon, an expert amateur photographer, a volcanologist and a keen mountaineer. He was a member – along with Dr John Smith Flett (1869–1947), after whom the lecture theatre at the former Geological Museum (now the Natural History Museum) is named – of the Geological Survey and of the Royal Society Commission, established to investigate the aftermath of the volcanic eruptions of Soufrière, St Vincent and Mont Pelée; some of his photographs were later published in his *Volcanic Studies in Many Lands* of 1903. In 1902 Anderson travelled to Mexico to attend the 10th Congrès Géologique International, and then sailed to Guatemala to study the after-effects of the 1902 earthquake and subsequent tidal wave.

In September 1831 the inaugural meeting of the British Association for the Advancement of Science (BA) was held at the Yorkshire Museum. The geologist John Phillips (1800–74) was one of the organisers. He became the BA's first Assistant Secretary in 1832, a post he held until 1859; he was its president in 1865. Phillips was employed in 1826 as the Yorkshire Society's first museum keeper. He is a key figure in the development of UK geology, through his scientific work and his 1844 biography of William Smith. By 1831 he was engaged by University College London, and in 1834 he accepted the professorship of geology at King's College London, but still kept his York post. In 1834 he was elected a Fellow of the Royal Society. In 1845 he was awarded the Geological Society's Wollaston Medal; he was the Society's president in 1859 and 1860. In 1840 he resigned from the Yorkshire Museum and was appointed to the Geological Survey of Great Britain under Sir Henry De la Beche (1796–1855). In 1844 he became professor of geology in the University of Dublin. He was appointed to a readership at Oxford in 1856, whereupon he took a leading role in the foundation and arrangement of the new museum, opened in 1859. He was keeper of the Ashmolean Museum from 1854 until 1870.

The Yorkshire Museum's geology collections began with the discovery in 1821 of the Kirkdale Caves' ancient hyena den and its excavation in 1822 by Dean William Buckland. Today, the museum incorporates collections made by around 200 individuals, mainly amassed in the 19th century and acquired from then and through the 20th century. Amongst these collections is that of William Reed (1811–92). Among his 60,000 specimens is material derived from the work of several other collectors; for example, Red Crag material collected by Whincoppe and Jas Baker (see Lankester 1867), John William Elwes's (1850–1918) Hampshire Tertiary material, and Edward Wood's Carboniferous fossils; the latter consists of 10,000 specimens purchased by William Reed and then donated to the museum (Keeping 1881). Part of William Bean's (1787–1866) collection was purchased in 1859, with further specimens purchased by John Leckerby (1814–77) and added later through William Reed; amongst the 5,000 purchased specimens are some figured by John Phillips, and also by the Reverend George Young (1777–1848) and John Bird (described on the title page as an 'artist') in their co-authored volume, *A Geological Survey of the Yorkshire Coast* (1822). Other significant material includes that from John Bainbridge (Yorkshire material) in 1851, J F Walker in 1863–1907, T P Barkas (Northumberland Coal Measures fishes) in 1872, W H Horne in 1879, and R S Herries (Yorkshire Jurassic) in 1939. The Baker Red Crag material (Pleistocene fossils, shells and mammals) exemplifies the complex history of museum collections. It was sold in 1873 to Edward Charlesworth, who then sold it to William Reed before it was acquired by the Yorkshire Museum (Sherborn 1940, 11). Charlesworth, an authority on the (Pleistocene) Crag deposits, was sometime editor of the *Magazine of Natural History*. He had held posts at the British Museum and the museum of the Zoological Society of London when he applied in 1842 for the position of assistant secretary and librarian at the Geological Society of London. His unsuccessful application seemingly had much to do with the Society's internal politics, and with the personalities of the protagonists (Davies 2007, 73). Meanwhile, William Reed is significant to the Yorkshire Museum today because his bequest of 1892 established a £600 fund, which by 2010 had accrued £18,000, for 'the purchase of British fossils and works on natural history and especially on geology and palaeontology to be added to the geological collection and to the "Reed Reference Library".' In January 2010 it was proposed to vary the original terms of the bequest (with the agreement of the Charity Commission) to include the conservation of fossils, with the aim of spending the monies in their entirety on conserving the museum's large ichthyosaur. Whilst perhaps a laudable aim, it might be argued – in the true spirit of the bequest – that the funds could be better employed for palaeontological fieldwork and research, and that the conservation work could be more appropriately resourced out of exhibition and education funding sources. The Young and Bird (1832) publication exemplifies the significance of provincial museums as repositories of regionally significant geological specimens that are important in the history of the published geoheritage.

The Yorkshire Museum's geology collections include more than 1,000 type and figured specimens, with a catalogue published in the mid-1970s. Type material is internationally significant because it is the physical basis on which species are described; access to such specimens is needed for future research purposes, such as taxonomic studies. Its 7,500 rock specimens include the Bernard Hobson (1860–1932) collection of 3,000 worldwide igneous rocks and 1,500 accompanying thin sections. There are over 5,000 mineral specimens from Britain and the rest of the world. A small meteorite collection contains the Middlesbrough meteorite, which fell to Earth in 1881. Finally, 5,000 negatives and glass lantern slides (with around 1,500 duplicates prepared for lectures) created by Tempest Anderson are incorporated within the collection. The collections, with some 112,500 specimens, are particularly strong in Lower Carboniferous, Mesozoic and Tertiary material, with especially good Yorkshire coverage. The supporting library has the original William Reed material at its core.

In 1960, the museum and gardens were given in trust to York City Council. In 1974, following local government reorganisation, North Yorkshire Council took over their running. Following yet another such reorganisation, the City of York Council set up the York Museums Trust in 2002 to manage the York Castle Museum, York Art Gallery, the Yorkshire Museum and the Museum Gardens. The museum was closed in November 2009 for a £2 million refurbishment. It reopened in August 2010 with three new major exhibition sections; the geology section ('Extinct: A Way of Life') was created to be a fun, family-oriented gallery featuring fossils, skeletons and animal specimens. Their chief UK geoheritage importance is that much of the material is associated with figures key to the development of both regional and international scientific geology; it is just a pity that a non-too-modest day entrance fee (£7.50 in 2015, albeit with concessions for the young) will undoubtedly deter many casual visitors.

Museo di Paleontologia e Geologia 'G Capellini', Bologna, Italy

Scientific natural history collecting in Italy originated in Bologna. In the 16th century, Ulisse Aldrovandi (1522–1605) formed a large and systematically arranged collection, exhibited to the public, of natural objects – mainland Europe's oldest natural history museum. Aldrovandi bequeathed the collection to his native Bologna in 1605. Its systematic arrangement established a scientific pattern followed by the city's later museums. In 1660 Senator Ferdinando Cospi (1606–86) donated his collection, the Museo Cospiano, to Bologna's Senate. In 1667 he printed at his own expense a five-volume catalogue; the first two volumes of this described the natural history specimens, and the last three described archaeological artefacts. The title of the catalogue prepared by Lorenzo Legati (d.1675) – *Museo Cospiano annesso a quello del famoso Ulisse Aldrovandi* [*The Museo Cospiano, to which is annexed that of the famous Ulisse Aldrovandi*] – indicates that it included the Aldrovandi collection. The catalogue illustrates Cospi's own 'Cabinet of Curiosities'; it includes true natural history specimens as well as some fictitious animals, such as a winged fish and a Hippocampus (a half-horse and half-fish mythical beast seen in ancient Greek art).

In 1743, the Academy of Sciences of the Institute of Bologna obtained the collection of natural objects that had been assembled by Senator Ferdinando Cospi in the 17th century – the 'Naturalia Museum', as it was by then called. The Museum of the Institute of Sciences was added in the 18th century; this had been established and enriched with thousands of specimens from all over the world by Count Luigi Ferdinando Marsigli (1658–1730). In 1796 Napoleon's troops (who were rampaging through Italy) sacked the museum; some, though not all, of the collections were returned in 1816, including many geological or palaeontological collections, although without their type specimens. All the collections were combined at the beginning of the 19th century into a single Natural History Museum. During the first half of the 19th century the museum was renovated, and in 1852 a new Museum of Natural History was officially inaugurated, but only lasted for eight years.

In 1860 Giovanni Capellini (1833–1922) was appointed by the University of Bologna to Italy's first chair of geology. He then established the Geological Museum of Palaeontology, on the basis of the historical collections of the Museum of Natural History. This date (1860) underlines the significance of politics in the museum's development; after the annexation of Emilia to the Kingdom of Italy, the Governor General of the Province of Emilia, Luigi Carlo Farini (1812–66), issued two decrees in early 1860 calling for the institution of three new chairs in mineralogy, zoology and geology, instead of retaining just the original chair of natural history; this also meant that the collections of the Museum of Natural History would be split into three new museums, corresponding to these chairs. Capellini enriched the collections of the newly established Geological Museum with worldwide geological and palaeontological material; he also developed its educational and exhibition facilities.

Whilst the Geological Museum was active from 1860, the 'G Capellini' epithet was officially inaugurated in 1881 at the Second International Geological Congress in Bologna, which he organised. Participating geologists brought rock and fossil collections to show at the Congress, and these were then given to the museum (Figure 4.1C). Its 15 galleries and the number and importance of its collections (with their one million specimens) make it Italy's largest palaeontological museum. It is divided into five sections: ancient collections, fossilised plants, fossilised vertebrates, rock and invertebrate collections, and fossils ordered according to their geographic origin in Italy and abroad. Within the exhibition galleries, vertebrates are dramatically displayed; these include the mastodon, Pliocene elephants, Eocene fish from Monte Bolca, Pliocene whales, marine reptiles such as ichthyosaurs, and terrestrial reptiles such as the life-sized model of the dinosaur Diplodocus. The newest inclusions are land birds from the Quaternary of New Zealand and large Pliocene insectivores from Argentina. The fossil plants on display include two large Eocene palm trees from Verona and the largest European collection of cycads. The museum is now part of the Sistema Museale di Ateneo (Bologna University Museum System), which brings together its various museums and archives to optimise their use and to recognise their historical significance and research potential; amongst the constituent museums is a separate Museum of Mineralogy. Within the incorporated 'Aldrovandi Tribune' are preserved parts of the 16th to 18th century collections, the museum's geoheritage core. Public admission is for a modest fee.

The Museum of Natural History, University of Florence

The Museum of Natural History in Florence was founded in 1775 by Grand Duke Pietro Leopoldo (1747–92); as the Imperial Regio Museo di Fisica e Storia Naturale, it consisted of several natural history collections housed within the Palazzo Torrigiani. Its origins lie in a complex series of events and the attentions of various key figures, some political and some scientific, dating from the 15th century when natural history collections began to be formed under the first Medici – particularly Lorenzo il Magnifico (1449–92) – in the family palace. Then, the specimens were mixed with art works, like the classic *Wunderkammern* ('cabinet of wonders') approach, but the only documented extant geological specimens are decorative items forming part of the Medici collection of the Mineralogy Section. In 1596 Ferdinando I de' Medici, Grand Duke of Tuscany (1549–1609), established the Museum of Natural History in Pisa, using part of the collections from Florence assembled by the Medici and housed in their palaces – particularly the Uffizi. This was the kernel of a collection later expanded to serve the University of Pisa's students.

However, in 1669 Nicholas Steno (1638–86), on the orders of Ferdinando II de' Medici, Grand Duke of Tuscany (1610–70), returned to Florence part of the mineralogical collection from Pisa; these specimens were combined with the ones he had collected in excursions in Italy, Hungary and Germany to form the first substantial core of the mineralogical and palaeontological collections. The physician Giovanni Targioni Tozzetti (1712–83), director of the Botanical Gardens, compiled in 1763 the *Catalogo delle produzioni naturali che si conservano nella Galleria Imperiale di Firenza disteso nell'anno 1763* [*Catalogue of the natural products that are kept in the Imperial Gallery of Florence, drawn up in the year 1763*] at the explicit request of the emperor Grand Duke Francis Stephen of Lorraine. In 1775 Grand Duke Pietro Leopoldo of Hapsburg-Lorraine (1747–92) established the Imperial Royal Museum of Physics and Natural History, transferring the natural science collections and scientific instruments of the Accademia del Cimento, housed in the Uffizi Gallery, to the Torrigiani Palace; Felice Fontana (1730–1805), professor at Pisa and court physicist, was installed as its director. Within a few years, the museum became the most important Italian centre of physical and natural science research and the site of a workshop producing wax anatomical models.

In 1807, by decree of Maria Louisa of Bourbon-Parma (1751–1819), regent of the Kingdom of Etruria, the museum became a university-level teaching centre, the Lyceum of Physics and Natural Sciences; six chairs (astronomy, physics, chemistry, mineralogy and zoology, botany, and comparative anatomy) were established. The Lyceum was abolished upon the return of Grand Duke Ferdinando III (1769–1824), but the teaching resumed with his successor and last Grand Duke, Leopoldo II (1797–1870). In 1859, after the expulsion of Lorraine, the provisional government of Tuscany, led by Bettino Ricasoli (1809–80), founded the Institute of Higher Practical Studies and Specialisation – something of a Tuscan super-university. Due to the scientific material and the existing expertise, the museum became the site of the 'Section of Natural Sciences', corresponding to the present-day Faculty of Sciences. By 1878 the marked development of scientific disciplines, and an increase to 15 chairs, enabled the entire Section of Natural Sciences to remain in the museum in Via Romana. Astronomy and later physics moved to the Arcetri Hill; chemistry moved to Via Gino Capponi; geology and mineralogy and then botany transferred to Piazza San Marco; while only zoology remained in Via Romana. Each teaching discipline reacquired its collections to create its departmental museum.

These moves led, after just over a century's existence, to the museum's end. In the first half of the 20th century, scientific specialisation reduced the significance of the various departmental museums, as the natural sciences generally found little need or use for specimens and displays. They lost their attraction and autonomy, as neglected appendages of the institutes, often being plundered of funds, space and personnel. Fortunately, the late-20th century revival of interest in natural history saw the University of Florence re-establish the autonomy of the natural history museums from their respective institutes, assigning funds, personnel and space in 1971. Then, in 1984, anticipating the establishment of a National Museum of Natural History proposed for Florence by the National Academy of the Lincei, the university reunited the departmental natural history museums into the Museum of Natural History of Florence University; it has six sections (housing eight million specimens) across four sites, of which two (with 250,000 specimens) – the Museo di Geologia e Paleontologia and the Museo di Mineralogia e Litologia – are of geoheritage interest.

Museo di Geologia e Paleontologia

This houses about 200,000 specimens from the collections of noted geologists and palaeontologists (Fucini, Dainelli, Marinelli, De Stefani, Stefanini, D'Ancona, Pecchioli). However, only part of the museum is open to the public. This public section of some 26,000 specimens is mostly Italian fossil mammals collected over two centuries, for which the museum has gained worldwide renown. There are also collections of invertebrates, rocks and plants (the palaeobotany collection includes some 8,000 specimens) occupying the second floor of the building, which is closed to the public.

Museo di Mineralogia e Litologia

This houses some 50,000 specimens spread across several collections. Of particular interest are the earliest items (about 500 specimens) from the Tribuna degli Uffizi, and the collection (about 5,000 items) of Giovanni Targioni Tozzetti. The large mineralogical collection, originating from the first half of the 16th century under the patronage of the Medici family, makes it Italy's most important such historical and scientific collection. It is Italy's finest natural history museum; its geoheritage interest is its significant early geological material.

Muséum National d'Histoire Naturelle, Paris, France

The Muséum National d'Histoire Naturelle (MNHN), the National Museum of Natural History, was founded during the French Revolution in June 1793. Its origins lie in the Jardin Royal des Plantes Médicinales (Royal Medicinal Plant Garden), created by King Louis XIII (1601–43) in 1635, which was directed and run by the royal physicians. The 1718 royal proclamation of Louis XV removed its purely medicinal function, enabling the garden, known as the Jardin du Roi (King's Garden), to focus on natural history. For much of the 18th century (1739–88), the garden was under the direction of Georges-Louis Leclerc, Comte de Buffon (1707–88), one of the leading naturalists of the Enlightenment. Remarkably for a royal institution, it survived the French Revolution by being reorganised in 1793 as the Republican Muséum National d'Histoire Naturelle, with the aims of instructing the public, amassing scientific collections and conducting scientific research. It was allotted 12 professorships of equal rank; amongst the early ones were the eminent comparative anatomist and founder of vertebrate palaeontology Baron Georges Léopold Chrétien Frédéric Dagobert Cuvier (1769–1832), and evolutionary pioneers Étienne Geoffroy Saint-Hilaire (1772–1844) and Jean-Baptiste de Lamarck (1744–1829); the first and last of these are particularly significant to French geoheritage.

Cuvier studied at Stuttgart's Caroline Academy, and after graduation was appointed in 1788 tutor to the son of the Comte d'Héricy at Fiquainville Château in Normandy. At the château, he began comparing fossils with extant forms in the early 1790s. In 1795 he became the assistant of Jean-Claude Mertrud (1728–1802), the newly created chair of comparative anatomy at the Jardin des Plantes. At the opening of the Institut de France in 1795 he read his first palaeontological paper (Cuvier 1799) and further published on the same topic of living and fossil elephants in 1806, as *Suite du mémoire sur les éléphants vivants et fossiles*; it was an analysis of the skeletal remains of African and Indian elephants, as well as mammoth fossils, plus a fossil skeleton then known as the 'Ohio animal', which he later named a mastodon. In 1796 he began lecturing at the École Centrale du Pantheon. In 1799 he was appointed professor of natural history in the Collège de France and in 1802 titular professor at the Jardin des Plantes. In 1800, after examining just a drawing, he was the first to identify a Bavarian fossil as a small flying reptile, the first known member of the pterosaurs, which he named the Ptero-Dactyle in 1809. In 1808 he identified a fossil found in Maastricht as a giant marine lizard, which, as the first known mosasaur, he named '*Mosasaurus*'. His *Théorie de la terre* [*Theory of the Earth*] of 1821 proposed that new species were created after periodic catastrophic floods, establishing himself as catastrophism's foremost proponent. The *Essais sur la géographie minéralogique des environs de Paris, avec une carte géognostique et des coupes de terrain* [*Essays on the mineralogical geography near Paris, with geognostic map and terrain sections*] of 1811 was co-authored with Alexandre Brongniart (1770–1847), an instructor at the École des Mines (Mining School) in Paris. Brongniart is particularly remembered today for his extensive studies on trilobites and his work on biostratigraphy. He published his *Traité élémentaire de minéralogie* [*Elementary treatise on minerals*] in 1807. He is less well known as having been director of the Sèvres Porcelain Factory (from 1800 until his death), where he applied his knowledge of chemistry and geology to the manufacturing process; he was also the founder of the Musée National de Céramique-Sèvres (the National Museum of Ceramics). His son was the botanist and palaeobotanist Adolphe-Théodore Brongniart (1801–76), who particularly worked on Carboniferous plant fossils; due to his much published work on the relationships between living and fossil plants, he has been called the 'Father of Palaeobotany'. Meanwhile, Cuvier was one of the first geologists to suggest that the Earth had been dominated in prehistoric times by reptiles and not mammals. He was a vehement opponent of the organic evolutionary theories of Lamarck and Saint-Hilaire, arguing that new species arose as a result of the opportunities made available after the extinctions caused by catastrophic floods.

Lamarck was a distinguished botanist and invertebrate (a term he coined) zoologist. He studied medicine and botany whilst working in Paris as a bank clerk; in 1778 he published to much acclaim his three-volume *Flore Française*, on the strength of which he was appointed an assistant botanist at the Jardin du Roi. In 1788 he was made keeper of its herbarium; in 1790, at the height of the French Revolution, Lamarck changed its name, so as to remove its close association with King Louis XVI, to the Jardin des Plantes. In 1793 he was appointed curator and professor of invertebrate zoology at the Muséum National d'Histoire Naturelle. In 1802 he published *Hydrogéologie*, in which he proposed a geological steady-state founded on a strict uniformitarianism. He argued that global currents tended to flow from east to west and continents eroded on their eastern borders, with the material carried across and deposited on the western borders; thus, the Earth's continents advanced steadily westward (in an early notion of continental drift) across the globe's surface. Also in 1802, he published *Recherches sur l'organisation des corps vivants* [*Research on the organisation of the living body*], in which he developed the flawed evolutionary theory for which he is perhaps best and unfairly remembered.

In the 19th century, particularly under the direction of the chemist Michel Eugène Chevreul (1786–1889), the museum continued to flourish, becoming a rival in scientific research to the University of Paris. An 1891 decree ended the museum's general physical scientific phase, returning its focus to natural history. After gaining financial autonomy in 1907, it began opening facilities throughout France during the 1920s and 1930s. In French public administration terms, the museum is classed as a 'grand établissement' of higher education. In recent decades, its research and education efforts have been focused on the effects of human exploitation on the natural environment. Today, its main site is in Paris with another 14 across France, with four in Paris including the original location at the Jardin des Plantes. The galleries open to the public are the Cabinet d'Histoire du Jardin des Plantes, the Galerie de Minéralogie et de Géologie, the Galerie de Paléontologie et d'Anatomie Comparée, and the famous Grande Galerie de l'Évolution. The Musée de l'Homme is also in Paris; it houses ethnographic and physical anthropology displays, including artefacts and fossils. Other parts of the museum in Paris include four scientific sites with one of some geoheritage interest, the Institut de Paléontologie Humaine.

The Galerie de Minéralogie et de Géologie displays a collection of 600,000 crystals, gemstones and minerals. It began in 1625, when minerals with medicinal properties were deposited in 'le droguier du jardin royal des plantes médicinales', thereafter, enriched on the orders of Louis XIV (1638–1715), it has been open to the public since 1745. The Galerie de Paléontologie et d'Anatomie Comparée, with a façade decorated with sculptures inspired by naturalists, was inaugurated in 1898 as part of the 1900 Paris Universal Exposition; it was created by Albert Gaudry (1827–1908), professor of palaeontology, and Georges Pouchet (1833–94), professor of comparative anatomy, to preserve and present to the public collections of great historical and scientific importance; its collections were amassed from 18th and 19th century traveller-naturalists' expeditions and the adjacent zoo. The palaeontology galleries, on two floors totalling some 2,500m^2, exhibit fossil vertebrates (especially dinosaurs and other extinct vertebrates) and invertebrates; the largest gallery is almost 80m long. The Grande Galerie de l'Évolution was established within the MNHN's former Zoology Gallery; opened in 1994, it covers the evolution of species and the world's biodiversity through the contemporary display of the old natural history collections. Overall, the museum's principal French geoheritage interest is the association of its collections with some of Europe's significant 19th century palaeontologists and evolutionary theorists.

Bergakademie Museum, Freiberg, Germany

There are two museums with geoheritage and historical significance in Freiberg: the Municipal and Mining Museum based in the Domherrenhof (Canons' Lodging) of 1484, a Late Gothic patrician house; and the mineral collection of the Bergakademie (Mining Academy). There is also a new mineral museum, the Terra Mineralia, opened in 2008. The reason for the foundation of these museums is that lead has been mined in the Freiberg area for around 1,000 years. With the discovery of silver around 1163, the government took an interest. In 1175 the Margrave of Meissen, Otto der Reiche (1125–90), founded the town with its castle, the Freudenstein. The Bergakademie Freiberg, the first school of mining engineering in the world, was founded in 1765 – on the suggestion of the General Mining Commissioner, Baron Friedrich Anton von Heynitz (1725–1802) – by Prince Xavier of Kursachsen (1730–1806). Its mineral museum was founded at the same time; its original core probably consisted of the collection formed by C E Gellert (1713–95), together with those of the Bergakademie's two founders – von Heynitz (1725–1802) and F W von Oppel (1720–69). Today the collections have some 80,000 specimens stored at the Mineralogical Institute (A-G-Werner-Bau) and at the Geological Institute (A-v-Humboldt-Bau). Most of the collections are open to the public for a modest entrance fee, and its stores can be used by researchers. The exhibition is subdivided into two main sets of collections. The systematic collection (with a noteworthy regional collection of minerals from eastern Germany) is arranged using Strunz's classification and Roesler's textbook, in which crystal chemistry data and genetic relationships are both considered. The important historical collections include the original mineral discoveries by Werner (and his archive), Breithaupt and Weisbach; Werner had sold his personal collection of more than 10,000 specimens to the Bergakademie for 40,000 talers.

The geological collections of the Bergakademie are probably the tenth oldest of the world's more than 450 important geoscientific collections, and have informed and underpinned the Bergakademie's teaching for some 250 years. The chief geoheritage interest of the museums and the Bergakademie is their association with the 'Father of German geology', Abraham Gottlob Werner (1749–1817). He was associated with them for more than 40 years and was the most respected geology professor of his day. The tuition provided by Werner and his immediate successors, in which the collections prominently featured, was unrivalled in its quality and relevance to 18th and 19th century mining.

Werner was born in German Prussian Silesia (now in Poland), into a family with a mining tradition. His youthful interest in rocks and minerals was indulged by his father. After private tuition he was educated at the Waisenschule at Bunzlau (now Boleslawiec). His father was inspector of the Duke of Solm's ironworks, and the teenage Werner clerked in the foundry; he began his managerial studies in Freiberg's new Bergakademie. He then joined the Saxon mining service instead of returning to the Duke of Solm's employment. He studied law at the University of Leipzig but left before graduating, and then became interested in the systematic identification and classification of minerals. Whilst a student at the Freiberg Bergakademie, he published the first modern textbook on descriptive mineralogy, *Von den äusserlichen Kennzeichen der Fossilien* [*On the external characters of fossils, including minerals*] in 1774; this contained a comprehensive colour scheme, of his own invention, for the description and classification of minerals. The work was translated into French by Claudine Guyton de Morveau (née Picardet) in 1790 and into English by Thomas Weaver in 1805. The international influence of Werner's scheme can be further judged from the work of Patrick Syme (1774–1845), painter to the Wernerian and Horticultural Societies of Edinburgh, who published in 1814 a revised version, entitled *Werner's Nomenclature of Colours, with Additions, arranged so as to render it useful to the Arts and Sciences*.

In 1775, based upon the promise shown in his first book, Werner was appointed professor of mineralogy at the Freiberg Bergakademie, serving out 42 years of professional life. It was K E Pabst von Ohain (fl. 1774), a former teacher and then friend, who got Werner his faculty job; he was also a major mineral collector. Werner actually published very little else, but what he did publish was often influential, including *Von den verschiedenerley Mineraliensammlungen* (1778); *Kurze Klassifikation und Beschreibung der verschiedenen Gebirgsarten* (1787); and *Abraham Gottlob Werner's letztes Mineral-System* (1817). He named eight new minerals, and wernerite (a variety of scapolite) is named in his honour. Werner recognised that, in his day, chemistry and crystallography were not sufficiently advanced to enable a sophisticated mineralogical system to be developed. However, he worked up simple descriptive standards of classification in the belief that such external characteristics were perhaps not unrelated to chemical composition.

Today, Werner is best known for his attempts to explain the stratified nature of rocks and their disposition. His major theory, 'neptunism', especially on the origin of basalts as a marine deposit, rivalled that of John Hutton's 'plutonism', in which they were formed by volcanic action. The subsequent controversy focused much geological discussion from the end of the 18th century well into the 19th century. Werner propounded that all rocks were once sediments or precipitates of a universal ocean. He is credited with coining the term 'geognosy' for the geological study of the Earth's structure, specifically its exterior and interior construction; he was the first geologist to work out a complete schema for the Earth's structure and the history of its formation. With no evidence that he ever travelled as an adult beyond Saxony, it must be assumed that, apart possibly from communications from ex-students, he based his theory on local fieldwork. Plagued by frail health for much of his adulthood, he abandoned fieldwork altogether in later life. Despite such limited mobility and few publications, he was elected to 22 major learned societies – including the Geological Society of London, Institut National de France, Imperial Society of Physics and Medicine of Moscow, Royal Prussian Academy of Sciences, and Royal Swedish Academy of Sciences.

Werner taught his students, especially through organised fieldwork (which was something of a first), to appreciate the broader implications and interrelations of geology. His teaching and the reputation of the Freiberg Bergakademie attracted students from across Europe, many of whom went on to become lecturers who spread his theories throughout Europe and North America. Amongst them were geologists and mineralogists like Christian Leopold von Buch (1774–1853), Alexander von Humboldt (1769–1859), Jean d'Aubuisson de Voisins (1769–1841), Robert Jameson (1774–1854), who became professor of geology at Edinburgh, and Carl Friedrich Christian Mohs (1773–1839). A key man at the Bergakademie in Werner's time was M H Klaproth (1743–1817), the founder of quantitative mineral analysis.

Apart from Werner, the most significant mineralogist amongst the early figures associated with the Bergakademie is Carl Friedrich Christian Mohs (1773–1839). After studying chemistry, mathematics and physics at the University of Halle, he went on to the Bergakademie before becoming a mine foreman in 1801. In 1802 he was employed in Austria to identify the minerals in a banker's private collection. In 1812 he moved to Graz, becoming a professor, and was employed by Archduke Johann in his newly established museum and science academy (subsequently split into the Joanneum and Graz University of Technology), but returned to Freiberg in 1818. In 1826 he established himself in Vienna. Whilst in Graz, Mohs began to classify minerals by their physical properties rather than by their chemical composition, as was traditional at the time. This approach became the basis of the hardness scale developed by Mohs, based on the relative hardness of ten minerals. Although minerals are classified today by chemical characteristics, their physical properties are still useful in the field and Mohs' hardness scale is still learned and applied by geologists – a lasting Bergakademie geoheritage.

Terra Mineralia, Freiberg, Germany

The Terra Mineralia museum opened in 2008. For Erika Pohl-Ströher (b.1919), a Swiss national born in Saxony, it seemed natural to locate some of her outstanding mineral collection in the home-town of scientific mineralogy. Her private collection – the world's largest – of more than 80,000 minerals, gemstones and meteorites was amassed over 60 years. Although mainly collected for aesthetic value, its specimens are scientifically significant. In 2004 she offered some 3,500 of her finest specimens on permanent loan to the Technische Universität Bergakademie Freiberg, with two major conditions: they would be properly curated and made available for public enjoyment, although an entrance fee could be charged. The museum, housed within Freiberg Castle, was designed around this large donation as the bulk of its displays. The original castle was built (1175–77) to protect the area's silver deposits, found a few years earlier; between 1566 and 1577 Elector Augustus (1526–86) had the present Renaissance-style Freudenstein Castle replacement built to reflect the growing status and importance of its inhabitants and of the area. Over time the castle lost its significance and had various uses, including a grain store and hospital. Eventually it fell into disrepair. Lying empty after its last business (a restaurant opened in 1986) had closed in 1991, the castle was given a new lease of life, partly related to its original significance. In 2003 the city council decided to grant joint usage to the Technische Universität Bergakademie Freiberg and the Saxon Archives of Mining and Metallurgy, and a four-year reconstruction and refurbishment programme began in 2004. Four museum 'galleries' over three storeys display specimens from Africa, America, Asia and Europe; a separate area is dedicated to Australian material. A particular feature is the 'Treasure Chamber', in which some of the larger and most magnificent specimens are displayed in tailor-made cases. Researchers and students have access to material not displayed. In many ways, Terra Mineralia epitomises the latest approach to geological museums, wherein public exhibition and the spectacular outweigh scientific investigation and publication; perhaps, and of German geoheritage significance, something of a return to the initial role of Europe's Renaissance cabinets of curiosities.

MUSEUMS OF GEOLOGY – SOME CLOSING THOUGHTS

Europe's geological museums are a repository of historically and scientifically significant specimens and archives. The oldest collections are associated with the leading museums of their country, as is seen in Cambridge and Bologna. Many were also the first to establish formal geology teaching, with their collections evolving to support students; as a consequence of individual geological studies and research programmes, these collections were greatly enlarged and now underpin the 'history of geology' element of geoheritage. Most museums and their incorporated collections have had complex histories of ownership and engagement with leading figures in the development of geology. Many of these individuals – through their published work, their collections, and their surviving papers and archives – have bequeathed to science a major geoheritage resource, one of significance now and into the future.

REFERENCES

Anderson, T, 1903 *Volcanic Studies in Many Lands*, John Murray, London

Arts Council England, 2011 *Culture, Knowledge and Understanding: Great Museums and Libraries for Everyone – A Companion Document to Achieving Great Art for Everyone*, Arts Council England, London

Bacon, F, 1594 *Gefta Grayorum: Or, The History of the High and Mighty Prince, Henry Prince of Purpoole*, W Canning, London [The Malone Society Reprints, 1914]

Brongniart, A, 1807 *Traité élémentaire de minéralogie* [*Elementary treatise on minerals*] [in 2 vols], Imprimerie de Crapelet, Paris

Burns Owen Partnership, 2005 *New Directions in Social Policy: Developing the Evidence Base for Museums, Libraries and Archives in England* [report by Burns Owen Partnership], Museums, Libraries and Archives Council, London

Cuvier, G, 1799 Mémoire sur les espèces d'éléphants vivants et fossiles [On the species of living and fossil elephants], *Mémoires de l'Institut National des Sciences et Arts: Sciences, Mathématiques et Physiques*, 2, Mem: 1–22

— 1806 Suite du mémoire sur les éléphans vivans et fossiles, *Annales du Muséum National d'Histoire Naturelle* 8: 1–58; 93–155; 249–69

— 1821 *Théorie de la terre* [*Theory of the Earth*], F D Pillot, Paris

Cuvier, G, and Brongniart, A, 1811 *Essais sur la géographie minéralogique des environs de Paris, avec une carte géognostique et des coupes de terrain* [*Essays on the mineralogical geography near Paris, with geognostic map and terrain sections*], Baudouin, Paris

Davies, G L H, 2007 *Whatever is under the Earth: The Geological Society of London (1807–2007)*, The Geological Society, London

Doughty, P, 1996 Foreword, in K W James, *Damned Nonsense! – The Geological Career of the Third Earl of Inniskillen*, Ulster Museum, Belfast

Farey, J, 1811 *General View of the Agriculture and Minerals of Derbyshire*, Vol 1, G and W Nicol, London

Fifield, R, 1985 A mine of information: treasures of the Earth, *New Scientist*, No 147 (10 October), 54

Gurian, E H, 1982 Museum's relationship to education, in *Museums and Education*, Danish-ICOM/CECA, Copenhagen, 17–20

Haan, J de, 2010 Museum statistics and cultural policy, paper presented at the 12th plenary meeting in the European Group on Museum Statistics (EGMUS): *Museum Statistics and Museum Policies – New Developments*, 21–22 October 2010, Oslo, http://www.egmus.eu/fileadmin/intern/Museum_statistics_and_cultural_policy_Jos_de_Haan_v3_incl_CV.pdf [accessed 05/09/2015]

Harris, J, 1704 and 1710 *Lexicon Technicum: or, An Universal English Dictionary of Arts and Sciences: Explaining not only the Terms of Art, but the Arts Themselves* [in 2 vols], printed for D Brown, T Goodwin, J Walthoe, T Newborough, J Nicholson, D Midwinter and F Coggan, London

Henry, C J, and Hose, T A, 2015 The contribution of maps to appreciating physical landscape: examples from Derbyshire's Peak District, in *Appreciating Physical Landscapes: Three Hundred Years of Geotourism*, *Special Publication No 417* (ed T A Hose), Geological Society, London

Hunt, R, and Rudler, F W, 1867 *A Descriptive Guide to the Museum of Practical Geology* (3 edn), HMSO, London

Keeping, W, 1881 Presentation of the Edward Wood Collection to the York Museum, *Geological Magazine* 8 (02), 96

Lamarck, J-B de, 1778 *Flore Française, ou, Descriptions succinctes de toutes les plantes qui croissent naturalle-ment en France* [*French flora, or, succinct descriptions of the plants that grow naturally in France*] [in 6 vols], Desray, Paris

— 1802a *Hydrogéologie*, Lamarck, Paris

— 1802b *Recherches sur l'organisation des corps vivants* [*Research on the organisation of the living body*], Lamarck, Paris

Lankester, E R, 1867 *On the* Structure *of the* Tooth *in* Ziphius Sowerbiensis (Micropteron Sowerbiensis, Eschricht), *and on some* Fossil Cetacean Teeth, by E Ray Lankester, FRS, of Christ Church, Oxford, read 8 May 1867, *Transactions of the Royal Microscopal Society*, New Series, XV, 55–64

Lapworth, C, 1879 On the tripartite classification of the Lower Paleozoic, *Geological Magazine*, 26, 1–15

Legati, L, 1677 *Museo Cospiano: annesso a quello del famoso Ulisse Aldrovandi e donato alla sua patria dall'illustrissimo signor Ferdinando Cospi* [*The Museo Cospiano, to which is annexed that of the famous Ulisse Aldrovandi, and donated to his homeland by the illustrious Mr Ferdinand Cospi*], Giacomo Monti, Bologna

Ligorio, P, 1571 *Rimedi contra terremoti per la sicurezza degli edifici* [*Remedies against earthquakes for the safety of buildings*], Turin

Lloyd, D, 1983 *The History of Ludlow Museum 1833–1983*, Ludlow Historical Research Group, Ludlow

Nudds, J R, 1994 *Directory of British Geological Museums: Miscellaneous Paper No 18*, The Geological Society, London

Ousby, I, 1990 *The Englishman's England: Taste, Travel and the Rise of Tourism*, Cambridge University Press, Cambridge

Phillips, J, 1844 *Memoirs of William Smith, LLD, author of the 'Map of the strata of England and Wales'*, John Murray, London

Plot, R, 1677 *The Natural History of Oxford-shire, being an essay towards the Natural History of England*, printed at the Theater, Oxford

— 1686 *Natural History of Stafford-shire*, printed at the Theater, Oxford

Potter, J, 2006 *Strange Blooms: The Curious Lives and Adventures of the John Tradescants*, Atlantic, London

Price, D, 1989 A life of dedication: A G Brighton (1900–1988) and the Sedgwick Museum, Cambridge, *Geological Curator* 5 (3), 95–9

Robinson, E, 1996 Earth Galleries open: Impression 1, *Geology Today* 12, 128–30

Sherborn, C D, 1940 *Where is the – collection? An account of the various natural history collections which have come under the notice of the compiler*, Cambridge University Press, Cambridge

Standing Commission on Museums and Galleries, 1977 *Report on University Museums*, HMSO, London

Syme, P, and Werner, A B, 1814 *Werner's Nomenclature of Colours, with Additions, arranged so as to render it highly useful to the Arts and Sciences; particularly Zoology, Botany, Chemistry, Mineralogy and Morbid Anatomy. Annexed to which are examples selected from well-known objects in the animal, vegetable, and mineral kingdoms*, William Blackwood, Edinburgh, John Murray, London, Robert Baldwin, London

Thomas, I, and Cooper, M, 2008 The geology of Chatsworth House, Derbyshire, *Mercian Geologist*, 17 (1), 27–42

Tozzetti, G T, 1763 *Catalogo delle produzioni naturali che si conservano nella Galleria Imperiale di Firenza disteso nell'anno 1763* [*Catalogue of the natural products that are kept in the Imperial Gallery of Florence, drawn up in the year 1763*], Florence

Tradescant, J, 1657 *Musaeum Tradescantianum, or a Collection of Rarities preserved at South Lambeth near London by John Tradescant*, London

Uglow, J, 2002 *The Lunar Men: The Friends who made the Future*, Faber and Faber, London

Watson, W, 1811 *The Strata of Derbyshire*, W Todd, Sheffield [1973 reprint, with a new introduction by Trevor D Ford, Moorland Reprints, Hartington]

Werner, A G, 1774 *Von den äusserlichen Kennzeichen der Fossilien* [*On the external characters of fossils, or of minerals*], Siegfried Lebrecht Crusius, Leipzig

— 1778 *Von den verschiedenerley Mineraliensammlungen, aus denen ein vollständiges Mineralienkabinet bestehen soll* [*Of the various mineral collections, which make up a complete mineral cabinet*], Leipzig

— 1787 *Kurze Klassifikation und Beschreibung der verschiedenen Gebirgsarten* [*Short classification and description of the different kinds of rock*], Walther, Dresden

— 1817 *Abraham Gottlob Werner's letztes Mineral-System* [*Abraham Gottlob Werner's last mineral system*], bey Craz und Gerlach und bey C Gerold, Freiberg and Vienna

Whewell, W, 1837 *History of the Inductive Sciences, From the Earliest to the Present Times* [in 3 vols], John W Parker, London, J and J Deighton, Cambridge

— 1840 *The Philosophy of the Inductive Sciences, Founded upon their History* [in 2 vols], John W Parker, London, J and J Deighton, Cambridge

Woodward, A S, 1889–1901 *Catalogue of Fossil Fishes in the British Museum (Natural History)* [in 4 vols], British Museum (Natural History), London

Woodward, J, 1695 *An Essay toward a Natural History of the Earth and Terreſtrial Bodies, especially Minerals: and alſo of the Seas, Rivers, and Springs. With an Account of the Universal Deluge: and of the effects it had on the Earth*, A Bettesworth and W Taylor, London

— 1696 *Brief Instructions for Making Observations in all Parts of the World*, Richard Wilkin, London [1973 reprint with an introduction by V A Eyles, Sherborn Fund Facsimili Number 4, Society for the Bibliography of Natural History, London]

— 1714 *Naturalis historia telluris illustrata et aucta*, Richard Wilkin, London

— 1728 *Fossils of all Kinds, Digested into a Method, Suitable to their mutual Relation and Affinity; with the Names by which they were known to the Antients, and those by which they are at this Day known: and Notes conducing to the setting forth the Natural History, and the main Uses, of some of the most considerable of them. As also several Papers tending to the further Advancement of the Knowledge of Minerals, of the Ores of Metalls, and of all other Subterraneous Productions. By John Woodward, MD, late Professor of Physick at Gresham College, Fellow of the College of Physicians, and the Royal Society*, William Innys, London

— 1728–29 *An Attempt towards a Natural Hiftory of the Fossils of England; in a Catalogue of the Englifh Fossils in the Collection of J Woodward, MD. Containing a Description and Historical Account of each; with Obfervations and Experiments, made in order to difcover, as well the Origin and Nature of them, as their Medicinal, Mechanical, and other Ufes* [in 2 vols], J Osborn and T Longman, London

Worm, O, 1655 *Museum Wormianum, seu, Historia rerum rariorum: tam naturalium, quam artificialium, tam domesticarum, quam exoticarum, quae Hafniae Danorum in aedibus authoris servantur* [*Museum Wormianum, or, History of rare objects, both natural and artificial, domestic and exotic, which are preserved in the house of the author, in Copenhagen*], Lugduni Batavorum/Apud Iohannem Elsevirium, Leiden

Young, G, and Bird, J, 1822 *A Geological Survey of the Yorkshire Coast, Describing the Strata and Fossils Occurring between the Humber and the Tees, from the German Ocean to the Plain of York*, George Clark, Whitby

Further reading

Alfrey, J, and Putnam, T 1992 *The Industrial Heritage – Managing Resources and Uses*, Routledge, London

Allen, D E, 1978 *The Naturalist in Britain: A Social History*, Allen Lane/Penguin Books, London

Barber, L, 1980 *The Heyday of Natural History, 1820–1870*, Jonathan Cape, London

Butler, S, 1992 *Science and Technology Museums*, Leicester University Press, Leicester

Carr, W, and Hartnett, J, 1996 *Education and the Struggle for Democracy: The Politics of Educational Ideas*, Open University Press, Milton Keynes

Cato, P S, and Jones, C (eds), 1991 *Natural History Museums: Directions for Growth*, Texas Tech University Press, Lubbock

Chamberlin, E R, 1979 *Preserving the Past*, J M Dent and Sons, London

Flowers, W H, 1898 *Essays on Museums and Other Subjects connected with Natural History*, Macmillan and Company, London

Hudson, K, 1981 *A Social History of Archaeology*, Macmillan, London

— 1987 *Museums of Influence*, Cambridge University Press, Cambridge

Lowenthal, D, 1996 *The Heritage Crusade and the Spoils of History*, Viking, London

Merriman, N, 1991 *Beyond the Glass Case*, Leicester University Press, Leicester

Miers, H A, 1928 *A Report on the Public Museums of the British Isles*, Carnegie United Kingdom Trustees, Dunfermline

Pearce, S M, 1999 *On Collecting – An Investigation into Collecting in the European Tradition*, Routledge, London

Purcell, R W, and Gould, S J, 1992 *Finders, Keepers: Eight Collectors*, Hutchison Radius, London

Pyenson, L, and Sheets-Pyenson, S, 1999 *Servants of Nature: A History of Scientific Institutions, Enterprises and Sensibilities*, Harper Collins, London

Schnapp, A, 1996 *The Discovery of the Past: The Origins of Archaeology*, British Museum Press, London

Tait, S, 1989 *Palaces of Discovery: The Changing World of Britain's Museums*, Quiller, London

Tinniswood, A, 1989 *A History of Country House Visiting*, National Trust/Basil Blackwell, London/Oxford

Towner, J, 1996 *An Historical Geography of Recreation and Tourism in the Western World 1540–1940*, John Wiley, London

Trench, R, 1990 *Travellers in Britain: Three Centuries of Discovery*, Arum Press, London

Vaccari, E, 2009 Mining academies as centers of geological research in Europe between the 18th and 19th centuries, *De Re Metallica*, 13, 35–41

Williams, R, 1963 *Culture and Society 1780–1950*, Penguin, Harmondsworth

Wilson, D M, 1989 *The British Museum: Purpose and Politics*, British Museum Publications, London

Woodward, H B, 1907 *The History of the Geological Society*, The Geological Society, London

Woolnough, F, 1916 The future of provincial museums, *Museums Journal* 16 (6), 125–31

5

Geoheritage for Sale:
Collectors, Dealers and Auction Houses.

Thomas A Hose

Acquiring geoheritage

For many natural history collectors, including geologists, there is something intrinsically rewarding about displaying specimens personally acquired in the field; for others, however, mere ownership of such material was and is enough. Historically, the first approach was typified by naturalists researching their collections for publication, whilst simple ownership was the purview of aristocratic and mercantile patrons building up cabinets of curiosity (see Chapter 4). A good example of a field collector is the amateur palaeontologist and artist Elizabeth Philpot (1780–1857), who spent most of her adult life in Lyme Regis where she built up an extensive collection of fossils, especially fish (now in the Oxford University Museum), from the local cliffs (Torrens 1995, 260). A good example of the aristocratic collector is Georgiana, Duchess of Devonshire (1757–1806), who purchased material from the mineral dealer White Watson (1760–1835); she also employed him to tutor her son, later the 6th Duke (1790–1858), in geology. Some patrons, such as the mineralogist Robert Ferguson (1767–1840) were also collectors in their own right; a founder member of the Geological Society of London, he was one of the 16 subscribers to the Comte de Bournon's (1715–1825) classic treatise on the mineral aragonite (Bournon 1808; Lewis 2009a) and, perhaps less gallantly, he was at the heart of one of the major divorce scandals (Nagel 2004) of the 19th century, between the diplomat Thomas Bruce, 7th Earl of Elgin (1766–1841) and his first wife Mary (née Nisbett), the Countess of Elgin (1778–1855). Patrons such as the Duchess of Devonshire bought from professional natural history preparators and dealers, such as J R Gregory & Company (see later); consideration of their role provides a background to the development of the natural history (including geological) material eventually sold in auction houses.

Auctioneers' and natural history dealerships' catalogues detailing items for sale are invaluable to anyone tracing the origins of material that has often been dispersed across several collections over time. However, because geological specimens in particular were often acquired directly from miners, quarrymen and other private collectors (with little or no surviving original correspondence), their origins are sometimes uncertain. Further, 'Recognition of the commercial as opposed to the amateur collector has, however, often been covert – their names frequently being omitted from museum records' (Rolfe, Milner and Hay 1988, 139), reflecting 'the common prejudice of museums against buying fossils' (Taylor 1987, 60). Public libraries have been less active in acquiring early geological texts and maps, especially when major private collections have been dispersed through auction.

FROM TAVERN TO AUCTION HOUSE

Auctions were relatively uncommonly used to sell books, art and natural history items until the 18th century. Before then, haggling, often through intermediaries, and set-price sales were usual. Europe's first recorded auction house, the 'Auktionsverk', was opened in Stockholm in 1674 by Baron Claes Rålamb (1622–98). Another early major Swedish auction house is the Uppsala 'Auktionskammare', established in 1731. Elsewhere, the 'Dorotheum' in Vienna, now the largest auction house in German-speaking and continental Europe, was established in 1707. In the late 17th century *The London Gazette* began reporting artwork auctions; by the 18th century's close, auctions of artworks, libraries and natural history collections were commonly held in taverns and coffee-houses – which was the genesis of the Sotheby's auction house (Lacey 1998, 20–2). In examining the various auction houses, some note of their history is useful in that it illuminates the challenge posed by numerous premises and name changes in keeping track of the origins of traded geoheritage material. Further, these changes and the individuals involved have not always been well documented, even by the major institutions. Interestingly, even their official histories and websites cannot be relied upon for their accuracy (Lacey 1998, 23–4).

EUROPEAN AUCTION HOUSES

Today, two large auction houses dominate the European auction scene: Christie's and Sotheby's. Christie's was founded by James Christie (1730–1803), who is recorded as renting auction rooms in 1762, although newspaper advertisements for his sales appeared from 1759; the earliest catalogue dates from 1766. He held sales in his 'Great Rooms' on London's Pall Mall. By 1778, James Christie had moved on to art auctions and he arranged the sale of Sir Robert Walpole's collection of pictures on behalf of his grandson, George Walpole; they were sold to Catherine the Great of Russia for £40,000 (£560,000 at today's prices). In 1795, Christie realised £25,000 (£280,000 at today's prices) on the auction over the course of five days of the contents of Sir Joshua Reynolds' studio. Christie's established a reputation as a leading auction house by taking advantage of London's status as the major centre of the international art trade, especially after the French Revolution. James Christie's son, another James, took over the business on his father's death in 1803. In 1823 Christie's moved to its present London headquarters in London's St James's district, where it has held various auctions of natural history and geological interest. In 2015 Christie's auctioned a fine (1915 x 1630mm) copy of *A Physical and Geological Map of England & Wales … (on the basis of the original map of Wm. Smith 1815), revised and improved. London: The Geological Society, July [but 1 August] 1865* by George Bellas Greenough (1778–1855) for just over £8,000. Christie's is presently the world's largest auction house, with headquarters in London and New York; it now holds regular auctions at its London showroom of 'Travel, Science and Natural History', which include geoheritage material.

Its major rival, Sotheby's, began in the 1730s when Samuel Baker (1713–78) started selling books at fixed prices; his first surviving catalogue for a fixed price sale, held at the Rose and Crown public house in London's Covent Garden district, was issued in February 1734. Baker held probably his first true auction sale, in which the price realised was the highest offered, when he presided in March 1744 over the disposal of several hundred scarce and valuable books from the library of the Rt Hon Sir John Stanley, 1st Baronet (1663–1744) of Grangegorman, Co. Dublin; he was an Irish politician who had held such posts as secretary to several Lords Chamberlains of the Household (1689–99), Commissioner of Stamp Duties (1698–1700) and Chief Secretary for Ireland (1713–14), and he was (from 1688) a Fellow of the Royal Society. Following the Stanley sale, Baker was approached to auction the books of several major Londoners, including those of Dr Richard Mead (1673–1754), physician to Queen Anne (1665–1714). Mead was a collector of books, classical sculpture, zoological specimens and gemstones – all of which he made freely available for study at his house in London's Bloomsbury district. The quality and variety of his non-book collection can be gauged from its sale catalogue, prepared by J Langford for his auction in March 1755, that realised just over £3,332 (approximately £600,000 at today's prices); it included geological material such as 'A model of Governor Pitts great diamond in crystal' (Langford 1855, 9) and 'Various specimens of petrified wood … Various pieces of crystal and amber, with flies and other matter inclos'd' (Langford 1855, 10), together with specimens of native silver and silver ores (Langford 1855, 11). His collection of some 100,000 books took 56 days to auction, and Baker realised (Lacey 1998, 22) over £5,508 (approximately £990,000 at today's prices). Baker had so much business that in 1767 he took on a partner, George Leigh, and the firm was renamed to Messrs Baker and Leigh – one of several changes of name for the business over the next 50 years.

After Baker's death in 1778 his estate, including the business, was divided between his nephew John Sotheby (1740–1807) and George Leigh; the business was consequently renamed to Messrs Leigh and Sotheby. By 1823, with John Sotheby and Leigh both dead, the business had passed to Samuel Sotheby (d.1842), John's son. In turn his son, Samuel Leigh Sotheby, joined the business and the two of them developed new markets – such as autographs in 1819 – and co-authored texts on early books. Under the Sotheby family's management, the auction house eventually extended its activities into auctioning prints, medals and coins, whilst continuing with library and some natural history sales. These sales included the significant library of the antiquarian Benjamin Heywood Bright (1787–1843) in 1844 to 1845. Several catalogues (Sotheby 1845) of the Bright sale survive, and it is known that some literary material was purchased by the British Museum (Murphy and O'Driscoll 2013, 73–4). Bright's non-book collections included some geological material amassed by himself and his father, Richard Bright (1754–1840); interestingly, he also donated a mineral collection with some meteorite material (Fletcher 1907) to the British Museum in 1873.

Sotheby's is presently the world's fourth oldest auction house in continuous operation. It operates today from 90 locations in 40 countries, with its headquarters now in New York, although it maintains London premises. It was reported in the 1970s that Sotheby's had held bi-monthly natural history auctions, 'selling mainly minerals and shells' (Tinker 1971). For the first time since the 19th century, Sotheby's auctioned a major mineral collection – that of the gem dealer Joseph A Freilich (b.1952) – as a single event over two days in January 2001 in New York. In February 2000, Freilich had exhibited 150 of the most important pieces of his 500-specimen collection, originally displayed in a basement gallery at his Long Island home, at the Tucson Gem and Mineral Show and in the *Mineralogical Record*, a bi-monthly publication for collectors and dealers. Freilich also exhibited the collection in seven cities in the USA and Europe, to generate interest before consigning it to Sotheby's in September 2000. Sotheby's engaged the designer of the permanent displays at the American Museum of Natural History in New York to prepare the auction pre-sale exhibition. Alongside the minerals, his library of some 600 volumes (ranging from 15th century texts on alchemy to 20th century mineralogical texts) was also auctioned. Sotheby's now holds regular high-value natural history specimen and literature sales. One Paris sale of 76 lots (including mounted mammoth, ichthyosaur and plesiosaur specimens) in September 2014 realised over €1 million. Another Paris sale in June 2015 realised the highest price, €1.3 million, ever paid in Europe for a mounted '*Allosaurus*' dinosaur skeleton.

Two other major auction houses were established in London in the late 18th century: Phillips in 1796 and Bonhams in 1793. Phillips was founded in London in 1796 by Harry Phillips (d.1840), formerly senior clerk at Christie's. In that year he conducted a dozen successful auctions. He introduced evening receptions prior to the auction as a means to promote business, a practice common today in the major auction houses. His prestigious clients included Beau Brummel (1778–1840) and Napoleon Bonaparte (1769–1821). Phillips' son, William Augustus, inherited the business and in 1879 changed the name to Messrs Phillips & Son. When he brought his son-in-law, Frederick Neale, into the business in 1882 it was renamed to Phillips, Son & Neale, under which name it traded into the 1970s when, with a regional network of auction houses, it became just Phillips and auctioned everything from furniture to art and estates. In 1999 Phillips was bought by Bernard Arnault and then merged with the private art dealership of Simon de Pury and Daniela Luxembourg. That dealership took majority control of the company in 2002; in 2008 Mercury Group acquired their majority holding and finally the whole company in 2013, whereupon the Phillips name was resurrected with a new state-of-the-art London headquarters in Berkeley Square. By then, however, it was firmly established as a purely fine art auctioneer with no interest in geoheritage material.

Bonhams is one of the world's oldest and largest privately owned auctioneers. Its Natural History Department holds three auctions annually, making it the field's international leader. As the firm's website indicates: 'An extraordinarily wide range of categories are offered, including mineral specimens; fossils, including dinosauria, petrified wood, amber and … meteorites.' Bonhams also sources new items, not previously offered for auction, from fossil preparators and mine and quarry owners. In May 2014, Bonhams in Los Angeles held a 431-lot 'Gems, Minerals, Lapidary Works of Art and Natural History' auction. Minerals from India, Mexico, Morocco and Japan, collected from the 1950s to the 1980s, were a significant, large and varied offering that included vanadinites, smithsonites, azurites, stibnites and zeolites. The largest ruby crystal (228,000 carats) ever auctioned was also in the auction. Other mineralogical highlights included a collection of 27 rhodochrosites, colloquially 'Red Gold', from South Africa's Kalahari Manganese Field (an area no longer producing minerals), and an exceptionally rare and large demantoid (unusually, a green) garnet. Other unmounted and rare faceted gemstones, amber, gold nuggets and meteorites were also in the auction, along with mineralogically based artworks and jewellery. Auctioned fossils included dinosaur, mammal and plant material such as petrified wood. Clearly, as the recent sales at Bonhams and Sotheby's suggest, natural history (including geological) material is back in vogue in the major auction houses. This material is generally purchased from specialist dealers, some of which have lengthy histories, although not on a par with the auction houses.

THREE OF ENGLAND'S MAJOR EARLY MINERAL DEALERSHIPS

Jacob Forster (1739–1806)

The 19th century witnessed the emergence and expansion of specialist natural history and geology (especially mineral) dealers. It was to these that most collectors turned for their acquisitions, rather than the major auction houses – a trend that has continued into the 21st century. Chief amongst such mineral dealers in London were Jacob Forster (1739–1806) and James Reynolds Gregory (1832–99). Forster had established his London mineral dealership around 1766. He was well known for always holding a considerable stock of the rarest and most precious mineral specimens. The mineral 'forsterite' was named in his honour (Frondel 1972) by Armand Lévy (1794–1841), who cited Forster's contribution 'to the advancement of mineralogy by his extensive connections in that branch of science in every part of the world, and by having laid the foundation of one of the finest private collections now in the possession of Mr Heuland' (see below). Forster counted among his clients many of the leading mineral and fossil collectors of his time. He was, in terms of the quality of his material and its widespread incorporation in contemporary collections, the most important London mineral dealer of the 18th century's second half; he is generally acknowledged to be the best dealer in Europe at the time.

Forster established sales offices in Paris (run by his brother) and St Petersburg sometime before 1769, and they were still open at the time of his death. His London business was managed by his wife Elizabeth (1735–1816) during his foreign buying trips. He regularly sold specimens to all of the important mineral collectors in Paris, and was also dealing in fine shells, corals and polished gems. Forster held four spectacular Paris mineral auctions between 1769 and 1783, with catalogues prepared by the famous French mineralogist Jean-Baptiste Louis Romé de l'Isle (1736–90). Considered one of the creators of modern crystallography, Romé de l'Isle authored *Essai de Cristallographie* in 1772, the second edition of which, published as *Cristallographie* (in three volumes and with an atlas) in 1783, was for many years an influential text.

In 1802 Forster sold a large collection of splendid minerals to the St Petersburg Mining Institute's museum. The Institute was under the auspices of Tsar Alexander I (1777–1825), and it is probably no coincidence that Forster spent his last ten years in Russia, actually dying in St Petersburg. John Henry Heuland (1738–1856), Forster's nephew, then moved to London to take over the business and help sort out Forster's personal collection. Russell (1950, 398–401) has provided a most useful list of the business's sales catalogues for the period 1808 to 1848. Heuland catalogued the sales stock bequeathed to Forster's wife, for sale at auction in some 5,860 lots; held over 45 days in 1808, it realised more than £3,000 (£185,300 at today's prices). Heuland engaged the mineralogist Armand Lévy (1795–1841) to catalogue his own collection, which was eventually sold in 1820 to Charles Hampden Turner. Levy's three-volume catalogue was published in 1838 as *Description d'une collection de minéraux, formée par M. Henri Heuland, et appartenant a M. Ch. Hampden Turner*. In honour of Turner, Lévy named a rare mineral 'turnerite', also known as pictite; however, in 1866 James Dwight Dana identified it as monazite. Turner's collection was later sold to the mineralogist Henry Ludlam (1824–80), whose collection in turn was eventually bequeathed to the Museum of Practical Geology in London. In 1808 Heuland took over the business, which he managed into the late 1840s (Russell 1950). This was clearly profitable, judging by the prices reported to have been realised in several auctions. However, the great variation in inflation rates (O'Donaghue, Goulding and Allen 2004) over the course of Heuland's ownership of the business makes any price comparison with those of today somewhat speculative.[1] An 1829

---

[1] The period 1800 to 1850 is marked by great variation in inflation and deflation rates (O'Donaghue, Goulding and Allen 2004). Inflation varied from its lowest in 1808 of 3.4% to its highest in 1825 of 17%. Deflation was at its lowest in 1847 at 11.3%. This level of variation makes strict price comparisons with today, especially when earnings are not factored in, rather speculative – but it is still useful in contextualising the relative worth of geoheritage sale items.

sale (Russell 1950, 399) of 1,020 lots over six days realised some £1,956 (£180,000 at today's prices), and one of the two 1847 sales (Russell 1950, 401) of 400 lots over two days realised just over £221 (£22,000 at today's prices); the indicated modern price comparisons are actually similar to those recorded at modern-day sales.

James Reynolds Gregory

James Reynolds Gregory started work as an assistant at a London silk and jewellery business. In 1858, aged just 26, he established his own mineral and fossil business. He amassed a superb personal collection of meteorites, authoring several associated papers. His competence is indicated by his election as a Fellow of the Geological Society and by his memberships of the Mineralogical Society, the Mineralogical Society of France, and the Royal Society of Arts. Gregory was one of London's most important mineral dealers in the second half of the 19th century. His company became one of the longest-running mineral dealerships in the world, second only in age and influence to the F Krantz Company of Bonn; this was founded in 1833 by Adam August Krantz (1809–72) when he was a student at the Freiberg Mining Academy. He moved the business to Berlin in 1836 and, finally, to Bonn in 1850. August Krantz became one of the foremost mineral dealers in Europe, and the company is still trading today as the world's oldest geological dealership, unlike Gregory's. However, Gregory's was one of the best London dealerships and exhibited at many major commercial shows in London and elsewhere; it won awards for the excellence of its specimens in London (1862, 1883 and 1884) and Paris (1867). Gregory did little field collecting, mainly purchasing at auction; he also bought and exchanged from other dealers and private collectors. He supplied many of the major contemporary collectors, together with scientists requiring research samples.

Around 1896 the business became J R Gregory & Company; it was still under James's management but he was by then assisted by his son Albert, who had been working for the business since he was a teenager. When James died in December 1899, the business passed to Albert. James's meteorite collection was broken up and sold by his sons, with a large percentage being incorporated into what would become the Natural History Museum. Over time the company absorbed several others, including Russell & Shaw, Samuel Henson and Francis H Butler. In 1931 Percy Bottley took over the company, renaming it (in deference to his predecessors) to 'Gregory, Bottley & Company'. Bottley added the stock of G H Richards in 1936. During World War II, the business carried on with several excellent private collections acquired and then sold on. After Percy Bottley's death in 1981 the business was sold to Brian Lloyd, who had previously operated a London mineral dealership, and it became Gregory, Bottley & Lloyd. Lloyd continued the company's specialisation in historic collections and in 1997 acquired that of Robert Ferguson (1767–1840). In 2008 the business moved to Walmer, Kent. Over 150 years, in its various guises, the dealership had ten different London addresses.

In 2014, with none of the Lloyd family wishing to continue trading, the business closed; the Canterbury Auction Galleries sold off the remaining stock, fixtures and fittings for a meagre £180,000 on 12 June 2014 (Sims 2014). All 532 lots were sold to both UK and international buyers. It took more than nine hours to complete the auction, with around 150 buyers packed into the saleroom and with strong interest from buyers from Australia, the Netherlands and Italy. The lots included a collection of 85 specimens of polished amber (many encasing preserved insects), which sold for £5,000, and a set of 17 tourmaline crystals from Afghanistan and Pakistan, which sold for £1,700. The James Gregory Educational Fossil Collection, comprising 200 groups of exhibits, sold for £4,000 – double its low estimate. There was also undoubtedly remnant material within the auction from the old Ferguson acquisition; interestingly, tourmaline crystals from the Ferguson collection appeared for sale on eBay in July 2014.

Isaiah Deck (1792–1853)

Isaiah Deck (1792–1853) established himself in 1825 as a pharmacist and mineral dealer in Cambridge (Green 1978). He perfected a method for casting high-quality fossil replicas, and supplied geological equipment ranging from hammers to goniometers (precision instruments used to measure the angles between mineral crystal faces). He advertised himself as a mineralogist who 'furnishes every description of Geological specimens from the various localities of England and the Continent – Fossils … volcanic rocks and minerals … single specimen to extensively fitted up cabinets' (Green 1978, 132). He published a guide to collecting and arranging a geological collection in 1841. In 1834 he was employed to arrange the huge mineral collection of Sir John St Aubyn (1758–1839). The collection was split into four portions: two small ones for St Aubyn's wife and daughter; a larger one for the Devonport Museum (the remains of which are now in the Plymouth City Museum); the remainder was auctioned and Deck bought a substantial portion of the lots for himself. His competence and social standing can be gauged from his election to Fellow of the Geological Society in 1838; he donated a few specimens to the Society's collections in 1841. Some of his specimens are now in the Sedgwick Museum, Cambridge, in the Yorkshire Museum, York (fossils and casts recorded as donated from 1837 to 1844) and in the collections of the British Geological Survey.

THE MARKET FOR HISTORICAL GEOHERITAGE: A PARTIAL PAST RECORD

Over time, many of the records of auctioneers and dealerships relating to 19th and early 20th century dispersals of natural history collections were lost, except for some sale catalogues. There are also few published histories of regional auction houses (but see Westgarth 2009) and dealerships, let alone natural history and geological (but see Cooper 2007) dealerships. Hence, when researching the origin of specimens in museum collections, apart from often incomplete acquisition registers, it is the material attached to or housed with the specimens that can best identify their individual origins. Unhelpfully, it was seemingly fashionable in the past for some curators to create new labels and dispose of the old ones – so losing important data – especially if they had significantly deteriorated due to age or poor storage. Consequently, tracking down the origins of individual specimens requires significant detective work, often tracing movements through a variety of agencies, as exemplified by the case study below of the Glasgow and Plymouth museums. The other case studies trace the history, fate and connections of specific collectors and their collections, and also indicate something of the financial viability of their associated dealerships; they particularly show how the fate of collections is closely linked to the roles and interconnectedness of auction houses, dealerships and collectors.

Glasgow and Plymouth Museum connections

The David Corse Glen Collection was purchased by the Glasgow Museum and Art Gallery in 1896 (Campbell 1976, 341). Glen was an engineer with an interest in geology – particularly mineralogy. An active member of the Geological Society of Glasgow, he published a few papers in their *Transactions*. Like many of its contemporaries, Glen's collection was amassed by incorporating material from other collections. Its major addition was the mineral specimens purchased from the executors of Archibald Thomas Frederick Fraser (1736–1815), in whose family ownership it had been since it was purchased at auction in London in 1805; the catalogue for that auction claimed that it contained specimens from the extensive collections formed in the mid-18th century (Eddy 2008, 113; Wilson 1994a, 1994b) by the 3rd Earl of Bute (1713–92) (Lewis 2009b; Wilson 1994b) and Abraham Gevers (1712–80) of Rotterdam, amongst others (Campbell 1976, 341). According to the auction catalogue, this early material was originally assembled by 'a Gentleman deceased, well known to the first Mineralogists in this country'. That gentleman has been identified from a note in a copy of the sales catalogue held by Sir Arthur Russell (1878–1964), written by Philip Rashleigh (1729–1811), which states: 'This collection is said to have been Mr Atkinson's and to have been purchased by Mr Fraser for £550.' The Atkinson Collection at the time of the 1805 sale comprised 2,701 lots, arranged according to Babington's 1799 system (based on a mineral's chemistry, physical and external characteristics). Sadly, as Campbell (1977) noted, there are 'no biographical details of Mr Atkinson'. After Glen's death in 1892, his entire collection – including minerals, fossils, rocks, shells and antiquities – was purchased from his executors by Glasgow Museum for £500.

Some of the distinctive labels on 18th century specimens within Glasgow's Glen Collection are identical in style to those on some specimens in Plymouth City Museum. The Plymouth specimens are in the Sir John St Aubyn mineral collection, which was transferred to the museum in 1924 in a series of cabinets. The value of this collection lies in its various associations, including Glasgow. Sir John St Aubyn, 5th Baronet (1758–1839) was elected a Fellow of the Royal Society in 1797. He collected a huge number of engravings and etchings; when they were sold at Phillips's Auction Rooms in April 1840, the collection was so vast that the sale took 17 days. Sir John's father (the 4th Baronet) was brought up by Dr William Borlase (1695–1772), a passionate mineral collector. It was perhaps this influence that generated Sir John's interest in minerals and natural history. Sir John amassed his own significant geological collection through personal collection, exchange, gift and acquisition. He purchased in 1799 two major mineral collections. The first was that of Dr William Babington (1756–1833), who had previously obtained the collection from John Stuart, the 3rd Earl of Bute (1713–92) (Wilson 1994b). The second was the fossil and mineral collection of Richard Greene (1716–93) of Lichfield. Sir John then incorporated these collections into his own. After his death, Sir John's collection was split by the mineral dealer Isaiah Deck (1792–1853), and although an extensive collection was arranged for the Civil Military Library at Devonport (now at Plymouth City Museum), the remaining minerals were auctioned.

The distinctive labels at the Glasgow and Plymouth museums are almost certainly those used by Jacques-Louis, Comte de Bournon (1771–1825), a refugee from the French Revolution who escaped to England in 1792. Because of his advanced expertise in mineralogy, he was immediately elected to the Royal Society; he soon became a leading figure within London's scientific circles. At that time mineral collecting was very much in vogue among the wealthy elite, who spent huge sums on their collections. Bournon was employed by some of these men to curate and enhance their collections. Unfortunately, his main patron, Sir Charles Greville, died in 1809 and his income was suddenly much reduced. Unsurprisingly, when the British Museum purchased Greville's mineral collection, Bournon, who had worked on the collection for 18 years, rather expected to be employed to catalogue it. However, he was offered a derisory salary and Sir Joseph Banks then offered the job to someone else. Bournon catalogued the St Aubyn

Collection in around 1806. The preliminary address of the 1805 sale catalogue, the basis of the Glen Collection, states that it was compiled by Dr William Babington 'and other intimate friends of the deceased'; it appears that the Comte de Bournon also compiled part of the 1805 catalogue. The fate of the Glen Collection exemplifies the changing fortunes of geoheritage collections, and the challenges to modern research and study posed by their break-up and dispersal.

Richard Greene of Lichfield and his geological cabinet

Richard Greene (1716–93) was born in Lichfield, where he eventually lived as a surgeon and apothecary. His portrait and its motto were described by James Boswell (1740–95) as 'truly characteristical of his disposition, *Nemo sibi vivat* [no man lives for himself]' (Nicols 1831, 319). It was engraved in his lifetime, and is inserted in Stebbing Shaw's (1762–1802) *History and Antiquities of Staffordshire* of 1801. Greene was an antiquary, a great collector of curiosities and a frequent contributor to the *Gentleman's Magazine*. Shaw was favoured by Greene's son with the loan of some valuable manuscripts and plates from Greene's museum, for use in his *History and Antiquities of Staffordshire*, and he embodied in his account of Lichfield a description of Greene's museum collection. Greene deposited his collection and arranged it as a museum of his day – a cabinet – in the ancient registry office of the bishops of Lichfield, which then stood almost opposite the cathedral's south door. The museum was well-placed because Lichfield was regarded as the cultural capital of the West Midlands. A view of one side of the room of this museum appeared in the *Gentleman's Magazine* for 1788; it was also reproduced in Stebbing Shaw's *History and Antiquities of Staffordshire*. The various natural history specimens in the museum directly influenced Erasmus Darwin's (1731–82) natural science interests in evolution and classification.

Erasmus Darwin lived in Lichfield from 1756 to 1781, and the spectacular shells in the museum might have added something to Darwin's inspiration for his famous motto: 'Everything from Shells'. Many of them were probably donated by Sir Ashton Lever (1729–88) of Manchester, a famous natural history collector. Interestingly, Lever, when financially embarrassed, hit upon the idea of raising funds by disposing of his collection via a lottery.[2] Greene's museum was open with free weekend admission. In 1773, *A Descriptive Catalogue of the Rarities in Mr Greene's Museum at Lichfield* was printed; this had a dedication to Ashton Lever, 'from whose noble repository some of the most curious of the rarities had been drawn'. A 'general syllabus of its contents' and a second edition of the catalogue were published in 1782; a third edition of the catalogue was issued in 1786. Although it was neither the first nor the only museum of curiosities in England, it was still a rare and unusual institution for its time. The only similar collection then open to the public was the Ashmolean in Oxford; although the collection of Sir Hans Sloane (1660–1753) was given to the nation in 1753, it only opened as the British Museum in 1759.

In the 1786 catalogue of Greene's museum, Daines Barrington (1727–1800), a lawyer, antiquary and naturalist, is mentioned among the benefactors; he was particularly known for his bird observations, some of which were described in a paper, 'Experiments and Observations on the Singing of Birds', presented to the Royal Society (Barrington 1773) and incorporated into Volume 2 of Thomas Pennant's (1726–98) *British Zoology* (Pennant 1776) as Appendix V

---

2 Lever reasoned that 'his life's work stood more chance of being preserved and perhaps built up further if it passed to a new owner as a single entity' (Allen 1976, 69). Fortuitously the winning ticket was held by James Parkinson, who then had one of the UK's finest zoological collections in England, noted for its malacological holdings. Some 20 years later, when Parkinson sold it, the bulk went to the Imperial Museum, Vienna.

(Barrington 1776a); the same volume as Appendix IV also has 'Of the Small Birds of Flight' (Barrington 1776b). In 1781 he attempted to explain the origin of fossils. On a copy of the second edition of 1782 of the catalogue, held at Stafford's William Salt Library, is the dedication in Greene's hand: 'From R Greene to his worthy and much esteem'd Friend and benefactor Mr J White, London'. It was through the Reverend John White that both Richard Greene and Ashton Lever knew his brother, the celebrated naturalist Gilbert White (1720–95), who authored *The Natural History and Antiquities of Selbourne* (1789); this includes letters written to Thomas Pennant and Daines Barrington. White also collected natural history specimens and donated some to Greene and Ashton Lever.

The third edition of Greene's catalogue (1786) was dedicated to Ashton Lever and 'Mr Pennant, immortalized by his various, faithful, and splendid publications, in Antiquities and Natural History'. Thomas Pennant (1726–98) was one of the 18th century's leading naturalists; he authored some 25 works, including *British Zoology* (1776–77), *A History of Quadrupeds* (1781) and *A Tour in Wales* (1778 and 1781). As a Fellow of the Royal Society, he was probably involved in the decision to send the *HMS Endeavour* to the Pacific in 1768. A close friend of Joseph Banks, he later included information received from him about the Australian fauna in his *A History of Quadrupeds* (1781) and in the fourth volume of his *Outlines of the Globe* (1798). All three of Greene's catalogues are dedicated to Sir Ashton Lever, 'from whose noble repository some of the most curious of the rarities were drawn'. A close friendship is suggested by the style of Greene's dedication, especially that of 1782: 'To you, my kind friend, by whose encouragement I was instigated, and by whose good offices I was enabled, to commence Virtuoso, I, once more, dedicate a descriptive catalogue of the Lichfield Museum'. The 1786 dedication described Ashton Lever as 'illustrious and generous Benefactor immortalized by his own matchless Museum'. Lever began his own collection, starting with seashells, in about 1760.[3]

By 1773, Greene's collection was particularly rich in coins, crucifixes, watches and natural history, including geological specimens. By 1786, before the only known illustration of (just a side room of) the collection had been published in the *Gentleman's Magazine* (White 1788), it had been augmented by new minerals, orreries, deeds and manuscripts, missals, muskets and armour. It also contained numerous curiosities from the South Sea Islands presented by David Samwell (1751–98), surgeon of the *HMS Discovery*, to the Lichfield poet Miss Anna Seward (1742–1809), who transferred them to Greene. In the five-page list of the collection's benefactors can be found the names of many of the most important late 18th century scientists – leading figures in the English Enlightenment – such as Matthew Boulton (1728–1809), Erasmus Darwin, Charles Darwin (1809–82), Dr Samuel Johnson, Thomas Pennant, Reverend Samuel Pegge (1704–96), Dr John Taylor (1740–84) and Dr William Withering (1741–99). Samuel Johnson (1709–84) was actually related to Greene. He and Henry Thrale (1724/30–1781) and family (who occasionally journeyed with Johnson, such as to Wales in 1774) visited and admired Greene's museum in July 1774. Two years later, Boswell and Johnson viewed it together. Boswell admired the 'wonderful

---

3   Lever's collection developed into one of the UK's richest private natural history collections and included a menagerie. Its first public display took place in Manchester in 1766. Following its success, Lever's museum opened to the public at Alkrington Hall near Manchester in 1771. In 1774, Lever moved to London and the following year opened his 'Holophusicon' in Leicester Square to the public. Captain James Cook was so impressed by Lever's collection that he gave his Australian items to the museum – the first place in Europe to display Australian specimens.

collection' with the neat labels, printed at Greene's own press, and the board with the 'names of contributors marked in gold letters' (Boswell 1824, 164).

It was the *Gentleman's Magazine* that widely broke the news of Greene's death in June 1793, mentioning that he was a 'surgeon and apothecary, and one of the aldermen of that city. He was the proprietor of a museum that merited and attracted the notice of the antiquary and curious of every denomination; to the collection of which he dedicated the principal part of his life, and which, free of charge, was open to the inspection of the curious.' A few years after Greene's death, the collection was broken up. Of geoheritage interest is the fact that, in 1799, Greene's son sold the fossils and minerals to Sir John St Aubyn for £100 (£11,000 at today's prices); both purchaser and price are indicative of the collection's geological, especially mineralogical, significance.

Robert Ferguson and his geological connections

Robert Cauford Ferguson (1767–1840) was a wealthy patron of science, perhaps best known in his day for eloping with Mary, the wife of Lord Elgin (of the Marbles). Ferguson was a keen and skilled mineral collector. He was a Fellow of the Royal Society (1806) and a founding vice-president (along with, amongst others, the Comte de Bournon and Sir John St Aubyn) of the Geological Society (1810). He travelled extensively through Europe between 1795 and 1805, collecting minerals and fossils. From 1786 to 1810 he wrote 30 journals, detailing his early life and travels in central Europe and his expanding mineralogical interests and acquaintances; the journals also included accounts of his philandering and political views. He made large purchases of minerals during his travels and compiled comprehensive lists. Most specimens are numbered and described, with purchase prices. Unfortunately, on returning to Scotland he seems to have renumbered the collection, making it difficult to match many of the specimens with the lists. His largest purchase, for £500 in 1801, was the collection of Baron Peter Block (1764–1818), which filled five packing cases. Various other purchases were made between 1796 and his departure from Paris in 1803. With a declared value of £815 (£90,000 at today's prices), and an accompanying master packing list with individual case values and other very detailed documentation, they were packed into a dozen other cases for shipment. Particularly noteworthy in Ferguson's collection was the main part of the meteorite from a shower that fell on the town of L'Aigle in April 1803; it alone was auctioned at Christie's for £30,000 (£45,000 at today's prices) in 1998.

Shortly after Ferguson's death, Alexander Rose (1781–1860), a mineral dealer and eventually lecturer in mineralogy at Queen's College, Edinburgh, compiled a report, presumably for the Ferguson family, on the mineral collection; Rose, after some of his students established it in 1834, was the Edinburgh Geological Society's president until 1846. His report highlighted the collection's major strengths and noted that the specimens were crowded and dusty, suggesting that Ferguson had rather lost interest in his collection in his later years. Some specimens were disposed of over the next few years, with a few going to the mineralogist Sir Arthur Edward Ian Montagu Russell, 6th Baronet (1878–1964). A superb specimen of lanarkite from Scotland is now in the Russell Collection at the Natural History Museum in London, together with three Scottish minerals illustrated or mentioned in James Sowerby's (1757–1822) *British Mineralogy* (1804–17): a millerite (plate 158) from Lord Elgin's lime quarry at Broomhall, Fifeshire; a water-worn blue topaz, mentioned but not illustrated (plate 163); and an etched beryl (plate 421). Murray's *Statistical Account of Scotland* (1845) notes that Ferguson's collection's 'richness and extent is surpassed by few private collections of this sort in the Kingdom'. For pretty much the next 150 years, little mention was made of the collection until 1997, when it was sold to Gregory, Bottley and Lloyd. Much of the collection (of mainly European minerals, fossils and lapidary items) by then needed considerable effort to sort and catalogue, with a large quantity of low-quality unlabelled rocks and minerals. Parts of the collection were then sold on to other major dealers and collectors based in the USA. The closure of Gregory, Bottley and Lloyd in 2014 perhaps also indicates that the market for such material now lies beyond the UK and probably also continental Europe. Ferguson's collection exemplifies the incompleteness of the record of such geoheritage assets, due to the changing fortunes of mineralogical inquiry and publishing once the Victorian passion for mineral collecting had passed.

BUILDING UP MODERN GEOLOGICAL COLLECTIONS

As any modern UK natural history curator will testify, the funds available for purchases of specimens for their collections are far less than those available to curators of archaeology, social history and fine art. This is because 'Museum geologists have been conditioned into spending very little of their museum's purchase grants which has led to low expectation, little financial demand ...' (Doughty 1984, 8). The preferred method of acquisition was seemingly by private and, less commonly, institutional sponsored fieldwork. This is because, for scientific research purposes, acquired specimens must be fully and accurately documented in terms of their find context and location. Commercial material often lacks this information, and fossils usually have their appearance enhanced for display during preparation; this often leads to the loss of key scientific information. However, this full documentation is often considered much less important for mineral specimens, for which often vague locations are supplied by mineral dealers. What is generally unrecognised by museum managers and curators, as well as many collectors, is the considerable time it takes to either collect or prepare specimens; the prices achieved by many dealers and their preparators often fail to fully reflect these efforts. It is also probably the case that many museum geological collections achieved their scientific significance (often by the donation and acquisition of major private and society collections) mainly in the 19th century, when much scientific endeavour was focused on taxonomy. Type and figured specimens are the most scientifically important legacy of that work, to which are being added the fruits of modern taxonomic research and revision.

A particularly notable exception to the emphasis on modern collection acquisition by field-work collection was the post-war rebuilding of the geology collections of Liverpool Museum. Much of the collections had been lost when the museum was damaged by fire during the World War II blitz of the city in May 1941. The museum's first appointed post-war keeper of geology focused on building up the collections, and 'Hundreds of specimens were bought using the War Damage Fund, and Geoff [Tresise] also participated in a Geologists' Association expedition to the Harz Mountains of Germany to collect rocks and minerals for the museum' (Edwards 2008, 287). This rebuilding had begun as early as 1942, with the purchase of some shells and geological specimens. A particular emphasis was on augmenting and replacing lost specimens for the mineral collections. One of the earliest original collections was the 1893 bequest of the 15th Earl of Derby's mineral collection. Through selective purchases from 1870 to 1892, it was one of the most magnificent collections of agates and allied materials ever assembled. Sadly, of the 782 specimens listed in the initial Derby catalogue, only 112 survived World War II. Today there are some 13,000 mineral specimens in the museum's collections. Its pre-war palaeontology collections were almost completely destroyed; fortuitously, some of the historic material had been removed from the museum and survived to be augmented after the war by acquisitions, purchase and donation, to some 36,000 specimens. Few other museums, except perhaps the nationals in London and Cardiff, spent as much on the purchase of natural history or geological specimens over the same period.

Taylor suggested that:

> There seems to be no risk of fossils becoming a commodity comparable to Old Masters with their record-breaking prices … the development of a market for fossils as collectables requires fundamental changes in taste and fashion, of which there is, so far, little sign. However, we should remember that taste varies, and the auction houses have the proven ability to push new classes of object into fashion.
>
> (Taylor 1987, 61)

Taylor also noted that the auction market for fossils was insignificant compared with that for antiquarian books and *objets d'art*. He made the prescient point that most museum managers falsely assumed that fossils had no real financial value. Hence, there was:

> little to lose by trumpeting the fact that fossils are worth good money. Unless we do this, no museum will want to spend money on its fossils. The only way to convince a committee to allocate £400 to conserving an ichthyosaur will, all too often, be to emphasise its value of £10,000, not just its scientific or historical importance.
>
> (Taylor 1987, 60)

Six papers in the section entitled 'The Financial Value of Collections' feature in 'The First International Conference on the Value and Valuation of Natural Science Collections' (Nudds and Pettitt 1997). None of them covered the auction price of such collections, or any history of geology dealers and museums. Perhaps surprisingly for many geology curators, the monetary value of mineral and fossil specimens, as witnessed by the sales at major auction houses such as Christie's, has risen dramatically in the last 30 years. Indeed, 'In the past few years vertebrate fossils have become highly sought-after items, and their legitimate and black-market values have soared' (Shelton 1997, 149); further, 'The monetary value of fossils has soared in a geometric progression, and theft from and vandalism of sites and collections have risen sharply' (Shelton 1997, 149). Naturally, the highest prices have been paid for spectacular items rather than run-of-the-mill small specimens that are the basis of taxonomic and scientific research, as well as local and regional studies, and which characterise the bulk of geological collections.

ANTIQUARIAN GEOHERITAGE PUBLICATIONS

Although the auction and retail markets for geology (excepting spectacular mineral and fossil specimens) have not generally realised significantly high prices, those achieved for the published fruits of historical geological research and enquiry are a different story. There has long been a thriving market in antiquarian books, and especially for maps; in relatively recent times this has acknowledged the existence of a specialist geology audience. This market has generally divided into collectors of geological maps and geology books. The former are catered for by a relatively

TABLE 5.1 AUCTION PRICES REALISED AT CHRISTIE'S FOR CHARLES LYELL PUBLICATIONS

Book	Note	Year	Place	Price realised	
Lyell, C. *Elements of Geology*		2000	New York	$1,763	
		2012	London	($15,460)	£10,000
Lyell, C. *Principles of Geology*		1993	London		£880
	3 volumes only	1994	London	($762)	£495
		1994	London		£418
		1994	New York	$3,450	
		1998	New York	$7,475	
		2001	New York	$5,288	
	3 volumes only	2001	New York	$529	
		2003	London	($4,995)	£3,055
		2009	London	($5,864)	£4,000

small number of specialist dealers, and the latter by a fairly large number of specialist and general antiquarian book dealers. Interestingly, a list (dated April 2011) on the Geological Society's website has just ten, including Christie's, who style themselves 'Antiquarian and second-hand booksellers specialising in geology and natural history'.

Christie's has auctioned a number of 19th century geology texts over the past 20 years, at prices that have generally shown a year-on-year increase for like-for-like volumes that is beyond the level of general inflation, allowing for the usual market variations, suggesting that they have been sold and purchased both for their intrinsic and investment values. Sales of copies of Lyell's two major general works (*Elements of Geology* and *Principles of Geology*) at Christie's (Table 5.1) show a marked variation in the prices realised, rather than a simple linear increase. In 1998 Christie's auction of the Haskell F Norman Library of Science and Medicine (see Norman 1991) saw an original offprint of *The Theory of the Earth*, from the *Transactions of the Royal Society of Edinburgh* by James Hutton, realise $36,800; the high price for the offprint was because it was a privately printed signed presentation copy. Christie's has also auctioned a number of William Smith maps and associated publications (Table 5.2).

The UK's major geology map dealer, London-based '19th Century Geological Maps', regularly holds some 3,000 items in stock; much of that stock relates to the output of the UK's Geological Survey through its various incarnations over two centuries, with many of the items being no longer available from official sources. There is also, with other dealers, a small but thriving market in the local and regional geology guides that appeared from the mid-19th century onwards. Areas much covered by these include the Isle of Wight and Yorkshire, with prices varying from £25 to £150 and generally comparable in real-term prices (although they were much less affordable when originally published) with their modern counterparts.

TABLE 5.2 AUCTION PRICES REALISED AT CHRISTIE'S FOR WILLIAM SMITH PUBLICATIONS

Book	Note	Year	Place	Price realised
A Delineation of the Strata of England & Wales	1st edition	1996	New York	$21,850
	1st edition with the memoir (from The Haskell F. Norman Library of Science and Medicine auction)	1998	New York	$118,000
	1st edition, 2nd issue	2002	New York	$53,775
	1st edition from the 5th and final unnumbered and unsigned series	2003	London	($28,458) £16,730
Strata Identified by Organised Fossils, Containing Prints on Coloured paper of the Most Characteristic Specimens in Each Stratum	(from The Haskell F. Norman Library of Science and Medicine auction)	1998	New York	$57,500
A New Geological Map of England	1st edition (76.7 x 62.9 cm) (from The Haskell F. Norman Library of Science and Medicine auction)	1998	New York	$14,950
	1st edition, 1st issue (77.2 x 64.2 cm)	2009	London	($13,089) £8,125
Geological Map of Durham; on which are delineated, by colours, the courses and width of the strata		2013	London	($6,858) £4,500

SUMMARY

This brief consideration of Europe's saleable geological material has indicated just how valuable geological specimens and publications are as a geoheritage resource – financially as well as for their historical and scientific value. The prices of antiquarian geology publications, usually with several copies available, are much easier to compare than the prices of mineral or fossil specimens, which by their very nature are individually unique. Recent auction prices for mineral specimens indicate their real-world worth at a time when most natural history and geology museums have underestimated the financial (or replacement) value of their holdings. This summary history of the UK's major auction houses contextualises the development of the overall marketplace for cultural items, into which the historical geoheritage resource has been incorporated. Consideration of specific material originating from one major private collector but now held in two provincial museums illustrates the interconnections between past dealers and collectors; the biographies of some of them help to explain how significant geological material has survived to the present and also, perhaps, why so much collected in the field, especially in the 19th century (see Chapter 6), has unfortunately been lost.

The only major guide to the changing financial values and fortunes of natural history and geology collections is *Natural History Auctions 1700–1972: A Register of Sales in the British Isles* (Chalmers-Hunt 1976). It has a rather metropolitan focus, yet the list itself, because of its bibliographic references, remains extremely useful. The incorporated essays by Rolfe and Embrey, on fossil and mineral sales respectively, and the introductory sections in Cleeveley (1983) provide additional information. For Scotland, one major volume (Stace, Pettitt and Waterston 1987) summarises its natural history collections; however, much of the volume's information is consigned to its included microfiche sheets, for which few modern libraries, let alone individuals, have the required reader, meaning that accessing the wealth of information is challenging. Most detailed published studies on collections have focused on their scientific or socio-cultural values as major aspects of geoheritage, rather than on their financial worth. Since the 'First International Conference on the Value and Valuation of Natural Science Collections' was held at Manchester Museum in April 1995, little has really followed it in the UK. Only the History of Geology Group's April 2011 'Conference on Geological Collectors and Collecting' addressed the issue. In the UK and Europe, 'the increase in literature on museums and collecting within the history of natural history has not often extended to natural history dealers' (Adelman 2012, 18), unlike the situation in the USA (Barrow 2000). This chapter has indicated the potential for such studies, and their usefulness to the geological community in understanding Europe's diverse cultural geoheritage.

REFERENCES

Adelman, J, 2012 An insight into commercial natural history: Richard Glennon, William Hinchy and the nineteenth-century trade in giant Irish deer remains, *Archives of Natural History* 39 (1), 16–26

Allen, D E, 1976 *The Naturalist in Britain: A Social History*, Allen Lane/Penguin Books, London/Harmondsworth

Babbington, W, 1799 *A New System of Mineralogy in the Form of [a] Catalogue, after the Manner of Baron Born's Systematic Catalogue of the Collection of Fossils of Mlle Eleonore de Raab*, printed by T Bensley, London

Barrington, D, 1773 Experiments and Observations on the Singing of Birds, by the Hon. Daines Barrington,

Vice Pres. RS, in a letter to Mathew Maty, MD, Sec. RS, *Philosophical Transactions* (1683–1775), 1773–01–01, 63: 249–91

— 1776a Experiments and Observations on the Singing of Birds, in T Pennant, *British Zoology* (4 edn), Vol II, Appendix V, Benjamin White, London, 660–708

— 1776b Of the Small Birds of Flight, in T Pennant, *British Zoology* (4 edn), Vol II, Appendix IV, Benjamin White, London, 649–59

Barrow, M V Jr, 2000 The specimen dealer: entrepreneurial natural history in America's gilded age, *Journal of the History of Biology* 33, 493–534

Boswell, J, 1824 *The Life of Samuel Johnson, LLD, Comprehending an Account of His Studies and Numerous Works, in Chronological Order: A Series of His Epistolary Correspondence and Conversations with Many Eminent Persons; and Various Original Pieces of His Composition Never Before Published: the Whole Exhibiting a View of Literature and Literary Men in Great Britain, for Near Half a Century During which He Flourished* [vol 3 of 4 vols], Charles Ewer and Timothy Bedlington, Boston

Bournon, J L de, 1808 *Traité complet de la chaux carbonatée et de l'arragonite* [*Complete treatise on lime carbonate and aragonite*] [in 3 vols], William Phillips, London

Buck, L, and Dodd, P, 1991 *Relative Values, or What's Art Worth?*, BBC Books, London

Campbell, E, 1976 Geological collections and collectors of note: Glasgow Museum and Art Gallery, *Newsletter of the Geological Curators' Group* 7, 336–45

— 1977 Geological collections and collectors of note: Glasgow Museum and Art Gallery, *Newsletter of the Geological Curators' Group* 10, 484

Chalmers-Hunt, J M, 1976 *Natural History Auctions 1700–1972: A Register of Sales in the British Isles*, Sotheby Parke Bernet, London

Cleeveley, R J, 1983 *World Palaeontological Collections*, British Museum (Natural History), London

Cooper, J, 1971 *Under the Hammer: The Auctions and Auctioneers of London*, Constable, London

Cooper, M P, 2007 Robbing the Sparry Garniture: A 200-year history of British mineral dealers, *Mineralogical Record*, Tucson

Courtney, W P, 2004 Sir John St Aubyn (1758–1839), in *Oxford Dictionary of National Biography*, Oxford University Press, Oxford

Curry, D A, 1975 The Plymouth City Museum Mineral Collection, *Newsletter of the Geological Curators' Group* 1 (3), 132–7

Deck, I, 1841 *A System of Geological Arrangement, intended as a Catalogue to facilitate the Classification of Rocks & Geological Specimens, to which is added Directions for Collecting Fossils & Geological Specimens*, I Deck, Ipswich

Doughty, P S, 1984 The next ten years, *Geological Curator* 4 (1), 5–9

Eddy, M, 2008 *The Language of Mineralogy: John Walker, Chemistry and the Edinburgh Medical School, 1750–1800*, Ashgate Publishing Ltd, Farnham

Edwards, M, 2008 Presentation of the A G Brighton Medal to Geoffrey Tresise: address by Mandy Edwards, Chairman of the GCG at the GCG AGM, National Museum of Ireland, Dublin, 3 December 2007, *Geological Curator* 8 (10), 487–8

Fletcher, L, 1907 Historical note relative to the meteoritic fragments labelled 'Cape of Good Hope' and 'Great Fish River', *Mineralogical Magazine* 14, 37–40

Frondel, C, 1972 Jacob Forster (1739–1806) and his connections with forsterite and palladium, *Mineralogical Magazine* 38, 545–50

Green, M, 1978 Isaiah Deck (1792–1853): Continuing the story of the Decks, chemists of Cambridge, *Newsletter of the Geological Curators' Group* 2 (3), 130–7

Hampden Turner, A M C, 1838 *Description d'une collection de minéraux, formée par M. Henri Heuland, et appartenant a M. Ch. Hampden Turner* [in 3 vols], F Richter et Haas, London

Lacey, R, 1998 *Sotheby's – Bidding for Class*, Little, Brown and Company, London

Langford, J, 1855 *A Catalogue of the Genuine and Entire Collection of Valuable Gems, Bronzes, Marble and other Busts and Antiquities of the Late Doctor Mead*, J Langford, London

Lewis, C L E, 2009a Doctoring geology: the medical origins of the Geological Society, in *The Making of the Geological Society of London, Special Publication No 317* (eds C L E Lewis and S J Knell), The Geological Society, London, 49–92

— 2009b Our favourite science: Lord Bute and James Parkinson searching for a theory of the earth, in *Geology and Religion: A History of Harmony and Hostility, Special Publication No 310* (ed M Kölbl-Ebert), The Geological Society, London, 111–26

Lloyd, B, 2000 The journals of Robert Ferguson (1767–1840), *Mineralogical Record* 31 (5), 425–42

Meyer, K E, 1979 *The Art Museum: Power, Money and Ethics*, William Morrow, New York

Murphy, K, and O'Driscoll, S (eds), 2013 *Studies in Ephemera: Text and Image in Eighteenth-Century Print*, Bucknell University Press, Lewisburg

Murray, 1845 *The New Statistical Account of Scotland*, Edinburgh

Nagel, S, 2004 *Mistress of Elgin: A Biography of Mary Nisbet, Countess of Elgin*, William Morrow, New York

Nicols, J, 1831 *Illustrations of the Literary History of the Eighteenth Century* Vol VI, J B Nichols and Son, London

Norman, H F, 1991 Introduction, in *The Haskell F Norman Library of Science & Medicine* (eds D H Hook and J M Norman), Jeremy Norman, San Francisco

Nudds, J R, and Pettitt, C W (eds), 1997 *Value and Valuation of Natural Science Collections*, The Geological Society, London

O'Donaghue, J, Goulding, L, and Allen, G, 2004 Consumer price inflation since 1750, *Economic Trends* 604, 38–46 [ISN 00130400]

Pennant, T, 1776 *British Zoology* (4 edn), Vol II, Benjamin White, London

— 1778 and 1781 *A Tour in Wales* [in 2 vols], Henry Hughs, London

— 1781 *A History of Quadrupeds*, Benjamin White, London

— 1798 *Outlines of the Globe* [in 4 vols], Henry Hughs, London

Price, M T, 1993 Museums and the mineral specimen market, in *Value and Valuation of Natural Science Collections* (eds J R Nudds and C W Pettitt), The Geological Society, London, 154–7

Rolfe, W D I, 1976 Fossil sales, in *Natural History Auctions 1700–1972* (ed J M Chalmers-Hunt), Sotheby Parke Bernet, London, 32–8

Rolfe, W D I, Milner, A C, and Hay, F G, 1988 The price of fossils, in *Special Papers in Palaeontology 40: The Use and Conservation of Palaeontological Sites* (eds P R Crowther and W A Wimbledon), Palaeontological Society, London, 139–71

Romé de l'Isle, J-B L, 1772 *Essai de Cristallographie*, Chez P Fr Didot le Jeune, Paris

— 1783 *Cristallographie, ou Description des formes propres a tous les corps du regne minéral, dans l'état de combinaison saline, pierreuse ou métallique* [in 3 vols and with an atlas], Imprimerie de Monsieur, Paris

Russell, A, 1950 John Henry Heuland, *Mineralogical Magazine* 29 (211), 395–405

Secord, A, 1994 Corresponding interests: artisans and gentlemen in nineteenth-century natural history, *The British Journal for the History of Science* 27 (4), 383–408

Shaw, S, 1801 *History and Antiquities of Staffordshire* [in 2 vols], J Nichols and Son, London

Shelton, S, 1997 The effect of market prices on the value and valuation of vertebrate fossil sites and specimens, in *Value and Valuation of Natural Science Collections* (eds J R Nudds and C W Pettitt), The Geological Society, London, 149–53

Sims, S, 2014 Walmer fossil dealers sell collection for £180k and retire in style, *Dover Express*, 26 June 2014

Sotheby, S L, 1845 *Catalogue of the Valuable Library of the Late Benjamin Heywood Bright, Esq. Containing a Most Extensive Collection of Valuable, Rare, and Curious Books, in All Classes of Literature, which Will be Sold by Auction, Etc. (Catalogue of the Concluding Part of the Twenty-fourth Day's Sale ... Containing the Books on Natural History, Geology, Mineralogy, Mining, &c)*, Compton and Ritchie, printers, London

Sowerby, J, 1804–17 *British Mineralogy: or Coloured Figures intended to elucidate the Mineralogy of Great Britain* [in 5 vols], printed by R Taylor (but Vol 5 by Arding & Merrett) and sold by the Author, J Sowerby, and by White, London

Stace, H E, Pettitt, C W A, and Waterston, C D, 1987 *Natural Science Collections in Scotland*, National Museums of Scotland, Edinburgh

Taylor, M A, 1987 'Fine fossils for sale': the professional collector and the museum, *Geological Curator* 5 (2), 55–64

Tinker, J, 1971 Auctions: extinct mounted birds, *New Scientist and Science Journal*, March 1971, 578

Torrens, H, 1995 Mary Anning (1799–1847) of Lyme: 'The greatest fossilist the world ever knew', *Journal for the History of Science* 28 (3), 257–84

Walley, G, 1997 The social history value of natural history collections, in *Value and Valuation of Natural Science Collections* (eds J R Nudds and C W Pettitt), The Geological Society, London, 49–60

Wells, J W, 1947 A list of books on the personalities of geology, *The Ohio Journal of Science* 47 (5), 192–9

Westgarth, M W, 2009 A biographical dictionary of nineteenth century antique and curiosity dealers, *Regional Furniture* XXIII, Regional Furniture Society, Glasgow

White, G, 1789 *The Natural History and Antiquities of Selbourne*, B White and Son, London

White, J, 1788 A folding plate of Mr Greene's Museum at Lichfield, *Gentleman's Magazine* LVIII (ii) 847

Wilson, W E, 1994a The history of mineral collecting 1530 to 1799, *The Mineralogical Record* 25, 6

— 1994b The history of mineral collecting: 1530–1599, The Earl of Bute (1713–1792), *The Mineralogical Record* 25, 69–70

FURTHER READING

Barber, L, 1980 *The Heyday of Natural History, 1820–1870*, Jonathan Cape, London

Elsner, J, and Cardinal, R (eds), 1994 *The Culture of Collecting*, Reaktion Books, London

Knight, D M, 1989 *Natural Science Books in English 1600–1900*, Portman Books, London

Muensterberger, W, 1994 *Collecting: An Unruly Passion – Psychological Perspectives*, Princeton University Press, Princeton

Nudds, J R, 1994 *Directory of British Geological Museums: Miscellaneous Paper No 18*, The Geological Society, London

Pearce, S M, 1999 *On Collecting – An Investigation into Collecting in the European Tradition*, Routledge, London

Stace, H E, Pettitt, C W A, and Waterston, C D, 1987 *Natural Science Collections in Scotland*, National Museums of Scotland, Edinburgh

Tinniswood, A, 1998 *The Polite Tourist: A History of Country Visiting*, National Trust Enterprises, London

6

Geoheritage in the Field

Thomas A Hose

This chapter provides a historical narrative of the development of field geology by both casual (amateur and leisure geologists) and dedicated (professional and academic geologists) geotourists (Hose 2000a, 136), and the publications and maps that informed and encouraged their travels. Geology field excursions for students date from late 18th century 'Wernerian times' in Germany, when they had the practical purpose of training mining engineers and managers. However, the acceptance of the hardships of fieldwork by other individuals, especially from the social elite, required a major shift in perceptions and customs to overcome prevalent restrictive social conventions. Fieldwork was a necessity for professional and academic geologists during the 'heroic age of geology' (Zitell 1901) from around 1790 to 1820. The published results of that and later fieldwork formed the historic literary geoheritage discussed in Chapter 5. Geological and natural history society journals, regional geology guides, local field guides, student textbooks, and populist books and pamphlets on geology appeared from the late 18th century onwards.

The geologists and their geological fieldwork and publishing activity (considered in this chapter) were preceded by leisure travellers, antiquarians and tourists (whose perambulations were especially promoted by various landscape aesthetic movements) and their dedicated publications. Indeed, without these earlier socio-cultural trends and publications that opened up Europe's wild and remote places to leisure travellers, it is unlikely that field geology would have advanced in late 18th and 19th century Britain and Europe. This is because they laid the foundations of acceptable outdoor practices for gentlemen, and just a few gentlewomen, of some financial means and social standing necessary to venture into increasingly wild landscapes. These various outdoor excursionists have left a written and published record of their activities that has been partly noted in Chapter 3 and is also partly examined in Chapter 7. Whilst dedicated geology fieldwork is very much associated with the late 18th and 19th centuries, and continues to the present day, its European origins are much earlier. These origins lie in the gradual opening up, often as transit places between civilised locations such as monasteries and towns, of Europe's wild places to travellers. This opening up was achieved through the journeys of working (albeit skilled) craftsmen, merchants and those on Church business, although it is only the latter who feature in the written record.

Journeymen and travellers

From the 16th century in Europe the travels of itinerant Christian scholars, usually from merchant and aristocratic families, became increasingly important in awakening interest in and knowledge of previously unknown and wild places. Journeys to study at the famous educational institutions of France (Paris and Montpellier), England (Cambridge and Oxford) and Italy (Bologna and

Florence) also later developed into a customary component of even secular education. Meanwhile, journeymen's travels to enhance the skills of trainee craftsmen were part of the traditional and precisely defined guild world of the apprenticed artisan. Journeying in England continued up to the 19th century; it still persists in German-speaking countries (especially for carpenters), and in France it continues in the form of the 'Compagnons du Tour de France', a French organisation offering young men and (latterly) women a craft-learning experience; for this, whilst apprenticed to a competent master, they journey around France and experience character-building community life by staying at some 80 Compagnon houses. The journeyman practice is perhaps not unlike the common assertion that a geologist is only as good as the number of rocks and places (s)he has seen. Some journeymen, especially from Germany, published detailed accounts which are of interest with regards to field excursions and field guides. By the close of the 18th century, travelling had changed from a means into an end in itself. People travelled to gain knowledge, particularly on the Grand Tour (Black 1992, Chapter 9; Dolan 2001, Chapter 3). The basis of the latter was the appreciation and collection of cultural heritage, usually with some personal or bought-in visual record and a written log, often with accounts illuminating contemporary thoughts on landscapes and society.

Early alpinism and pedestrianism

Beer (1930) still provides a useful introduction to early modern alpinism, but whilst his account is restricted to 16th to 18th century travellers, there are accounts from at least the 14th century of alpine ascents. For example, one of the earliest recorded instances of leisure, rather than pilgrimage or business, mountain travel is that by the cleric and poet Francesco Petrarca (1304–74). In 1336 he ascended Mont Ventoux (1,912m) (see Ward 1891, Chapter 5). A century later, when Michault Taillevent (c.1390/1395–c.1448/1458), poet to the Duke of Burgundy (Deschaux 1975), traversed the Jura Mountains in 1430, he was quite horrified by the sheer rock faces and the terrifying, thunderous mountain cascades he encountered; in this he was somewhat akin to those 16th century travellers who ventured into Britain's rather lesser mountains and high hills. The Englishman Thomas Coryat (c.1577–1617) undertook a European tour in 1608, around half of which he walked; he published *Coryat's Crudities Hastily gobled up in five Moneths travells in France, &c* in 1611 (Coryat 1905). He is often cited as the first Englishman to undertake the European 'Grand Tour', which became a mainstay of the education of the British upper class in the 18th century, and he is recognised as one of the earliest Englishmen to traverse the Alps (Beer 1930, 60–3) – the first recorded being Sir Edward Unton (1538–82) in 1563 to 1564 (Beer 1930, 57–8). *Coryat's Crudities* is the first dedicated travel account by an Englishman, and as such might well qualify Coryat as the first recorded tourist. However, he was not really an alpinist, being 'most attracted to fertile flat or rolling farmlands. He was indifferent to mountains and their vistas. He regarded the Alps not as a destination but as an impediment to his travels' (Christensen 2012, 347). When the French courtier Antoine de la Sale (c.1388–c.1462) climbed the crater of a volcano in the Lipari Islands in 1407, he became perhaps the earliest recorded European volcanic geotourist. Sale also journeyed, recording his adventures in a chapter of *La Salade* (1440–44) that also has a map of the ascent from Montemonaco, from Norcia to the Monti Sibillini.

English equestrianism – Celia Fiennes

The early travellers and geologists in the UK up to the 19th century who have left published accounts mainly travelled by horse, or horse and carriage (see Burke 1942), occasionally by boat, and exceptionally on foot. The geologists James Hutton (1726–97) and the Reverend William Buckland (1784–1856) undertook their fieldwork on horseback, whilst Roderick Impey Murchison (1792–1871) and Gideon Algernon Mantell (1790–1852) preferred horse and carriage (Wyse Jackson 2007, 135); however, Murchison was not averse to strenuous pedestrianism for his detailed field-mapping studies in the Welsh Borderland from 1831. During the 17th century, people commonly travelled 'not for the sake of travelling, as we do today, but in a spirit of inquiry, observing and noting, in the way of Leland and Camden' (Burke 1942, 26). Celia Fiennes (1662–1741), who left one of the earliest sets of English travel journals, was an elite horseback traveller seeking her health by a 'variety and change of aire and exercise' (Morris 1982, 32) between 1684 and about 1703. She was keen on domestic tourism and had much patriotic interest and intent in travel, for if 'Ladies, much more Gentlemen, would spend some of their tyme in Journeys to visit their native land … it would also form such an Idea of England, add much to its glory and esteem in our minds and cure the evil Itch of over-valueing [*sic*] foreign parts' (Morris 1982, 32). She was interested in the natural world and wild places only when their resources promised a commercial return, as in mining districts; she is the earliest recorded geotourist in England (Hose 2008), following her recorded visit in particular to the copper mines at Ashbourne in Derbyshire, of which she gave a brief description (Morris 1982, 111). Fiennes worked up her journal notes in 1702 into a travel memoir that was never actually intended for publication; it was passed down through the family before its transcription and eventual publication in 1888, with an edited scholarly edition in 1949 (Morris 1949), after which it has never been out of print.

A continental pedestrian in England – Karl Moritz

Surprisingly few continental travellers, given its relative political stability compared with the rest of Europe, journeyed around England in the 17th and 18th centuries. Karl Philipp Moritz (1756–93) was therefore an exception when he undertook a walking tour. His published account of 1783 (a second edition followed in 1785), *Reisen eines Deutschen in England im Jahre 1782*, initially translated into English in 1795 as *Travels, chiefly on Foot, through several parts of England in 1782, described in Letters to a Friend* (Moritz 1797), covers his June and July pedestrian ramblings through the East Midlands. He visited and wrote a full account of Peak Cavern, Castleton, illustrated on the title page of the German editions. His keen observations on Derbyshire geology and landscape probably make him England's first recorded foreign geotourist. Within ten years of Moritz's tour, a fashion for walking tours was developing in Britain (see Marples 1959) and Europe; in part this was a consequence of the general increase in travel for its own sake, made easier due to improvements in roads and accommodation, which meant that people of even modest means could emulate, at least at home, the Grand Tour. That domestic Grand Tour specifically included a visit to the Peak District and its natural and historical 'wonders'.

TOURISTS INSTEAD OF TRAVELLERS

After 1750 an increasing number of visitors recorded their travels to 'wild' Wales, the Peak District, the Lake District and the Scottish Highlands. In the late 18th century, 'traveller' as a descriptor began to be replaced by 'tourist'; it was incorporated within the title of probably the UK's first national guidebook, William Fordyce Mavor's (1798–1800) six-volume *The British Tourists; or Traveller's Pocket Companion, through England, Wales, Scotland, and Ireland*. This volume incorporated accounts by other well-known travellers, such as William Bray (1736–1832); for example, Bray's 1773 recollection, in his *Tour through some of the Midland Counties, into Derbyshire and Yorkshire by William Bray, FAS. Performed in 1777*, of the overt commercialisation of Peak Cavern, where 'on the guide's back, you enter Roger Rain's House ... Here you are entertained by a company of singers' (Bray 1783, in Mavor, Vol 2, 338–9). Within Mavor there is also the 'Journal of a Three Weeks Tour, in 1797, through Derbyshire to the Lakes. By a Gentleman of the University of Oxford', in which there is an account of another visit to Peak Cavern. The 'Gentleman' – seemingly J Grant, of whom nothing is known (Ousby 1990, note 133, 230) – was not so much angered by the guide and the singers as by the fees and tips he was expected to pay on his departure from the Cavern (Anon 1798, in Mavor, Vol 5, 233). This journal is particularly useful in any study of the early history of Peak District geotourism, because of its accounts and comments on the scenery, caves, mines, spas and accommodation around Matlock as well as Castleton; for example, he thought Eldon Hole was 'a horrid, fathomless chasm' (Anon 1798, in Mavor, Vol 5, 228) and reported that:

> Saxton's Inn, at Matlock, though I believe not so fashionable as Mason's, we found excellent and cheap. After paying our bill for a night, and bathing in the pellucid cistern of the Matlock water, we walked next morning, July 5th, to Smedley's shop of spar and petrifactions, some in the natural state, and some ingeniously manufactured into vases, cups, seals, and all varieties of furniture. In company with this virtuoso, went to explore a subterraneous, deep recess and cavern, on the top of an adjacent hill.
>
> (Anon 1798, in Mavor, Vol 5, 220–1)

Interestingly, 60 years later, White's 1857 *Directory of Derbyshire* has two Smedleys listed as mineral dealers in Matlock Bath.

Three major aesthetic movements, emerging around the time that Bray and the 'Gentleman' visited Peak Cavern, underpinned tourists' changing perceptions of, and relationship with, the wild landscapes and helped to make natural history – including geological fieldwork – an acceptable pursuit for the social elite. The first was the pursuit of the 'sublime', which was overlapped and followed by the 'picturesque', whilst the 'Romantic' was a dominant movement from about 1780 to 1850. The Romantic poet William Wordsworth's (1770–1850) 1835 edition of his *Guide to the Lakes* noted that: 'Sublimity is the result of Nature's first great dealings with the superficies of the earth; but the general tendency of her subsequent operations is towards the production of beauty; by a multiplicity of symmetrical parts uniting in a consistent whole' (Wordsworth 1835, 35). The 'picturesque', hinted at in Wordsworth's *Guide*, revelled in the appreciation of softer effects stemming from nature's subsequent operations – producing, for example, the appearance of differently coloured zones in the leaves of trees, and the harmony expressed by the curve of a river's meander or a lake shore, the grouping of the rocks and trees flanking them, the interplay of light and shade over these features, and the subtle colour gradations that seemingly meld the scene. The 'Romantic' was a movement that saw the expression of the feeling of landscape and its evocation in art and literature, especially poetry.

Essentially, landscapes could be framed like a picture, especially with the establishment of scenic 'stations' or viewpoints, some of which had purpose-built structures placed at them, as pioneered by Thomas West (1720–79) in the Lake District. West had travelled widely throughout Europe and, after accompanying various parties visiting the region, he wrote a detailed account of its scenery and landscape, particularly useful to artists: 'To encourage the taste of visiting the lakes by furnishing the traveller with a Guide; and for that purpose, the writer has here collected and laid before him, all the select stations and points of view' (West 1778, Introduction). As the Lake District's first dedicated guidebook, it was a huge success and ran to several editions. West was amongst the first writers to challenge the long-held notion of many travellers of a wild, barren and savage northern Britain, as promulgated by earlier writers such as Daniel Defoe (1660–1731); hence, he was instrumental in generating interest in the region and, in so doing, preparing the ground for its antiquarian and later geological exploration.

SCIENTIFIC INQUIRY RATHER THAN MERE LEISURE TRAVEL

Antiquarian antecedents

From the 17th century onwards antiquarians studied history, paying increasing attention to artefacts, historic and archaeological sites. Their approach is appositely encapsulated in the motto, 'We speak from facts not theory', adopted by the 18th century antiquarian Sir Richard Colt Hoare (1758–1838) and published as the first line of his two-volume *The Ancient History of Wiltshire* of 1812 and 1821. William Camden (1551–1623), a founding member of the Society of Antiquaries, published *Britannia* in Latin in 1586; this was the first comprehensive topographical survey of England and an essential introduction, despite its inconvenient format, for anyone considering serious cross-country journeys. The 1607 edition was the first to include a complete set of English county maps; these were based upon the mapping of Christopher Saxton (c.1640–c.1510) and John Norden (c.1547–1625). The first edition in English appeared in 1610; it had been translated and expanded (probably with Camden's collaboration) from the earlier Latin edition by Philemon Holland (1552–1637). The 1675 edition incorporated route or strip maps, pioneered by John Ogilby (1600–76), for the post roads of England and Wales. Although still a less than portable atlas of 300 pages, and measuring 13.5 x 18 inches (34cm x 46cm), Ogilby's accurate and detailed maps, at a scale of one inch to one mile (1:63,360), were centred upon the stage coach route corridor; stately homes visible from the road and towns, villages and staging posts are pictorially depicted and measured distances given. Ogilby's maps greatly assisted travellers' navigation on their selected linear routes by showing the main sights nearby. The development of topographic mapping (see later) over the next 300 years influenced and assisted the activities of travellers, antiquarians and the later geology field excursionists.

The most notable 17th century antiquarian was John Aubrey (1626–97); between 1656 and 1691 he prepared a manuscript of *The Natural History of Wiltshire* that was only widely published much later (Britton 1847). As well as describing such items as 'Springs Medicinal' and 'Reptiles and Insects', it included a section on 'Architecture', with accounts of prehistoric sites and ruins alongside great houses. Aubrey's manuscript is contemporaneous with the significant published antiquarian volumes of Robert Plot (1640–96), Keeper of the Ashmolean Museum – *The Natural History of Oxford-shire* (1677) and *The Natural History of Stafford-shire* (1686) – which were notable for their inclusion of geological material. The former is of significant geoheritage interest because, for the first time anywhere, it illustrated (Plot 1686, Tab viii) a bone – now lost – of a '*Megalosaurus*' dinosaur. The latter is significant in geoheritage terms because its Chapter 3 ('Of the Earth') is a topographical and geological account – the earliest record of Staffordshire's coals and clays.

During the early 18th century, antiquarians such as William Stukeley (1687–1765) determined relative dates for archaeological sites without historical records, with a growing emphasis on excavation. William Cunnington (1754–1810) and his patron Richard Colt Hoare (1758–1838) undertook some of the most competent excavations of the ancient barrows, earthworks and Roman sites around Salisbury Plain, systematically recording their findings (Hoare 1812 and 1821), using stratigraphy to distinguish between primary and secondary burials. Between 1785 and 1790, Hoare undertook several continental tours and sometime afterwards published accounts, some of which were illustrated by his own travel sketches. He toured Wales (in 1800, 1805 and 1806) and Ireland (1806), preparing numerous sketches, and published in 1807 *Journal of a Tour in Ireland*. He makes, for field excursions, the telling point that: 'The English are regarded by foreigners as a <u>rambling</u> nation; but I am proud to think, that this <u>vagabond</u> spirit arises, not from any dissatisfaction with our own home, or country … but from a laudable desire of research and information' (Hoare 1807, iv–v). It is commonly assumed today that antiquarian subject matter is somewhat esoteric and of little interest to the public; it might be argued that this assumption chimes with a common view of today's geologists and their field guides!

Officers, gentlemen and geologists

Many of the early geologists, often through military service at home and sometimes abroad, could draw upon both practical training and their experience of the rigours of travel and accommodation in wild and remote landscapes. Indeed, some of the founding members of the Geological Society of London had a military background (Rose 2009). By the mid-19th century, geology's relevance to the military was sufficiently recognised so that between 1858 and 1882 it was a discrete part of the curriculum at the Royal Military College, Sandhurst, and later at the Staff College, Camberley. Before its inclusion in the curriculum, several early pioneering geologists had passed through the Royal Military Academy at both Marlowe and Sandhurst – most notably Roderick Impey Murchison (1792–1871). Murchison had been an army officer without a future after the cessation of Napoleonic hostilities, when he entered upon his geological studies. He made good use of horse-carriages, conducting a very rapid survey of Russia's geology by such conveyances in 1841. From 1828 to 1831 he explored the Auvergne volcanic region, parts of southern France, northern Italy, the Tyrol and Switzerland; from May 1828 to February 1829 he travelled with Charles Lyell (1797–1875) to the Auvergne and Italy. In 1831 he began the Welsh Borderland fieldwork, with arduous excursions on foot which led to his best-known publication, *The Silurian System* of 1839. In 1855 he was appointed the second Director-General of the British Geological Survey and Director of the Royal School of Mines as well as the Museum of Practical Geology.

John MacCulloch (1773–1835) (see Rose 1996), who produced the first geological map of Scotland (Bowden 2009), was also a former army man. His army career began in 1795 when, after completing his medical studies at Edinburgh University, he was appointed Surgeon's Mate in the Royal Artillery, simultaneously becoming Assistant Ordnance Chemist and Assayist to the Board of Ordnance. Promoted in 1804 to the post of Ordnance Chemist and Assayist, he consequently lectured at the Royal Military Academy in Woolwich, training officers for the Artillery and Royal Engineers. From 1814 he was lecturer in chemistry, and from 1819 until his death also lecturer in geology, at Addiscombe – the East India Company's military school. Hence, he was one of the UK's earliest practical geology lecturers. He was made redundant on a pension in early 1826 with the recommendation that he transferred his attentions and services to the Geological Survey; from 1826 to 1832 the Treasury actually paid for fieldwork until his major map was finished – posthumously published in 1836. MacCulloch authored two – possibly the first – English engineering geology textbooks, *A Geological Classification of Rocks* (1821) and *A System of Geology* (1831), for use at the East India Company's military school.

Other notable ex-army 19th century geologists include George Bellas Greenough (1778–1855) and Henry Thomas De la Beche (1796–1855). The former (see Kölbl-Ebert 2009), served part-time with the Light Horse Volunteers and was the Geological Society's first president. The latter, like Murchison, entered the army after training at the Military College in Great Marlow and fought in the Peninsular War – field service that was undoubtedly excellent preparation for life as a professional geologist. He became the first Director-General of the Geological Survey. Long before these British officers, the Italian Count Luigi Ferdinando Marsigli (1658–1730) was a military engineer commissioned into the army of the Austro-Hungarian Empire (see Stoye 1994). Whilst in that role he made copious topographical observations across its territories, from which resulted the lavishly illustrated six-volume *Danubius Pannonico-Mysicus* of 1726; it includes the first description and illustration of loess (a wind-blown silty deposit) in Europe as well as other geological information and maps, such as for Romania's mining regions. He established the Accademia delle Scienze dell'Istituto di Bologna (the Academy of Sciences of the Institute of Bologna) with associated museums in Bologna in 1715.

VENTURING INTO THE FIELD

Excursions and trips

From the late 18th century onwards, students have been venturing outdoors to examine geology in the field. Such travel has been variously termed a field excursion, field visit, field trip or field meeting. In the UK, the interpretation of charity law forced the Geologists' Association to rename its organised outdoor geological sessions. Green (1989, 18) notes that when it sought to 'establish its charitable status in the 1920s, the impression that the Association's excursions were essentially recreational in character led to a dispute with the Inland Revenue and the consequent introduction of the "field meeting" to replace the "excursion" in its circulars and *Proceedings*'. In the UK, the Geologists' Association has played a major role in promoting and reporting geological fieldwork. It was formed in 1858, just as the railways were emerging as the prime means to convey and accommodate large numbers of travellers. However, the multiplicity of railway companies and their independent timetables and ticketing arrangements initially presented the cross-country traveller with some route-planning challenges. For geologists based in the south, the railway network strongly influenced the localities selected for day excursions so that they were within a few hours of the capital. Prior to the advent of the railways, the most reliable and quickest transport to the south coast was coastal steamers. Coaches were also available to the same coast, and northwards from London there were the great coaching routes to Holyhead (the port for Ireland) and Newcastle (respectively the modern A5 and A1); however, they were slow and uncomfortable. Unsurprisingly, the Geologists' Association's first recorded excursion, on 9 April 1860, was by train to Folkestone; it left London at 8.35am and reached Dover just after midday. It was also the first recorded use of a printed field guide, in the form of a pamphlet distributed to the excursionists that reprinted an article from the *Geologist*.

Transport into the field

For much of the early history of geological fieldwork, and tourist travel in general, the means of getting to the preferred location had been horse, horse and carriage, and then the railway – with pedestrianism being a generally on-site activity. However, by the 1870s the ordinary (or 'penny farthing') bicycle was available and helped to spread the cycling habit. Although hardly suitable for fieldwork, at least one bicycle and a tricycle made an extended European geological journey in the late 19th century (Cole 1894). From the 1880s the chain-driven single-speed safety bicycle, coupled with pneumatic tyres by the 1890s, was ideal for geological fieldwork. The Geologists' Association experimented with cycling excursions at the 19th century's close; on 4 April 1899 such an excursion, from Winchfield to Wokingham, was led by H W Monckton (1857–1931). Up to 1905 at least another seven followed, but only two were reported in its *Proceedings* as 'Cycling Excursions'. Participation was minimal, with the Excursion Secretary noting that the three 1889 cycling excursions had only five, three and nine members respectively; he recommended that just one should be included in the 1900 programme (Green 1989). By 1907 the Annual Report recorded that: 'No special cycling excursions have been arranged, they not having been well attended in former seasons' – effectively ending them. Some geologists promoted cycling field excursions thereafter, most notably Grenville Arthur James Cole (1859–1924), who led quite lengthy student cycling trips in Ireland – the accounts of some of which still survive in Trinity College, Dublin (Wyse Jackson 2007). On the eve of the Great War, the Geologists' Association (having already employed a tram and a boat for its 1914 Lyme Regis excursion) was trialling the motor car for a Surrey excursion. Its first recorded use of motor vehicles was the 13 April 1907 excursion to Tonbridge. After the Great War the Association variously employed charabancs, motor-buses and even a canal boat for part of its long excursion to Llangollen. It continued to use the railway and motor-buses until widespread car ownership led to the inevitable instruction by the 1970s to meet at a suitable car-park.

MAPS, GUIDEBOOKS AND FIELD INSTRUCTIONS

Topographic maps

Field excursions require, apart from transport and accommodation, the services of either a competent guide or guidebooks and maps (see Henry and Hose 2016). Whilst today's guidebooks invariably have topographic maps, this was not usual until the late 19th century. Before then travellers had no way, apart from text descriptions, to appreciate the nature of the physical land-scapes they intended to visit. In the early 19th century, fairly accurate small-scale road maps were in many trade directories and some guidebooks. By the 1820s, half-inch-to-the-mile (1:126,720) topographic mapping became available. The most useful and widely used topographic maps, due to their reasonable price, were produced by John Bartholomew (1831–93). Initially he supplied other publishers, such as Ward Lock, with maps for their guidebooks; these commonly 'quarter-inch' (1:253,440) scale maps first appeared with hachures, then hachures with contours, before finally just contours to show relief; the introduction of colour layering in 1880 improved the perceptibility of high ground. In the 1880s the firm of Bartholomew published maps in their own right. Bartholomew adapted the accurate Ordnance Survey series into the more useful and popular national map series, commonly known as 'half-inch' (1:126,720) scale,maps; they were used by walkers, then cyclists, and eventually motorists up to the 1980s.

By the early 1900s, with the improvements in topographic representation, travellers' aesthetic appreciation of landscape from maps was considerably easier than in the late 1700s. At least by the 1820s they could get a broad overview of an area's surface geology in map form. One of the earliest field geology manuals had noted, despite their then scarcity, that: 'It is of the very first importance that the geologist should, before he proceeds to the examination of a country, be provided with the best physical map that can be procured, so that his observations may be recorded on that which will not deceive him' (De la Beche 1833, 600). However, good quality official maps were markedly expensive in real terms up to the mid-20th century. In 19th century England, the first Ordnance Survey maps cost three guineas (£3.30), 20 days wages for a craftsman (Hewitt 2010, 166–7). Whilst such a price restricted their purchase to the social elite, their use was somewhat more egalitarian when they could be borrowed from the developing public libraries and mechanics' institutes to which the Ordnance Survey was prepared to donate copies. Those same institutions and the emerging natural science societies also afforded lower middle and working class readers the opportunity to read the latest geology texts.

Geological maps and atlases

The improvements in topographic maps and the developing understanding of stratigraphy facilitated the systematic mapping of Britain's rocks in the early 1800s (see Henry and Hose 2016). Elsewhere in Europe, especially Germany, maps of specific mining districts had been prepared in the late 18th century. Official geological surveying in the UK only began in the 1830s as the Ordnance Geological Survey, and as the Geological Survey of England and Wales from 1845; the Survey used the Ordnance Survey one-inch topographic series as base-maps. However, when the 'Father of English geology', William Smith (1769–1839) was searching for a base-map on which to show his strata, John Cary's county-based road map was the obvious choice because of its accuracy and simplicity; it was actually simplified even further to serve as the base of William Smith's geology map. Cary's atlas, a *New Map of England and Wales and part of Scotland* (1794), was first used by William Smith for his *A Delineation of the Strata of England and Wales and part of Scotland*. Cary and Smith actually worked together to produce a suitable base-map for presenting geology. Cary re-engraved his 1794 atlas version, simplifying it for clarity by removing administrative boundaries and many place names. Cary, at Smith's request, also enhanced detail in the drainage systems.

After Smith had published his (financially unsuccessful) geological map in 1815, he started working on two related projects – geological cross-sections and county geological maps. Both were expensive, major undertakings that stretched his resources and were never completed. The county maps were based upon those in Cary's 1809 *New and Correct English Atlas*. The coloured maps, with Smith's geological information, were published as six atlases of four county maps each (just 21 counties, because Yorkshire needed four maps) between 1819 and 1824. Cary published three further county geology maps, but they were uncoloured. As well as Smith's national geological map, and before the Geological Survey maps in the mid-1850s, several small-scale national and regional geological maps appeared in geology texts such as Phillips (1818), Conybeare and Phillips (1822), Johnston (1850) and Woodward (1876).

Commercial publishers recognised the popular interest and need for more accessible and affordable geological information, and combined the county atlas format with geological and railway information. *The Geological Atlas of Great Britain*, published by James Reynolds in 1860 and 1889, was a response to the public's interest in geology and ready access to travel via the railway network. In Reynolds's editions, the maps were accompanied by very brief text explanations. The hand-coloured maps were of the English counties – singly and in pairs or small groups – with just two maps for Wales and a folded map of Scotland, partly reflecting market demographics. The 1860 edition was published before the Geological Survey was completed and relied on information from commercial sources. In the 1889 edition, the text was expanded and the maps substantially revised for geological boundary changes, with annotations of geological features and fossil sites, plus an updated railway network.

Edward Stanford (1827–1904) eventually acquired Reynolds's map plates. His first edition of 1904 retained the pocket-sized format, at 7.5 x 5.25 inches (18.75 x 13.13 cm), kept and updated the railway lines, introduced colour printing and 50 monochrome plates of fossils, and expanded the text to include the geology encountered along main railway routes; its text included small woodcut illustrations of scenery and geological cross-sections. In the second edition of 1907, *The Geological Atlas of Great Britain and Ireland* added maps of Ireland with an accompanying text. The 1914 third edition added the Channel Islands. In 1913, *A Photographic Supplement to Stanford's Geological Atlas of Great Britain and Ireland* was published as a companion volume to the *Atlas*; its remarkably clear and finely printed 2.5 x 3.5 inches (6.25 x 8.75 cm) images are in geo-chronological order and are described with references to relevant geological memoirs and texts. The *Supplement* was cross-referenced to the second edition and referred to in the third edition. The last, and commercially unsuccessful, edition with an uncoloured version of the British Geological Survey map was published in 1964; despite having an up-to-date map and descriptions, the maps were difficult to read. The popularity and commercial success of the Reynolds and Stanford atlases up to the 1960s is evident, however, since they went to numerous editions. With their original pocketable size, copies were intended to be carried on journeys – especially railway journeys. They were much used and consequently few survive intact, as the low-cost binding was not durable. Sadly, there is no modern equivalent.

Guidebooks

The tourist guidebook was primarily a 19th century development, for British and German travellers who had both the money and intellectual curiosity to travel in any numbers (Sillitoe 1995). John Murray's 'handbooks', published from 1820 onwards, were written for the educated elite traveller; he produced a compendium focused on visitors' perspectives of what was important, including where to eat, stay and bank. John Murray's of Edinburgh also produced British county guidebooks. By the mid-19th century elite travellers were a less important market than the burgeoning middle-classes, with more modest means and education, and from 1854 the *Ward Lock Guides* were intended to meet their (supposedly) less demanding needs. Karl Baedeker's *Handbuchlein* adopted Murray's term, rather than guidebook, and were the first to employ asterisks as commendations promoting significant sites and sights to tourists. His first guide to England, in German, was published in 1862. By the late 19th century, tourists were well provided with good-quality affordable guidebooks, many of which mention geology; however, such mentions were not always accurate, as shown by: 'took the ferry to Northern Ireland for a day trip to the Giant's Causeway. Formations among the huge granite [they are actually basalt] blocks seemed like pipes of a great church organ' (Hindley 1983, 52–3).

Field instruction texts and field guides

To promote geological fieldwork skills, a few specialist texts were published from the late 17th century onwards. The earliest was John Woodward's (1665–1728) *Brief Instructions for Making Observations in all Parts of the World and also for Collecting, Preserving and Sending over Natural Things* of 1696. Woodward also included *Brief Directions for making Observations and Collections, and for composing a travelling Register of all Sorts of Fossils* in his *Fossils of all Kinds, Digested into a Method, Suitable to their Mutual Relation and Affinity* (1728). Principal amongst the 19th century British field skills texts were De La Beche's *Geological Manual* (1831), *How to Observe Geology* (1835) and *The Geological Observer* (1851). An early Italian field observation guide, originally compiled by Antonio Vallisneri (1661–1730) to help a student, was *Indice di osservazioni* in the final pages (404–19) of G Perrucchini's *Continuazione dell'Estratto*, published in 1726. Giovanni Targioni Tozzetti (1712–83), a Florentine naturalist, sent instructions to his collaborators on geological collecting and observation (Vaccari 2007); these were published with specific regard to Tuscany as *Lista di Notizie d'Istoria Naturale della Toscana, che si ricercano* in his *Prodromo della Corografia e della Topografia Fisica della Toscana* of 1754 (see Vaccari 2007). In Germany the demands of mining education produced several texts, amongst which (in translation from the 1747 Swedish edition) was *Mineralogia* of 1750 by Johann Gottschalk Wallerias (1709–85), subsequently translated into French (1753), Russian (1753) and Latin (1772), with revisions up to the 1780s. Such texts, the beginnings of the modern geology field guide, aided the development and support of European scientific travel (see Stafford 1984).

A field guide has been defined as: 'An illustrated pocketable publication that specifically describes, and might well explain, the physical attributes of an area by reference to located sites and what can be found at them' (Hose 2006, 126); just substituting 'geology and/or geomorphology' for 'physical attributes' in that definition would create a workable definition for a geology field guide. Focused on the occurrence of rocks and their inclusions, geology field guides are akin to botanical floras that, from the mid-19th century onwards, guided botanists to specific plant locations; floras usually included a geological account of their area. Pocketability is somewhat arbitrary, depending on the size of users' pockets, although around A5 size can be considered a practical limit. Their development in Europe roughly coincides with that of tourist guidebooks. The field guides selected for inclusion below date from 1800 to 1960, coinciding with the period from the emergence of modern scientific geology to the beginning of the higher education geology boom; the drivers for the latter were the emergence of the plate tectonic concept, the discovery of North Sea oil and the promotion of environmental education.

EARLY UK REGIONAL GEOLOGY GUIDES

Some early UK regional geology texts

Today an internet search or perusal of specialist bookshops' shelves reveals a varied range (in terms of price, accuracy, modernity of geological information and communicative competence) of local, regional and national geology accounts and field guides. Indeed, few regions of the UK and Europe lack any such coverage. Such publications were initially developed in the early 19th century for amateur geologists (or casual geotourists), mainly wealthy and leisured gentlemen, rather than students and professional geologists. Some of the country's then burgeoning local and regional natural history societies, such as the Chester Society of Natural Science, Literature and

Art and the Woolhope Naturalists' Field Club (Burek and Hose 2016), also published accounts of field excursions in their transactions that could be read as field guides.

The earliest national texts on English and Welsh geology were by William Phillips (1773–1828) – *A Selection of Facts from the Best Authorities, arranged so as to form an Outline of the Geology of England and Wales* of 1818; Robert Bakewell (1767–1843) – *Introduction to Geology* of 1813 (with four further editions up to 1838); and William Conybeare and William Phillips (1787–1857) – *Outlines of the Geology of England and Wales* of 1822 (Figure 6.1B). The first and last of these volumes were reference works, necessarily with locality descriptions summarising what could be seen in the field. The first volume compiled all that was then known about stratigraphy. They all incorporated as a small fold-out a reduced version of Smith's 1815 map. The last of these classic 19th century volumes was Horace Bolingbrooke Woodward's (1848–1914) *The Geology of England and Wales* of 1876. De la Beche's 1830 *Sections and Views Illustrative of Geological Phenomena*, most of the plates of which had been published elsewhere, is a compilation with little specific to the British Isles and much on alpine Europe and exotic locations such as Jamaica, making it perhaps the geological equivalent of the 'Grand Tour' guide. The most significant and widely read 19th century geological text book was Charles Lyell's (1797–1875) *Principles of Geology* of 1830, which had a more or less global coverage.

Concomitantly with the national overviews, regional accounts were published; for example, *A Geological Survey of the Yorkshire Coast* (Young and Bird 1822). Jonathan Otley's 1823 Lake District guide, the first English geology field guide, was also the first to be tourist-focused. However, Algernon Gideon Mantell's (1790–1852) 1847 *Geological Excursion Round the Isle of Wight and the Adjacent Coast of Dorsetshire* (Figure 6.1A) was the first pocketable modern-format illustrated geology field guide. The greatest regional account was *Siluria* (Murchison 1854), developed from his two-volume masterpiece, *The Silurian System* (Murchison 1839). County-based geology accounts appeared in some late 19th century trade directories. The *Geology of the Counties of England and of North and South Wales* (Harrison 1882) was such a set of accounts bound into a single volume; Anderson's *Field Geology in the British Isles* (1983), published a century later, is its only modern counterpart. The Geologists Association's excursions were twice published as discrete volumes. Its compendium of field excursions between 1860 and 1890 'described many of the spots most important from their geological sections, or for the views they afford of the structure of the surrounding countryside; while the intending visitor to any of them may learn from it both how to employ his time most profitably, and how much time will be necessary for his purpose' (Holmes and Sherborn 1891, vi – vii). Its jubilee volume (Monckton and Herries 1909), covering its first 50 years, is something of a county geology guide, with limited locality information. Its influential and most widely used UK field guides, the *Centennial Guides*, appeared from 1958 onwards. The development and nature of field guides, reflecting the progressive change in them and their readership, can best be assessed by critiquing their content and examining the backgrounds of their authors for a specific region, such as southern England's 'Mantell Country'.

'Mantell Country' and its personalities

The 19th century classic geological study region, herein referred to as 'Mantell Country', comprises the counties of Hampshire and the Isle of Wight, Dorset and Sussex (Hose 2006, Figure 3). Much of its popularity was – and still is – due to students' exposure to its classic sites *in situ* on fieldwork, and through field guides and in textbooks since the mid-19th century. Key figures in the early development of field excursions to the area are Thomas Webster (1773–1884) and Gideon

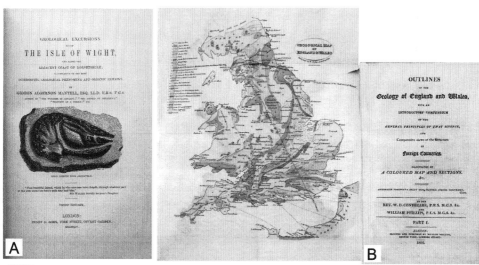

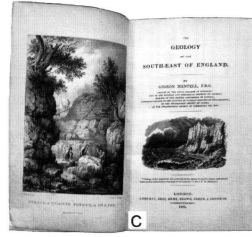

FIGURE 6.1 THE TITLE PAGES OF SOME 19TH CENTURY GEOLOGICAL GUIDES

A. *GEOLOGICAL EXCURSION ROUND THE ISLE OF WIGHT AND THE ADJACENT COAST OF DORSETSHIRE* (MANTELL 1854, 3 EDN).

B. *OUTLINES OF THE GEOLOGY OF ENGLAND AND WALES* (CONYBEARE AND PHILLIPS 1822), SHOWING THE FOLD-OUT GEOLOGICAL MAP.

C. *GEOLOGY OF THE SOUTH-EAST OF ENGLAND* (MANTELL 1833), SHOWING THE FRONTISPIECE TILGATE FOREST ILLUSTRATION.

D. *A DESCRIPTION OF THE PRINCIPAL PICTURESQUE BEAUTIES, ANTIQUITIES AND GEOLOGICAL PHENOMENA, OF THE ISLE OF WIGHT. WITH ADDITIONAL OBSERVATIONS ON THE STRATA OF THE ISLAND, AND THEIR CONTINUATION IN THE ADJACENT PARTS OF DORSETSHIRE BY THOMAS WEBSTER* (ENGLEFIELD 1816). NOTE THAT THE VOLUMES HAVE NOT BEEN REPRODUCED TO THE SAME SCALE.

Algernon Mantell (1790–1852). Webster studied architecture, his training as a draughtsperson underpinning his later geological work; he came to geological prominence through his work on the Isle of Wight and his engagement to write the geology section of the antiquarian Sir Henry C Englefield's (1752–1822) ground-breaking publication on the island's topography and antiquities, *A Description of the Principal Picturesque Beauties, Antiquities, and Geological Phenomena, of the Isle of Wight* (Figure 6.1D) of 1816 (see Heringman 2009). He was appointed the Geological Society's house-secretary and curator in 1826; for many years he also edited and illustrated its *Transactions*.

Mantell secured a five-year apprenticeship at the age of 15 with a surgeon in Lewes, Sussex. On his father's death in 1807, he inherited sufficient funds to pursue a medical career as his apprenticeship ended. After formal medical education in London, in 1811 he was awarded a diploma as a Member of the Royal College of Surgeons and became a certificated obstetrician. Back in Lewes he formed a busy and lucrative medical partnership, but also made time to study the local fossils. In 1813 he began a correspondence with James Sowerby (1757–1822), the naturalist and illustrator, who catalogued fossil shells; Sowerby named *Ammonites mantelli* in his honour. After marrying in 1816, Mantell purchased his own medical practice and took up an appointment at the Royal Artillery Hospital near Lewes. After collecting marine fossils from the local Chalk he began acquiring freshwater and terrestrial fossils from a quarry in Tilgate Forest (Figure 6.1C). Shortly before his first major geological book, *The Fossils of South Downs* (Mantell 1822), was published, his wife found several large teeth in Tilgate Forest in 1822; he eventually discovered that they were similar to those of the modern iguana lizard, but much larger, and hence he assigned the genus name '*Iguanodon*' to the dinosaur.

When in 1821 Mantell planned his book on the geology of Sussex, it was an immediate success with 200 subscribers, including King George IV (1762–1830). Mantell was a great populariser of geology through his often lavishly illustrated books, including southern England's first regional geology guide, *Geology of the South-East of England* of 1833 (Figure 6.1C). In 1833 he relocated to Brighton, but his medical practice suffered financially – probably because he was neglecting it for fossil studies. Fortunately, when he was almost destitute, Brighton Town Council transformed his house into a museum where he gave a series of lectures that were published in 1838 as *The wonders of geology, or, A familiar exposition of geological phenomena: being the substance of a course of lectures delivered at Brighton*. The museum, possibly since he often waived the entrance fee, eventually failed. Destitute, in 1838 he offered to sell his collection to the British Museum for £5,000, but accepted £4,000. He then moved to London's Clapham Common and continued in medical practice, moving to Pimlico in 1844. In 1847 he published the first modern-format geology field guide, *Geological Excursion Round the Isle of Wight and the Adjacent Coast of Dorsetshire* (Figure 6.1A), which went to two further editions. His death in 1852 pre-dated the establishment of a body that would really promote geology excursions, as he had done for at least an elite audience. That post-Mantell impetus to geological excursions was the establishment of the Geologists' Association in 1858 and the publication of its field meeting reports. Such reports in the Geologists' Association's *Proceedings* sometimes outnumber the research papers; for example, those for 1900 had 44 excursion reports and just 15 papers (Middlemiss 1989). Surprisingly, it delayed its first visit to the region until 1879; thereafter, its 19th century visits can be listed as: Barton Cliffs, 1880, 1888 and 1894; Bournemouth, 1880; the Isle of Purbeck, 1882; the Isle of Wight, 1881; Lyme Regis, 1889; Sherbourne and Bridport, 1885; and Weymouth and Portland, 1879. The area remained popular with the Association in the 20th century and four of its *Centennial Guides* were based there.

'Mantell Country': Sussex geo-publications

Sussex was the first county described in any detail, especially by Mantell, because of its proximity and easy access to London's active geological community, centred on the Geological Society. Peter John Martin's (1786–1860) *A Geological Memoir on a Part of Sussex*, published in 1828, was based upon a presentation to the Society in March 1827. It is a comprehensive regional geology account with a theoretical geomorphological discourse on the area's denudational history, and it has a fine coloured geological section. Its 'Advertisement to the General Reader' stated that 'a few words may be addressed to the less eager inquirer … For although "all the fine folk now-a-days, run up and down amongst the hills, knapping stones to find out how the world was made", yet the subject may not be so familiar' (Martin 1828, v), indicating that it was intended for both casual and dedicated geological inquirers (or geotourists).

Mantell particularly popularised the region. His first attempts, *The Fossils of the South Downs* of 1822 and *Illustrations of the Geology of Sussex* of 1827, were costly illustrated library volumes with locality-specific information; most of the former's 42 plates (but not the map) and its substantial text were incorporated within the latter volume. His *Geology of the South-East of England* of 1833 is an almost pocketable volume with some geotourism information – for example, at Rottingdean: 'The beach near this place contains semi-translucent pebbles of agate, and chalcedony … collected by visitors and when cut and polished are used for bracelets and other ornamental purposes' (Mantell 1833, 40–1). His *A Day's Ramble in and around the Ancient Town of Lewes* of 1846 had a geological chapter and his usual self-promotion: 'The large chalk-pit, now disused … was my principal field of research, when I began to investigate the organic remains of Sussex; and from it I obtained the first fossil fish discovered in the chalk of the South Downs' (Mantell 1846, 127–8).

Frederick Dixon's (1799–1849) posthumous 1850 text, *The Geology of Sussex; or the Geology and Fossils of the Tertiary and Cretaceous Formations of Sussex*, was a worthy successor to Mantell's publications. Several major illustrators and palaeontologists were involved in preparing its first edition. The preface, written by Dixon, noted:

> An apology may, perhaps, be expected for blending antiquarian notices with the description of a geological work. A few years ago there were no Archaeological associations or journals; but a local geologist, whose immediate researches were into the antiquities of remoter epochs of his district, could hardly fail to have his interest excited by the analogous evidence of the past history of his own race. I was therefore led to add the facts that came to my notice, and so attempt to fill up the hiatus between the last geological change and the existing period.
>
> (Dixon 1850, ix–x)

Like some of the earlier pioneers of geology, such as the Reverend William Buckland (1784–1856) and the Reverend William Daniel Conybeare (1787–1857), he overtly promoted his religious conviction: 'In the following pages I have entered into no speculative inquiry. Geology is not at variance with the sacred truths of Scripture; and it must be borne in mind that every fossil, as well as recent being, is the record of the will of God' (Dixon 1850, x). In 1878 it went to a second edition, which incorporated reproductions of the plates from Mantell's *The Fossils of the South Downs*. Little of merit followed these 19th century Sussex field guides for some time, although there were several centred on London that partly covered the county. The preface to one by

George MacDonald Davies, with a highly descriptive text, indicated that many modern authors aimed 'to provide a handy guide to geological fieldwork ... useful to many who are interested in the natural sciences and desire a field-acquaintance with the geological formations which occur in their district. To those who take pleasure in country walks the route here described will prove attractive' (Davies 1914). One field guide (Davies 1939) reversed the usual practice (instigated by Lyell in the mid-19th century) of beginning with the older rocks; hence, its users started with rocks that had fossils similar to modern forms before progressing onto the less familiar older fossils. It also promoted Geologists' Association membership and the benefits of museums and public libraries. In the 20th century's second half, the Geologists' Association published several Sussex guides in its *Centennial Guides*. Its *Geology of the Central Weald: The Hastings Beds* (Allen 1960) adopted a technical approach: 'An attempt is made to spotlight those sedimentary features of the Hastings cyclothems most important in interpreting the environments of deposition within the basin and in inferring the geology and geography of the uplands surrounding it' (Allen 1960, 1); its itineraries are well-detailed routes with some attempt at interpretation, as is the case with the much revised *The Weald* (Kirkaldy 1976). *The Weald* (Gibbons 1981) was one of a new format of highly pocketable field guides, with rounded corners and laminated covers designed to meet the needs of amateur geologists; its 'Invited Forward' is noteworthy for its inclusion of the geoconservation message: 'To observe and record, collecting any necessary specimens only from fallen rock, will in general give a better understanding of geology than will a physical assault on selected portions of the rock face. The indiscriminate use of hammers is thus as profitless as it is damaging.' To reinforce the conservation message, the volume's back cover alluded to the Geologists' Association's geological code of conduct. Some guides, such as Brooks (2001), focus on fossil-collecting from a limited area, Hastings; that guide also continues the 19th century tradition of privately printed geo-publications.

'Mantell Country': Isle of Wight geo-publications

The Isle of Wight attracted numerous early geological publications because: 'In the splendid sections exposed along the line of coast we are enabled to examine the strata and trace their relationships to each other in a manner which is impossible in inland districts, while the beauty of the scenery ... lends an additional charm to the investigation' (Harrison 1882, 103). The aesthetic landscape emphasis is repeated in many of today's field guides. Englefield's *A Description of the Principal Picturesque Beauties, Antiquities, and Geological Phenomena, of the Isle of Wight* of 1816 was a large seven-chapter volume (with much detail on woodlands and trees) with 48 plates, two fold-out maps and 12 detailed geological 'letters' by Thomas Webster (1773–1844). Its geoheritage significance is the seminal approach that, its preface indicated, was the result of field observations made between 1799 and 1801, which involved 'visiting repeatedly almost every part which contained anything worthy of notice, making copious notes and numerous sketches on the spot' (Englefield 1816, i). Its planned 1802 publication was delayed due to the author's personal (and possibly scandalous) private difficulties; this facilitated his requesting 'Mr Webster, who was not unacquainted with natural history, and who, to the talents of an expert draftsman, added the more valuable qualities of most scrupulous accuracy and patient investigation' (Englefield 1816, ii). Webster's command of geology can be gleaned from his twelfth letter, in which 'the north side of the island, consisting of a series of horizontal strata, had been deposited in an immense basin; and that it was not an improbable circumstance ... that they might be found to correspond with some of those in the basin of Paris lately described' (Webster, in Englefield 1816, 226–7). He then describes the rock succession in some analytical detail. His researches were aided by referring collected fossils for identification to James Parkinson (1755–1824), an eminent palaeontologist and a founder member of the Geological Society.

The area covered in Englefield's volume is similar to that of Mantell's *Geological Excursion Round the Isle of Wight and the Adjacent Coast of Dorsetshire* of 1847. Mantell's book was cheap enough to be taken into the field, probably explaining the poor condition of many non-library copies. Its preface justified the volume in words redolent of modern field guides: 'the Geology of the Island is but little known or regarded by the majority of the intelligent persons who every season flock by thousands to its shores … and take their departure without suspecting that they have been travelling over a country rich with the spoils of nature' (Mantell 1847, viii). The volume's nature can be gauged from its account of Alum Bay, which fully quotes Englefield, whilst for Brook Bay it notes that:

> As we descend to the sea-shore near Compton Chine, the top of the cliff … is seen to be composed of drifted gravel, clay and loam; and this alluvial covering continues for several miles along the coast … In this accumulation of drifted materials are numerous trunks of trees and quantities of hazel-nuts in the usual condition of peat or bog-wood … I picked up … teeth of the horse and deer.
>
> (Mantell 1854, 199–200)

Mantell's self-promotion is evident, as is his local knowledge gleaned from personal observation. Focused on fossil collecting, the volume has some 20 plates, 45 illustrations and a geology map.

Dr Ernest P Wilkins' *A Concise Exposition of the Geology, Antiquities, and Topography of the Isle of Wight* (Wilkins 1861) was seemingly an attempt to fill the tourist market niche; oddly, however, it was a subscription volume. In the opening sentence of its preface, he admitted that he did 'not pretend to have produced it from personal observation alone, but rather to have verified the investigations of others by my own researches' (Wilkins 1861, Preface). Its text is descriptive, with some attempt at explanation, as in:

> The Bembridge oyster beds are characterised by green marls and sandy bands, in many places abounding in species of Ostrea … The importance of these beds depends on the evident influx of salt water during their deposition, as marked by the abundance of oysters, accompanied by marine shells of various genera.
>
> (Wilkins 1861, 13)

The later Mark W Norman's *A Popular Guide to the Geology of the Isle of Wight* (1887), with 22 plates of fossils and topographic views (many more than its modern equivalents) and a map, is a stratigraphical account focused on fossil collecting that also covers Newport Museum and the Ventnor Collection. It is illustrated by a map, various sections, and 15 plates of fossils. Its genesis was a series of newspaper articles, originally published in the *Isle of Wight Advertiser*. Its preface justified the volume's publication on the grounds that Norman had been asked by friends and readers of the newspaper articles to produce a populist guide, as by then Mantell's volume had been published 40 years earlier. With its numerous lengthy quotes from the Geological Survey's publications, its success as a popular or even populist guide is perhaps doubtful. However, its emphasis on fossil collecting is obvious, for:

> The collector, if not fastidious as to the quality of his specimens, will not be troubled on the score of quantity, the organic remains being most abundant … The limestone yields some fair specimens of *Planorbis, Cyrena obtusa, Melania striata, Paludina, Limnea* … the fossil remains are of no great value for the cabinet.
>
> (Norman 1887, 149)

A populist field guide from 1922, *The Geological Story of the Isle of Wight* by J C Hughes, opened with the sentence:

> No better district could be chosen to begin the study of geology … abundant and interesting fossils to be found in the rocks, awaken in numbers of those who live in the Island, or visit its shores, a desire to know something of the story written in the rocks.
>
> (Hughes 1922, iii)

Its populist approach is obvious from the opening sentence of Chapter 1:

> Walking along the sea shore, with all its varied interest, many must from time to time have had their attention attracted by the shells to be seen, not lying on the sands, or in the pools, but firmly embedded in the solid rock of the cliffs and of the rock ledges which run out on to the shore, and have, it may be, wondered sometimes how they got there.
>
> (Hughes 1922, 1)

Likewise, the title of Chapter 3, 'The Wealden Strata: Land of the Iguanodon', employs the populist tag-line about dinosaurs; the chapter interestingly describes that landscape with, for example:

> What kind of trees grew in the country the river came from? Well, there were no oaks or beeches, no flowering chestnuts or apples or mays. But there were great forests of coniferous trees; that is trees like our pines and firs, cedars and yews, and araucarias; and there were cycads – a very different kind of tree, but also bearing cones – which you may see in a greenhouse in botanical gardens … With the logs and trunks of trees, which the river brought down, came floating down also the bodies of animals, which had lived in the country the river flowed through.
>
> (Hughes 1922, 18–19)

The volume's emphasis on collecting, although precise locations are not given, can be noted from: 'West of Hamstead Ledge the whole of the beds crop out on the shore, where beautifully preserved fossils may be collected' (Hughes 1922, 67). Its routes, such as 'The Lower *Crioceras* bed (16 ft.) follows, and crosses the bottom of Whale Chine' (Hughes 1922, 33–4) are technically descriptive.

The island has long attracted specialist geology groups' excursions. The Quaternary Research Association's guidebook (Barber 1987) is noteworthy for its profusion of maps and diagrams and for its concise, technical writing; for example: 'The geology … is dominated by the Cretaceous outlier of the Southern Downs. Truncations of the southern side of this by marine erosion has resulted in the formation of a large coastal landslip complex, known as the Undercliff

(Hutchinson, in Barber 1987, 123). The latest Geologists' Association guide (Insole *et al* 1998) has some 20 brief itineraries with numerous diagrams and route maps. The Geological Survey has published an intended populist large-format fold-out field guide (Gallois 1999), lacking in specific geosite detail. Geoconservation concerns at the close of the 20th century prompted the late 1980s *Guidelines for Collecting Fossils on the Isle of Wight*, probably the first such regional geoconservation publication.

'Mantell Country': Dorset geo-publications

Dorset's first dedicated populist field guide was that for Weymouth (Damon 1860); its preface suggested, with modern redolence, that a guide was needed, written in a 'scientific yet elementary and popular form … pointing out … where the various formations can be best examined, and … more easily identified by means of views, sections and other illustrations'. A second edition appeared almost a quarter of a century later. In the 20th century, Dorset has attracted a steady output of field guides. George MacDonald Davies published his first dedicated guide for the county (although see Davies 1914) in 1935; it was revised in 1939 and 1956, when its introduction noted:

> To the student of geology Dorset is well known, at least from books … But book-reading is a poor substitute for hammering the actual rocks and seeing how they fit into the structural fabric of the country … there are few areas where a beginner can get so clear an insight into geological structures and the work of the agents of erosion.
>
> (Davies 1956, 1)

Its introduction also noted – and how things have changed since then – that:

> Dorset is not so famed for the beauty of its coast line as its neighbour, Devon: hence it is not so over-run with visitors … Dorset is still unspoiled, and it will be an evil day when Poole Harbour is bridged and when the twisty lanes that lead to Tyneham and Kimmeridge are made more attractive to motorists. Only Lulworth and, latterly, Portland Bill attract the crowd. Even the towns – Lyme Regis, Bridport, Weymouth and Swanage – all have long-established industries of their own and are not primarily visitor traps.
>
> (Davies 1956, 1)

It is also rather noteworthy for its complete lack of geoconservation concern, with its suggestion that fieldwork is best undertaken with a hammer. Meanwhile, the Geologists' Association published several of its pocketable *Centennial Guides* (House 1958) and later volumes (Allison 1992; House 1989) for Dorset, with one (Ager and Smith 1973) specifically focused on the coastal strip of Dorset and adjacent Devon; they exemplify the evolution from the text-rich and sparsely illustrated field guide to the fully illustrated volume with limited text.

From a 1990s' populist pocketable field guide to its geology and scenery comes the rather descriptive locality text:

> Monmouth Beach … widens towards The Cobb because of the accretion of cobbles that have drifted eastwards. During storms cobbles are thrown from Monmouth Beach over the Cobb and onto the harbour. As on other beaches along the coast of Lyme Bay there is a tendency

for lateral grading from fine shingle in the west to large pebbles and cobbles to the east. This is probably because the larger storm waves produced by easterly and south-easterly winds take back only the finer material.

(Bird 1995, 153)

Contemporaneously, the British Geological Survey published what was intended as a populist large-format concertina-fold field guide to the Lulworth Cove area (Gallois 1995) that generally avoids jargon; however, apart from its inconvenient field use format, it employs quite technical text such as: 'The oldest strata exposed at Lulworth Cove are massive shelly limestones of the Portland Beds … The ammonites in the limestones, up to 60 centimetres in diameter and known locally as 'Portland giants', are latest Jurassic in age' (Gallois 1995); 'massive' in a technical sense, alongside 'strata' and 'shelly', have little meaning for its intended readership, presumably casual geotourists. The guidebook (Pfaff and Simcox 1998) produced by the local and privately owned visitor centre was a much more pocketable and readable volume.

'Mantell Country': geo-publications summarised

Over the course of some 200 years, the nature of the intended audience and the format and style of the region's field guides have changed. The audience has shifted from the social elite of considerable means to a mass one of moderate means. The field guides have moved away from library volumes (such as Englefield 1816) to truly pocketable ones (such as Gibbons 1981). These changes can be analysed by considering a number of intertwined facets of their content, typography and audience. Such an analysis involves considering their format (size, orientation, binding and materials); typography (font family, size and area taken; illustrations type and area taken; white space area); illustrations (number, type and area); content (facts and interpretations); approach (descriptive – *what can be seen*; explanative – *why it is there*; interpretative – *user hypothesis*); textual style and readability (intellectual access); intended audience; communicative competence (comprehensibility and readability); and affordability. Such an analysis cannot be wholly empirical, and necessarily relies for some facets upon the informed judgement of the assessor. Readability (in terms of reading age) can be gauged from a number of standard empirical tests, the most widely used being the Fry and Flesch tests.[1] Whilst these are generally reliable for modern texts, the older styles of writing found within 18th and 19th century texts, and past levels of individual literacy, might well impact upon their accuracy – but they are still a useful guide. An analysis of 'Mantell Country' geo-publications has been published (Hose 2007), along with several works that cover in some detail the various methods of analysis (Hose 1997, 1998a, 1998b, 2000b). Overall, the examined volumes noted within this consideration show high reading age requirements (Figure 6.2A), outside the ideal range and somewhat at odds with the claims of

---

1   Fry's Readability Formula requires the selection of three 100-word passages in the beginning, middle and end of a book, using no proper nouns. The number of sentences is counted in each passage and averaged among the three selections. Syllables are then counted and again averaged among the three selections. These two averages are then plotted as values on a published curve graph that is split into readability columns. The Flesch Reading Ease formula at its simplest requires counting the number of words per sentence (in a 100-word sample) and the number of syllables in those 100 words. Then the following formula is applied: Reading Ease = $206.835 - 0.846wI - 1.015sl$ (where wI = the number of syllables per 100 words and sl = the average number of words per sentence). Some word processing software packages include these and other tests; there are also online calculators (even for the Fry test, which does not display a graph) for such tests.

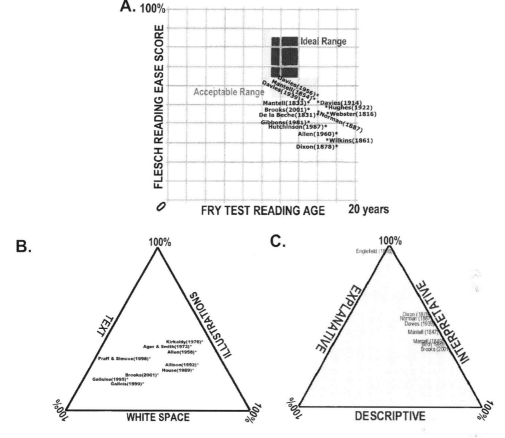

Figure 6.2 Analyses of 'Mantell Country' geo-publications
A. Readability of selected texts.
B. Typography of selected texts.
C. Nature of selected texts.
Note that all texts are discussed in this chapter and can be found in the references.

some to address the needs of casual geotourists. The volumes have become reduced in format size, and are relatively cheaper and more populist in their approach; the latter is reflected in the nature of their texts (Figure 6.2B) and the increase in the quality and number of illustrations employed (Figure 6.2C).

Conclusion

When major published geological discoveries are now seemingly the domain of the laboratory researcher, whose only contact with the real world is via remote images and data sets, the usefulness of classic field excursions has sometimes been called into question, not least because of their cost and associated safety issues. The latter, in particular, was not an issue that worried the early excursionists or their precursor travellers into wild landscapes. However, it is worth noting that 'many scientific discoveries of tomorrow will still be dependent on natural science fieldwork … Students need to have a framework of basic information and practical experience upon which to build their own particular expertise' (Gladfelter 2002, 411). That fieldwork has a long history and ancestry, the origins of which – the author suggests – can be traced back to the early journeymen, scholars and alpinists, and through the antiquarians and aesthetic landscape tourists to the scientific geologists. The travels of each of these groups were made, and to some extent limited by, the usual transport mode of their times. Their routes were directed and informed with varying accuracy by the available travelogues, field guides and maps; these popularised particular areas such as England's Peak District (Henry and Hose 2016), the Lake District (Hose 2008) and the south coast. Field excursions were popularised by regional natural history societies (see Burek and Hose 2016) and nationally in the UK by the Geologists' Association. The latter's excursions were particularly numerous from the late 19th to the early 20th centuries.

Geological inquiry began as an outdoor observational science and its pioneer authors fostered this approach. Maps and guides are an enduring vehicle, whether in hard-copy or digital format, introducing landscapes and geosites to dedicated and casual geotourists alike. Historic field guides, and the sources from which they were compiled, should be celebrated and made available to a wider public than the financially endowed antiquarian collector; embracing digital technology for their reproduction and dissemination is essential. Whilst the development of general tourist guides, including some with a topographical emphasis, has been examined (Buzzard 1993; Sillitoe 1995) in some detail, geology field guides have been very much neglected (but see Hose 2007). The study of geology's field excursions and the associated field literature reveals the scientific worth of past endeavours and their cultural significance to modern geoheritage. The lives and stories of travellers (as geotourists), geologists and authors can enrich and help to make the geological narrative interesting and accessible to today's casual audiences at geosites and geomorphosites.

REFERENCES

Ager, D V, and Smith, W E, 1973 *No 23: The Coast of South Devon and Dorset between Branscombe and Burton Bradstock*, Geologists' Association, London

Allen, P, 1960 *No 24: Geology of the Central Weald: The Hastings Beds*, Geologists' Association, London

Allison, R J, 1992 *No 47: The Coastal Landforms of Dorset*, Geologists' Association, London

Anderson, J G C, 1983 *Field Geology in the British Isles: A Guide to Regional Excursions*, Pergamon Press, Oxford

Arkell, W J, 1934 Whitsun field meeting: the Isle of Purbeck, *Proceedings of the Geologists' Association* 45, 412–19

Bakewell, R, 1813 *An Introduction to Geology, Illustrative of the General Structure of the Earth: Comprising the Elements of the Science, and an Outline of the Geology and Mineral Geography of England*, Richard Taylor and Company, London [4 further editions were published up to 1838]

— 1838 *An Introduction to Geology: Intended to Convey a Practical Knowledge of the Science, and Comprising the Most Important Recent Discoveries: With Explanations of the Facts and Phenomena which Serve to Confirm or Invalidate Various Geological Theories* (4 edn), Longmans, London

Barber, K E (ed), 1987 *Wessex and the Isle of Wight Field Guide (prepared to accompany the Annual Field Meeting held at Southampton and Cowes, 21–25 April 1987)*, Quaternary Research Association, Cambridge

Barnard, T, Capewell, J G, and Lang, W D, 1950 Whitsun field meeting at Lyme Regis, 1948, *Proceedings of the Geologists' Association*, 61, 156–8

Barton, H A, 1998 *Northern Arcadia: Foreign Travellers in Scandinavia, 1765–1815*, Southern Illinois University Press, Carbondale and Edwardsville

Beer, G R de, 1930 *Early Travellers in the Alps*, Sidgwick & Jackson, London

Bird, E, 1995 *Geology and Scenery of Dorset*, Ex Libris Press, Bradford on Avon

Black, J, 1992 *The British Abroad: The Grand Tour in the Eighteenth Century*, Allan Sutton, Stroud

Bowden, A J, 2009 Geology at the crossroads: aspects of the geological career of Dr John MacCullogh, in *The Making of the Geological Society of London, Special Publication No 317* (eds C L E Lewis and S J Knell), The Geological Society, London, 255–78

Britton, J, 1847 *The Natural History of Wiltshire by John Aubrey FRS (written between 1656 and 1691). Edited and Elucidated by Notes by John Britton FRS*, Wiltshire Topographical Society/J B Nicholls and Son, Salisbury/London

Brooks, K, 2001 *Geology and Fossils of the Hastings Area*, Ken Brooks, Bexhill-on-Sea

Burek, C V, and Hose, T A, 2016 The role of local societies in early modern geotourism: a case study of the Chester Society of Natural Science and the Woolhope Naturalists' Field Club, in *Appreciating Physical Landscapes: Three Hundred Years of Geotourism, Special Publication No 417* (ed T A Hose), The Geological Society, London

Burke, T, 1942 *Travel in England: From Pilgrim and Pack-Horse to Car and Plane*, B T Batsford Limited, London

Buzzard, J, 1993 *The Beaten Track: European Tourism, Literature, and the Ways to 'Culture' 1800–1913*, Clarendon Press, Oxford

Camden, W, 1586 *Britannia, sive Florentissimorum Regnorum Angliae, Scotiae, Hiberniae et insularum adjacentium*, Ralph Newberry, London

Cary, J, 1809 *Cary's New and Correct English Atlas: Being a New Set of County Maps from Actual Surveys, Exhibiting All the Direct & Principal Cross Roads, Cities, Towns, and most considerable Villages, Parks, Rivers, Navigable Canals &c*, John Cary, London

Christensen, T, 2012 *1616: The World in Motion*, Counterpoint Press, Berkeley

Cole, G A J, 1894 *The Gypsy Road: A Journey from Krakow to Coblentz*, Macmillan and Company, London and New York

Conybeare, W D, and Phillips, W, 1822 *Outlines of the Geology of England and Wales*, William Phillips, London

Coryat, T, 1905 *Coryat's Crudities Hastily gobled up in five Moneths travells in France, Savoy, Italy, Rhetia commonly called the Orisons country, Helvetia alias Switzerland, some parts of high Germany and the Netherlands; Newly digested in the hungry aire of Odcombe in the County of Somerset, and now dispersed to the nourishment of the travelling Members of this Kingdome* [of 1611] (in 2 vols), James MacLehose and Sons, Glasgow

Curry, D, and Wisden, D E, 1958 *No 14: Geology of some British Coastal Areas: The Southampton District including Barton (Hampshire) and Bracklesham (Sussex) Coastal Sections*, Geologists' Association, London

Damon, R, 1860 *Handbook of the Geology of Weymouth and the Isle of Portland*, E Stanford, London

Davies, G M, 1914 *Geological Excursions Round London*, Thomas Murby, London

— 1939 *Geology of London & South-East England*, Thomas Murby, London

— 1956 *The Dorset Coast: A Geological Guide* (2 edn), A and C Black, London

De la Beche, H T, 1830 *Sections and Views Illustrative of Geological Phenomena*, Treutel and Würtz, London

— 1831 *A Geological Manual*, Treutel and Würtz, London

— 1833 *A Geological Manual* (3 edn), Charles Knight, London

— 1835 *How to Observe Geology*, Charles Knight, London

— 1851 *The Geological Observer*, Longman, Brown, Green & Longmans, London

Defoe, D, 1724–27 *A Tour through the Whole Island of Great Britain* (in 3 vols), G Strahan, London

Deschaux, R, 1975 *Un poète bourguignon du XVe siècle, Michault Taillevent: édition et étude*, Librairie Droz, Geneva, 31–2

Dixon, F, 1850 *The Geology and Fossils of the Cretaceous and Tertiary Formations of Sussex*, Longman, Brown, Green and Longmans, London

— (revised by T R Jones), 1878 *The Geology of Sussex or The Geology and Fossils of the Cretaceous and Tertiary Formations of Sussex* (2 edn), William J Smith, Brighton

Dolan, B, 2001 *Ladies of the Grand Tour*, Harper Collins, London

Englefield, H C, 1816 *A Description of the Principal Picturesque Beauties, Antiquities and Geological Phenomena, of the Isle of Wight. With Additional Observations on the Strata of the Island, and their Continuation in the Adjacent Parts of Dorsetshire by Thomas Webster*, T Webster, London

Gallois, R, 1995 *Holiday Geology Guide: Lulworth Cove Area*, British Geological Survey, Keyworth

— 1999 *Holiday Geology Guide: Isle of Wight*, British Geological Survey, Keyworth

Gibbons, W, 1981 *The Weald: Rocks & Fossils – A Geological Field Guide*, Unwin, London

Gladfelter, E H, 2002 *Agassiz's Legacy: Scientists' Reflections on the Value of Field Experience*, Oxford University Press, Oxford and New York

Green, C P, 1989 Excursions in the past: a review of the Field Meeting Reports in the first one hundred volumes of the *Proceedings, Proceedings of the Geologists' Association* 100 (1), 17–29

Harrison, W J, 1882 *Geology of the Counties of England and of North and South Wales*, Kelly and Company/ Simpkin, Marshall and Company, London

Henry, C J, and Hose, T A, 2016 The contribution of maps to appreciating physical landscape: examples from Derbyshire's Peak District, in *Appreciating Physical Landscapes: Three Hundred Years of Geotourism, Special Publication No 417* (ed T A Hose), The Geological Society, London

Heringman, N, 2009 Picturesque ruins and geological antiquity: Thomas Webster and Sir Henry Englefield on the Isle of Wight, in *The Making of the Geological Society of London, Special Publication No 317* (eds C L E Lewis and S J Knell), The Geological Society, London, 299–318

Hewitt, R, 2010 *Map of a Nation: A Biography of the Ordnance Survey*, Granta, London

Hindley, G, 1983 *Tourists, Travellers and Pilgrims*, Hutchinson, London

Hoare, R C, 1807 *Journal of a Tour in Ireland, AD1806*, William Miller, London

— 1812 and 1821 *The Ancient History of Wiltshire* (in 2 vols), William Miller, London

Holmes, T V, and Sherborn, C D, 1891 *Geologists' Association: A Record of Excursions Made Between 1860 and 1890*, Stanford, London

Hose, T A, 1997 Geotourism – selling the earth to Europe, in *Engineering Geology and the Environment* (eds P G Marinos, G C Koukis, G C Tsiambaos and G C Stournas), Balkema, Rotterdam, 2955–60

— 1998a How was it for you? – matching geologic site media to audiences, in *Proceedings of the First UK RIGS Conference* (ed P Oliver), Worcester University College, Worcester, 117–44

— 1998b Mountains of fire from the present to the past – or effectively communicating the wonder of geologists to tourists, *Geologica Balcania*, 28 (3–4), 77–85

— 2000a European geotourism – geological interpretation and geoconservation promotion for tourists, in *Geological Heritage: Its Conservation and Management* (eds D Barretino, W A P Wimbledon and F Gallego), Sociedad Geológica de España/Instituto Tecnológico Geominero de España/ProGEO, Madrid, 127–46

— 2000b Rocks, rudists and writing: an examination of populist geosite literature, in *Proceedings of the Third UK RIGS Conference* (ed K Addison), Newton Rigg, Penrith, 39–62

— 2007 Leading the field: a contextual analysis of the field-excursion and the field-guide in England, in *Critical Issues in Leisure and Tourism Education: Current Trends and Developments in Leisure and Tourism Education* (eds E W Wickens, T A Hose and B Humberstone), Buckinghamshire Chilterns University College, High Wycombe, 115–32

— 2008 Towards a history of geotourism: definitions, antecedents and the future, in *The History of Geoconservation, Special Publication No 300* (eds C V Burek and C Prosser), The Geological Society, London, 37–60

— 2011 The English origins of geotourism (as a vehicle for geoconservation) and their relevance to current studies, *Acta geographica Slovenica* 51 (2), 343–60

House, M, 1958 *No 22: The Dorset Coast from Poole to the Chesil Beach*, Geologists' Association, London

— 1989 *Geology of the Dorset Coast*, Geologists' Association, London

Hughes, J C, 1922 *The Geological Story of the Isle of Wight*, Stanford, London

Insole, A, Daley, B, and Gale, A, 1998 *No 60: The Isle of Wight*, Geologists' Association, London

Johnston, A K, 1850 *The Physical Atlas of Natural Phenomena*, William Blackwood and Sons, Edinburgh and London

Kirkaldy, J F, 1976 *Guide No 29: The Weald* (3 edn) (eds F A Middlemiss, L J Allchin and H G Owen), Geologists' Association, London

Kölbl-Ebert, 2002 M George Bellas Greenough's 'Theory of the Earth' and its impact on the early Geological Society, in *The Making of the Geological Society of London, Special Publication No 317* (eds C L E Lewis and S J Knell), The Geological Society, London, 115–28

Lang, W D, 1906 Excursion to Lyme Regis, Easter 1906, *Proceedings of the Geologists' Association* (eds H B Woodward and G W Young), 19 (9), 323–4, 328–9

Lyell, C, 1830 *Principles of Geology* (in 3 volumes), John Murray, London

MacCulloch, J, 1821 *A Geological Classification of Rocks with Descriptive Synopses of the Species and Varieties, Comprising the Elements of Practical Geology*, Longman, London

— 1831 *A System of Geology, with a Theory of the Earth and an Explanation of its Connections with the Sacred Records* (in 2 vols), Longman, London

Mantell, G A, 1822 *The Fossils of the South Downs or Illustrations of Sussex Geology*, Lupton Relfe, London

— 1827 *Illustrations of the Geology of Sussex*, Lupton Relfe, London

— 1833 *The Geology of the South-East of England*, Longman, Rees, Orme, &c, London

— 1838 *The Wonders of Geology, Or, A Familiar Exposition of Geological Phenomena: Being the Substance of a Course of Lectures Delivered at Brighton* (in 2 vols), Relf and Fletcher, London

— 1846 *A Day's Ramble in and around the Ancient Town of Lewes*, Henry Bohn, London

— 1847 *Geological Excursion Round the Isle of Wight and Along the Adjacent Coast of Dorsetshire*, Henry G Bohn, London

— 1849 *Thoughts on a Pebble*, Reeves, Benham and Reeve, London

— 1854 *Geological Excursion Round the Isle of Wight and Along the Adjacent Coast of Dorsetshire* (3 edn), Henry G Bohn, London

Marples, M, 1959 *Shank's Pony*, J M Dent and Sons Ltd, London

Marsigli, L F, 1726 *Danubius Pannonico-Mysicus* (in 6 vols), P Gosse, R C Alberts, P de Hondt/Herm Uttwerf & Franc Changuion, The Hague/Amsterdam

Martin, P J, 1828 *A Geological Memoir on a Part of Sussex with some Observations upon Chalk-Basins, The Weald-Denudation and Outliers-By-Protrusion*, John Booth, London

Mavor, W, 1798–1800 *The British Tourists; or Traveller's Pocket Companion, through England, Wales, Scotland, and Ireland. Comprehending the Most Celebrated Tours in the British Islands* (in 6 volumes), E Newbury, London

Middlemiss, F A, 1989 One hundred volumes of geology: a personal review of the *Proceedings* since 1859, *Proceedings of the Geologists' Association* 100 (1), 55–72

Monckton, H W, and Herries, R S (eds), 1909 *Geology in the Field: The Jubilee Volume of the Geologists' Association, 1858–1908*, Stanford, London

Moritz, C P, 1797 *Travels, chiefly on Foot, through several parts of England in 1782, described in Letters to a Friend, by Charles P Moritz, a Literary Gentleman of Berlin, translated by A Lady* (2 edn), C G and J Robinson, London

Morris, C (ed), 1949 *The Journeys of Celia Fiennes*, Cresset Press, London

— 1982 *The Illustrated Journeys of Celia Fiennes 1685–1712*, Macdonald and Co, London

Murchison, R I, 1839 *The Silurian System founded on Geological Researches* (in 3 vols), John Murray, London

— 1854 *Siluria: The History of the Oldest Known Rocks, containing Organic Remains, with a Brief Sketch of the Distribution of Gold over the Earth*, John Murray, London

Norman, M W, 1887 *A popular guide to the geology of the Isle of Wight, with a note on its relation to that of the Isle of Purbeck*, Knight's Library, Ventnor

Ousby, I, 1990 *The Englishman's England*, Cambridge University Press, Cambridge

Perrucchini, G, 1726 Continuazione dell'Estratto d'alcune Notizie intorno alla Garfagnana, cavate dal primo Viaggio Montano del Signor Antonio Vallisnieri, *Supplementi al Giornale de' Letterati d'Italia (Venezia)* 3, 376–428

Pfaff, M, and Simcox, D, 1998 *Lulworth Rocks*, Weld Estate, Lulworth

Phillips, W, 1818 *A Selection of Facts from the Best Authorities, Arranged so as to Form an Outline of the Geology of England and Wales*, William Philips, London

Plot, R, 1677 *The Natural History of Oxford-shire, being an essay towards the Natural History of England*, printed at the Theater, Oxford

— 1686 *Natural History of Stafford-shire*, printed at the Theater, Oxford

Rose, E P F, 1996 Geologists and the army in nineteenth century Britain: a scientific and educational symbiosis?, *Proceedings of the Geologists' Association* 107, 129–41

— 2009 Military men: Napoleonic warfare and early members of the Geological Society, in *The Making of the Geological Society of London, Special Publication No 317* (eds C L E Lewis and S J Knell), The Geological Society, London, 219–47

Sillitoe, A, 1995 *Leading the Blind: A Century of Guidebook Travel 1815–1911*, Macmillan, London

Smith, W, 1815 *Delineation of the Strata of England with Part of Scotland; exhibiting the Collieries and Mines;*

the Marshes and Fen Lands originally overflowed by the Sea, and the Varieties of Soil according to the Variations in the Substrata, John Cary and J Wyld, London

Stafford, B M, 1984 *Voyage into Substance: Art, Science, Nature and the Illustrated Travel Account 1760–1840*, MIT Press, Cambridge

Stoye, J, 1994 *Marsigli's Europe 1680–1730: The Life and Times of Luigi Ferdinando Marsigli, Soldier and Virtuoso*, Yale University Press, New Haven and London

Tozzetti, G T, 1754 *Prodromo della Corografia e della Topografia Fisica della Toscana*, Stamperia Imperiale, Florence

Vaccari, E, 1907 The organized traveler: scientific instructions for geological travels in Italy and Europe during the eighteenth and nineteenth centuries, in *Four Centuries of Geological Travel: The Search for Knowledge on Foot, Bicycle, Sledge and Camel – Special Publication 287* (ed P N Wyse Jackson), The Geological Society, London, 7–17

Wallerias, J G, 1747 *Mineralogia eller Mineral Ricket indelt och beskrifvet*, Stockholm [Swedish edn in 1750, French in 1753, Russian in 1753 and Latin in 1772 – with revisions up to the 1780s]

Ward, M A, 1891 *Petrarch: A Sketch of his Life and Works*, Roberts Brothers, Boston

West, T, 1778 *A Guide to the Lakes: Dedicated to the Lovers of Landscape Studies, and to All Who Have Visited, or Intend to Visit the Lakes in Cumberland, Westmorland and Lancashire*, B Law, etc, London

Wilkins, E P, 1861 *A Concise Exposition of the Geology, Antiquities, and Topography of the Isle of Wight*, T Kentfield, Newport

Woodward, H B, 1876 *The Geology of England and Wales*, Phillip and Sons, London

Woodward, J, 1696 *Brief Instructions for making Observations in all Parts of the World*, Richard Wilkin, London [1973 reprint with an introduction by V A Eyles, Sherborn Fund Facsimili Number 4, Society for the Bibliography of Natural History, London]

— 1728 *Fossils of all Kinds, Digested into a Method, Suitable to their mutual Relation and Affinity; with the Names by which they were known to the Antients, and those by which they are at this Day known: and Notes conducing to the setting forth the Natural History, and the main Uses, of some of the most considerable of them. As also several Papers tending to the further Advancement of the Knowledge of Minerals, of the Ores of Metalls, and of all other Subterraneous Productions. By John Woodward, MD, late Professor of Physick at Gresham College, Fellow of the College of Physicians, and the Royal Society*, William Innys, London

Wordsworth, W, 1835 *The River Duddon, A Series of Sonnets: Vaudracour and Julia: and Other Poems. To which is annexed, a Topographical Description of the Country of the Lakes, in the North of England*, Hudson and Nicholson, Kendal

Wright, C W, and Curry, D, 1958 *Guide No 25: Geology of Some British Coastal Areas: The Isle of Wight. I The Cretaceous, II The Tertiary*, Geologists' Association, London

Wyse Jackson, P N, 2007 Grenville Arthur James Cole (1859–1924): the cycling geologist, in *Four Centuries of Geological Travel: The Search for Knowledge on Foot, Bicycle, Sledge and Camel – Special Publication 287* (ed P N Wyse Jackson), The Geological Society, London, 135–47

Young, G A, and Bird, J, 1822 *A Geological Survey of the Yorkshire Coast*, Longman, London

Zitell, K A von, 1913 *History of Geology and Palaeontology*, Walter Scott, London

7

Geoconservation: An Introduction to European Principles and Practices

Jonathan G Larwood

This chapter explores the origins of geoconservation in Europe, setting out the value placed on geodiversity and the geoconservation effort that is present at local, national and Europe-wide levels today. From the 17th century, geoconservation has grown from the *ad hoc* protection of individual geodiversity sites to a situation where geoconservation happens in every country in Europe; a growing network of 64 Global Geoparks and some 18 of the region's 39 natural World Heritage Sites are wholly or partly inscribed on the basis of their geodiversity. This geoconservation progress is illustrated with examples drawn from across Europe; different approaches are compared and the development of European geoconservation cooperation is described.

Defining geoconservation

Geoconservation, shorthand for geological conservation, is a relatively new term which warrants defining at the outset, before its development in Europe is considered. A number of authors have considered the scope of geoconservation. Burek and Prosser (2008) provide a summary of this discussion and conclude with the following statement: 'geoconservation can be defined as action taken with the intent of conserving and enhancing geological and geomorphological features, processes, sites and specimens, including associated promotional and awareness raising activities and the recording and rescue of data or specimens from features and sites threatened with loss or damage' (Burek and Prosser 2008, 2 and updated Prosser 2013a, 1). This is now a widely quoted definition for geoconservation. The basic principle of conservation needs to be examined in more detail and the question asked – what factors lead to the action of conservation? Whether undertaking conservation of natural or cultural heritage, there is a basic pattern that is followed. First, some value must be placed on the object of interest. It could be a painting, a sculpture, a building, a landscape, a habitat, an animal or a plant that has associated values, perhaps linked to beauty, historical record, rarity, an experience of the natural world and the functioning of the natural environment, or a resource that is critical to our well-being.

Having established a value, if a threat emerges, often through a risk of loss or deterioration, then the response is to counter that threat, to maintain and often enhance the value that is threatened; this response is the action of conservation. Gray (2013) examines this principle for geoconservation, setting out a simple conservation equation ('value + threat = conservation need') and considering in detail its components. In the context of geoconservation, what are we trying to conserve? The term geodiversity, again recent but in increasingly common use in the field of geoconservation, now needs to be defined. Many are familiar with the term 'biodiversity', shorthand for 'biological diversity', which emerged with the proposal to establish an International Convention on Biological Diversity (Wilson 1988). Gray (2013) identifies the first published use of the term geodiversity in Germany in 1993 and tracks its subsequent development. He concludes that the term emerged partly in response to the public and professional familiarity with 'biodiversity', and its wide use in environmental conservation as a parallel and shorthand for 'geological diversity'.

Gray's widely accepted definition states that geodiversity is 'the natural range (diversity) of geological (rocks, minerals, fossils), geomorphological (landforms, topography, physical processes), soil and hydrological features. It includes their assemblages, structures, systems and contributions to landscapes' (Gray 2013, 12). Geoconservation is therefore concerned with the conservation of geodiversity in all its facets. Gray (2013) provides the most comprehensive analysis of geodiversity, whilst Stace and Larwood (2006) provide a well-illustrated discussion of the relationship between geodiversity, people, places and nature. Gray, Gordon and Brown (2013) present a detailed analysis of the ecosystem role of geodiversity. Geodiversity clearly overlaps with geoheritage, but the distinction is that geodiversity is a wider term, whilst geoheritage is more focused on aspects of geodiversity with particular 'special' values (Wimbledon and Smith-Meyer 2012) – although such values can be wide ranging and both qualitative and quantitative in nature.

Returning to the conservation equation, many values are attached to geodiversity. These include the contribution it makes to our understanding of the Earth's evolution, the role of geodiversity in ecosystems, as a natural resource contributing to the well-being and economic wealth of society, a resource that defines the world's landscapes and environments, and one that is intimately linked to people and their diverse cultures and histories. However, despite the often-held belief that geodiversity is robust – rocks are there for ever – the threat to geodiversity can come from many quarters.

Society's growth, development and demand for mineral resources can remove, permanently conceal and restrict access to geodiversity, and can equally have an impact on natural processes which are fragile and susceptible to environmental change. Overvaluing geodiversity can lead to its deterioration through over-use, and to potential damage through activities such as specimen-collecting. Equally, limited understanding of the value of geodiversity can lead to neglect through lack of action and can cause damage, sometimes as a consequence of conserving other heritage values. This last point is important, as understanding and therefore developing a value for geodiversity is the critical first step in realising a need for conservation when threatened. As a consequence, geoconservation not only involves restoring and enhancing geodiversity that has been damaged, and taking action to counter, reduce or remove threats, but also promoting and widening awareness of the importance and relevance of geodiversity (Burek and Prosser 2008). Gordon and Macfadyen (2001), Prosser, Murphy and Larwood (2006) and Gray (2013) each examine in detail the values and the potential threats to geodiversity and consequent geoconservation responses. A number of examples, illustrating the development of geoconservation across Europe, are examined in this chapter and elsewhere within this volume.

ESTABLISHING VALUE: A FIRST STEP IN GEOCONSERVATION

Establishing a value for geodiversity is a critical first step in undertaking geoconservation; so, how has this value emerged in Europe? The late 18th and early 19th centuries witnessed the early development of geological science in Europe, as part of the ever growing and widening interest in natural history and associated disciplines. Three European scientists were at the heart of the early debate on the formation and age of the Earth. The German mineralogist and mining engineer Abraham Gottlob Werner (1749–1817) proposed that sedimentary rocks were deposited from a retreating flood, and that igneous rocks were the result of aqueous precipitation (for the established churches, this neatly matched the 'truth' of the biblical flood) – the Neptunism theory. James Hutton (1726–97), a Scottish geologist, developed the theory of Plutonism (in his 1788 *Theory of the Earth*), which considered that rocks were volcanic in origin and eroded to produce sedimentary rocks. He examined the relationship between igneous and sedimentary rocks, proving their separate origin.

Hutton further developed the theory of uniformitarianism: the geological processes we see operating today have been active throughout the Earth's history, with the intrusion and eruption of igneous rocks and repeated cycles of deposition, uplift, tilt and erosion over vast periods of time. Hutton's and Werner's 'discovery' of the scale of geological time should not be underestimated and can be compared to the Copernican shift (in a world in which we orbit the Sun) in the way it has influenced our cultural and scientific understanding (Cervato and Frodeman 2012). Lastly, Georges Cuvier (1769–1832), a French naturalist, a zoologist and the founder of comparative anatomy, observed repeated faunal change through his detailed study of fossils of the Paris Basin, and established the existence of extinction linked to repeated catastrophic events; that is, the catastrophism theory.

Hutton's uniformitarianism remains central to geological science today, and he is regarded as the 'Father of Modern Geology'. Cuvier's theories of extinction and catastrophism were important steps in the development of biostratigraphy and evolutionary theory and, although Werner's Neptunist ideas were superseded, his theories were central to early debate and he remains an important early mineralogist. What is shared by Werner, Hutton and Cuvier is that much of their work was based on field observation, sampling and interpretation. This approach lies at the heart of both geological science and geoconservation. That is, through geoconservation we aim to maintain the ability to see, sample, reinterpret and experience geodiversity, whether as scientists or simply by appreciating geodiversity in the world around us. It was also this immediacy of geology that led to its growing popularity as a science that cross-cut both the cultured and wealthy classes and the general public. Knell (1996) notes the tangible, comprehensible and romantic view of geology (compared to the more abstract sciences of mathematics, astronomy, physics and chemistry) and its presence (and relevance) wherever one is.

This popularisation, fascination and accessibility is epitomised by the visit of Sir Roderick Murchison (author of the 1835 *Silurian System*), as part of the 1849 British Association meeting, to the limestone mines of Dudley (West Midlands, England). Murchison's speech was attended by an estimated 15,000 people who arrived by canal boat and lined the underground limestone caverns, which were illuminated by gaslight and flares (Worton 2008). The event was reported and depicted in the *Illustrated London News*, and an anonymous ballad proclaimed Murchison as the 'King of Siluria'.

Museums further added to this growing interest, both as centres of learning and as places to present collections of rocks, fossils and minerals to the general public. The Natural History Museum in London, which opened in 1881 as the British Museum (Natural History), featured many of the fossil marine reptiles collected by Mary Anning (1799–1847) from the Dorset coast in the first half of the 19th century. The Muséum National d'Histoire Naturelle in Paris (founded in 1793 by, among others, Georges Cuvier) constructed the Gallery of Palaeontology and Comparative Anatomy in 1898; this presented the accumulated collections of the 18th and 19th centuries, including many dinosaurs and other extinct vertebrates. Mineral collections were also important, reflecting their link to economic benefit, and early collections are found in many national museums; for example, the Museum für Naturkunde (founded in 1810) in Berlin not only houses the Berlin specimen of *Archaeopteryx* (added in 1880) but also includes a mineral collection dating back to the Prussian Academy of 1700.

Local museums were equally important. In the UK, the Rotunda Museum in Scarborough, North Yorkshire, opened in 1829, was 'a building constructed in the round so as to provide the best opportunities for the stratigraphical arrangement' of William Smith's fossil collection (Knell 1996, 108). William Smith published the first large scale geological map of England, Wales and parts of Scotland in 1815, and is commonly referred to as the 'Father of English Geology'. Returning to Dudley, it was Murchison's earlier visits to the area that encouraged the establishment of a permanent museum. Founded, with the support of other eminent geologists, to house the fossil collections of local miners (Worton 2008), a geological society and field club were also established in 1841; this pattern of the development of geological museums alongside societal change is seen throughout Europe in the 19th century.

Geology was also part of parkland and country estate design. Since the 18th century, the use of rock to create romantic rocky landscapes and grottoes brought geology into these designed, manicured landscapes and to a wide, and often urban, public (Doyle *et al* 1996; Taylor 1995). Most remarkable is the Crystal Palace Park in south London (Doyle 2008). It was designed by Joseph Paxton in 1854 as a complex of Pleasure Grounds incorporating the reconstructed Crystal Palace of the 1851 Great Exhibition. Uniquely, it included a series of geological environments depicting the successive ages of Britain's geology in groups of rock strata tableaux (including a Derbyshire lead mine), accompanied by full size reconstructions of the dinosaurs, marine reptiles and extinct mammals that once lived in those environments. This was a visual spectacle, an outdoor laboratory and museum presenting the most up-to-date science and interpretation to the general public. Hence, by the end of the 19th century a value for geodiversity was well established in the field, through museums and learned societies and the everyday life of people in cities, towns and the countryside; but at what point does this value lead to sufficient concern to act in the light of threat, that is, to undertake geoconservation?

EARLY THREATS AND THE FIRST ACTS OF GEOCONSERVATION

The growth of interest in geodiversity encouraged more people to seek out geological phenomena near to where they lived, and increasingly further afield, with the consequent development of geotourism (Hose 2008; Hose in press). It is the pressure brought by increasing use and the potential impact of the visitor that led to the first examples of geoconservation. Much of the early attention given to geodiversity focused on the spectacular and the unusual in the landscape: caves, landforms such as tors and glacial erratics, fossil trees and igneous eruptions featured significantly.

For example, Erikstad (2008) identifies the Baumannshöle Cave in Germany as the first recorded example of geoconservation in Europe. Documented since the 15th century, there is a specific record of organised tours of this show cave since 1646 (Duckeck 2014). In 1668 it was subject to a nature conservation decree by Rudolf August, Duke of Braunschweig and Lüneburg,

FIGURE 7.1 THE LANDSCAPES OF FOUR UK WORLD HERITAGE SITES
A. HADRIAN'S WALL, PERCHED ON THE HIGH ESCARPMENT OF THE WHIN SILL, IN NORTHUMBERLAND, ENGLAND.
B. THE JURASSIC COAST, CHARMOUTH BEACH LOOKING TOWARDS GOLDEN CAP, IN DORSET, ENGLAND.
C. THE GIANT'S CAUSEWAY ON THE ANTRIM COAST OF NORTHERN IRELAND.
D. STUDLEY ROYAL PARK, INCLUDING THE RUINS OF FOUNTAINS ABBEY, IN NORTH YORKSHIRE, ENGLAND.

controlling access and establishing the first (in the world) official cave tour guide. In contrast, Doughty (2008) considers the growing interest in the iconic igneous Giant's Causeway (Figure 7.1C) on the Antrim Coast of Northern Ireland. This lava flow of (some 40,000) hexagonal basaltic columns, which erupted 50–60 million years ago, was first mentioned in literature in 1693 (see Doughty 2008), with an early account by Foley (1694). The Causeway drew intense curiosity, fuelled by the commissioned illustrations (due to its remote location) which were circulated widely throughout Europe, and reignited debate on the origins of basalt between Plutonists and Neptunists. By the middle of the 18th century it was an established tourist destination for the wealthy, but it wasn't until 1986 that it was afforded protection as a World Heritage Site.

Further early examples include the German volcanic Drachenfels (Dragon Rock) near Bonn, which was bought by the Prussian Crown Prince Frederick Wilhelm in 1832 to prevent quarrying of the hill, and was protected in 1836 as a 'Natural Monument' (considered the first nature reserve in Germany). In 1840 the Weltenburg Danube Cut (a narrow gorge cut by the Danube through Jurassic limestone) was protected by the Bavarian king, and in 1844 the 'Totenstein' granodiorite (a natural outcrop of a silica-rich igneous intrusion) in Saxony was purchased by the Royal Dynasty of Prussia and bequeathed to the people (Erikstad 2008; Wiedenbein 1994a, 1994b; Röhling *et al* 2012). There are examples of early erratic boulder protection from a number of countries (Gray 2013), including the 1838 protection of an erratic boulder in Neuchâtel in Switzerland (Reynard 2012), and Fritz Muhlberg's 1870s campaign to protect giant erratic boulders from being exploited as kerbstones (Stürm 2012). Also in the 1870s, the 'Boulder Committee' was established to identify and conserve remarkable erratics in Scotland (Milne Home 1872). In 1839 local regulation prevented the destruction and collection of dripstones from the Hungarian Baradla Cave and later, in 1866, F Kubinyi senior (a founder of the Hungarian Geological Society) had the lithified tree trunk of Iplytarnóc (preserved by volcanic eruption approximately 20 million years ago) covered by a protective roof (Bolner-Takács *et al* 2012) – an early example of *in situ* fossil preservation; today it is part of the Iplytarnóc Fossil Nature Reserve.

Other early examples of the protection of lithified trees include the Carboniferous Coal Measures at Wadsley Fossil Forest in Sheffield, England, discovered by Henry Clifton Sorby (1875), and the 'Fossil Grove' in Glasgow, Scotland, discovered in 1887. Both were protected *in situ* from deterioration by the construction of a covering building; that for 'Fossil Grove' still survives in Victoria Park (Thomas 2005; Thomas and Warren 2008), making it the longest continuously open and managed geotourism location in the UK (Hose 2008). Other examples include the purchase in 1856, by the Assembly of German Scientists and Doctors, of exotic blocks of granite in Upper Austria to protect them from quarrying activities (Hoffman and Schönlaub 2012), and in Czechoslovakia the protection, in 1884, of folded Palaeozoic limestones in Prague – named 'Barrande's Rock' in memory of the famous French palaeontologist Joachim Barrande (1799–1883) (Kriz 1994).

Such examples demonstrate the concerns that were emerging in Europe during the mid-19th century over the impact of people, particularly the potential damage caused by ever increasing visitor numbers, and clear concern over the destructive impact of quarrying. Responses were reactive and *ad hoc*, and it was not until the early 20th century that a more systematic approach to geoconservation began to emerge.

A STRUCTURED APPROACH TO GEOCONSERVATION

The establishment of a structured approach to geoconservation (and conservation, for that matter) has varied across Europe. The development of legislative frameworks and a systematic approach to identifying and protecting geodiversity have largely seen their origins in the early to mid 20th century. Here a more detailed account of the development of a structured approach to geoconservation in the UK is given; the approach in England, and in the wider UK, is well documented and is considered the first systematic approach to geoconservation, driven by a geoconservation audit and supported by legislation (Thomas and Cleal 2005; Prosser 2008) and by the growing experience in the practicalities of undertaking geoconservation. This is then compared to the wider development of structured geoconservation in Europe, and the comparison is summarised in Table 7.1. Prosser (2008) provides a detailed summary of the legislative, policy and administrative milestones for geoconservation in England, and reflects on the origin of geoconservation legislation (Prosser 2013b). Thomas and Cleal (2005) provide a review of the development of geoconservation in the UK, and this is further reviewed by Gray (2013).

TABLE 7.1 TABLE SHOWING THE CHRONOLOGY OF KEY DEVELOPMENTS IN EUROPEAN GEOCONSERVATION AS DESCRIBED IN THE CHAPTER

	Local geoconservation	National geoconservation	Systematic audits	International geoconservation
Pre 19th Century	**1668** – Baumannshole Cave, Germany - access control established **1694** – first written account of Giant's Causeway			
19th Century	**1830s** – protection of Hutton's Rock, Salisbury Crags, Edinburgh, Scotland **1836** – Drachenfels (Dragon Rock) purchase (1832) protects against quarrying becomes Germany's first 'Nature Reserve' **1838** – Neuchâtel erratic boulder protected, Switzerland **1844** – 'Totenstein' granodiorite purchased and bequeathed to the people, Germany **1856** – purchase of exotic granite blocks to protect from quarrying, Austria **1870s** – Muhlberg's 1870s campaign to protect giant erratics, Switzerland **1875** – Wadsley Fossil Forest, Sheffield, England protected by building **1884** – Barrande's Rock, folded limestones protected in Prague, Czechoslovakia **1887** – 'Fossil Grove', Glasgow, Scotland protected by building		**1870s** – 'Boulder Committee' to identify and conserve erratics, Scotland	

(CONT.)

TABLE 7.1 (CONT.)

	Local geoconservation	National geoconservation	Systematic audits	International geoconservation
20th Century		**1900** – cave conservation law enacted, Croatia **1909** – nature protection act conserves erratic boulders and establishment of first national Parks in Europe, Sweden **1910** – nature conservation act protects geological and mineralogical sites, Norway **1949** – National Parks and Access to the Countryside Act provides legal provision for the protection of 'geological or physiographical' features, United Kingdom	**1906** – systematic inventory of nature conservation sites, Germany **1940** – systematic survey of erratics (901 listed, 210 protected), Estonia **1945** – 390 important geological sites listed by the Nature Reserves Investigation Committee, England and Wales (Chubb, 1945)	**1948** – International Geological Congress (IGC) meets in London and encourages governments to list and protect geodiversity sites

Local geoconservation	National geoconservation	Systematic audits	International geoconservation
		1960 – over 600 geological and geomorphological Sites of Special Scientific Interest established in England, Scotland and Wales **1969** – Gea Project inventory of important geological, geomorphological and pedological sites, Netherlands	**1972** – UNESCO World Heritage Convention established, criterion viii specifically encompasses geodiversity World Heritage Sites
		1977 – Geological Conservation Review (GCR) initiated, systematic audit of geodiversity sites in England, Scotland and Wales	**1979** – first European geological World Heritage Site inscribed, Plitvice Lakes National Park, Croatia
			1988 – origin of the European Association for the Conservation of Geological Heritage (ProGEO)
1989 – formal establishment of Regionally Important Geological/Geomorphological Sites Groups (local and voluntary geoconservation) in England, Scotland and Wales	**1990** – *Earth science conservation in Great Britain* (Nature Conservancy Council) published; the first comprehensive guide to geoconservation		**1991** – first international geoconservation symposium, Digne, France and the Digne Declaration – *'memories' of the past* **1993** – UK Malvern International Conference
	1994 – *Earth Heritage* magazine initiated	**1994** – IUGS global geosites project and inventory initiated	

20th Century continued

(CONT.)

(CONT.)

TABLE 7.1 (CONT.)

Local geoconservation	National geoconservation	Systematic audits	International geoconservation
			2000 – European Geopark Network (now Global Geopark network) established **2004** – Council of Europe Rec(2004)3 on geoconservation adopted
	2008 – *History of Geoconservation* published		**2008** – International Union for the Conservation of Nature (IUCN) adopt resolution 4.040 on geoconservation **2009** – '*Geoheritage*', first international journal encompassing geoconservation launched
	2011 – UK Geodiversity Action Plan launched **2012** – Scottish Geodiversity Forum established **2013** – *Geoconservation for science and society* published **2014** – English Geodiversity Forum established		**2012** – *Geoheritage in Europe and its conservation* published by ProGEO and publication of ProGEO geoconservation protocol
		2015 – at least 15 European countries with an active geodiversity audit and inventory process	**2015** – 64 Global Geoparks in Europe and18 geodiversity World Heritage Sites in Europe **2015** – Global Geoparks awarded programme status by UNESCO

21ˢᵗ Century

GEOCONSERVATION: THE UK APPROACH

For its size, the UK is amongst the most geodiverse locations in the world, with more than 2,800 million years of geological time represented in its varied landscapes from mountain to coast. It has been at the centre of the development of geological sciences and, in part reflecting this, has been at the forefront of the development of geoconservation practice. In the UK, the earliest noted example of geoconservation is the successful 1830s legal action to save Salisbury Crags below Arthur's Seat in Edinburgh (McMillan 2008; Thomas and Warren 2008). Salisbury Crags includes 'Hutton's Rock' and is one of the key locations that helped shape James Hutton's Plutonist theory, exposing the junction between the igneous dolerite of Salisbury Crags and the underlying sedimentary rocks. Due to local concerns about the threat from quarrying, the consequent impact on the city landscape and the loss of 'Hutton's Rock', extraction was prevented and Hutton's locality saved; it remains one of the most used educational sites in Britain.

The previously noted *in situ* conservation of the Sheffield and Glasgow fossil forests provides two further examples of late 19th century *ad hoc* geoconservation. The first more systematic approach appears to have been the work of the Scottish 'Boulder Committee' (Milne Home 1872), which set out to identify glacial erratic boulders and to develop a scheme for their conservation. Importantly, this is a recognition of the need for a systematic and scientific approach to identifying and understanding a geological resource (here, erratic boulders) and then establishing a way of conserving the resource. The committee reported (Milne Home 1884) with a comprehensive list of boulders and a discussion of their origin, but there is no evidence of what further conservation action, if any, was taken.

Evans (1992) explores the wider origins of nature conservation in Britain, proposing that its origins go back to the 11th century management of Royal Forests and 'conserving' their hunting resource. In the late 19th century, concern over the decline in bird species led to demand for and development of wildlife conservation. Sheail (1976) explores in detail the development of modern nature conservation in Britain from the 19th century onwards, and especially considers the influence of naturalists' societies and their individual members. He makes the point that the establishment of nature reserves, with the immense changes in agricultural practice of the early and middle 20th century, was seen as the most effective way of protecting wildlife and conserving habitats; if these happened to have some feature of geological interest, it was a welcome – if unintended – consequence of nature conservation focused on the biotic world. It was not until the 1940s, however, that there was any serious intent to undertake geoconservation at a more systematic national level. In 1942 the Nature Reserves Investigation Committee was formed and a sub-committee was instructed to identify potential 'geological parks'. Chubb (1945) reports the outcome, indicating the different challenges of conserving geological sites, highlighting their national and international significance and potential vulnerability, and noting the successful protection of geological monuments in the USA. The report also lists 390 sites across England and Wales, falling into various categories of conservation areas, geological monuments, controlled sections and registered sections.

It is in this context that the first formal and encompassing legislation for nature conservation (the 1949 National Parks and Access to the Countryside Act) was enacted, and the nature conservation body, the Nature Conservancy, was established by Royal Charter. Importantly, the 1949 Act stated that: 'Where the Nature Conservancy is of opinion that an area of land, not being land for the time being managed as a nature reserve, is of special interest by reason of its flora, fauna, or <u>geological or physiographical features,</u> it shall be the duty of the Conservancy to notify that fact to the local planning authority in whose area the land is situated.' Since then, a number of modifications have been made to the 1949 Act – notably the Wildlife and Countryside Act (1981) and the Countryside and Rights of Way Act (2000), which are detailed by Prosser (2008).

The basic principle of a science-led and site-based approach to conservation, however, has remained largely the focus of UK geoconservation to the present day. The Nature Conservancy and its subsequent bodies (today Natural England, Scottish Natural Heritage and Natural Resources Wales) have been responsible for designating Sites of Special Scientific Interest (SSSIs), largely in private ownership, on the basis of their scientific value; this includes both biological and, importantly, geological SSSIs. They also manage a selection of these areas as National Nature Reserves (NNRs), either directly by acquisition or by establishing an approved body to take responsibility for management, primarily for their nature conservation interest. In 1950 William Macfadyen was employed as the first geologist of the Nature Conservancy (Prosser 2012). It was his job to review the lists of geological and geomorphological sites that were compiled by the Nature Reserves Investigation Committee – and an additional 60 sites in Scotland (Gordon 1992) – and to recommend the sites as either SSSIs or NNRs. Through to 1960, Macfadyen visited and successfully identified more than 600 sites across England, Wales and Scotland which were subsequently notified as geological and geomorphological SSSIs.

The systematic approach to geological site selection in the UK has been critical to successful and consistent geoconservation. In 1977 the Geological Conservation Review (GCR) was initiated to establish a more rigorous, robust and scientifically systematic approach than that inherited by Macfadyen to the identification of nationally important sites (Ellis *et al* 1996; Ellis 2008, 2011). The GCR was undertaken on a national scale (England, Scotland and Wales) by specialists from all geological fields. The aim of the GCR was to systematically assess the geological heritage of Great Britain and to identify, for conservation, localities of at least national value to earth science that should 'represent comprehensively the geological history of Britain and demonstrate the range and diversity of the best Earth science sites' (Ellis 2008, 124). Today, the geological SSSI network is underpinned by the GCR series, with every geological or geomorphological SSSI having to qualify as one or more GCR site.

Whilst the 1949 Act established a statutory approach to geoconservation at a national level, it was not until 1989 that a more systematic approach to the conservation of locally and regionally important geological sites was initiated (Prosser and King 1998). This was driven by the lack of recognition and protection for local geological sites (in part highlighted by the GCR), and by the need to support local geoconservation more coherently (as with an already established network of local wildlife sites and groups), and it built on already established (but geographically limited) local geoconservation activity, including the Black Country (in the West Midlands) and the counties of Avon and Shropshire. The concept of a Regionally Important Geological/ Geomorphological Site (RIGS) was established by the Nature Conservancy Council, and RIGS Groups formed as local voluntary groups to select RIGS (today known as Local Geological Sites) for each county in England, Scotland and Wales. Importantly, Local Geological Sites are a material consideration in making local planning and development decisions.

Prosser (2013a) reviewed the achievement in national and local site selection in the UK, listing the total number of sites selected. Across England, Scotland and Wales in 2013 there were 2,010 geological SSSIs (made up of more than 3,200 GCR sites), 13 geological NNRs and approximately 4,400 Local Geological Sites. In Northern Ireland there is a parallel system which has identified 121 Areas of Special Scientific Interest (ASSIs), equivalent to SSSIs. It is important to note that, despite the intentions to integrate biological and geological conservation, the 1949 Act has had a strong biological bias, also noted by Sheail (1976); consequently, many of the resources available for nature conservation are still being targeted at flora and fauna at national and local levels. As a result, geoconservation – whilst well advanced – has yet to fulfil its potential, particularly in terms of the relative profile, importance and resourcing it is afforded in nature conservation and environmental protection.

Beyond specifically designated sites there is less protection provided for geodiversity, although this is gradually being addressed. Geodiversity is protected more broadly through wider landscape designations, which include National Parks and Areas of Outstanding Natural Beauty (National Scenic Areas in Scotland), and is encompassed through means such as the European Landscape Convention (for example, Natural England 2010) and approaches such as National Character Areas in England and Landscape Character Areas in Wales. Recent developments have also seen the establishment of a UK-wide action plan for geodiversity – the UK Geodiversity Action Plan (UKGAP)[1] – which provides a framework for geodiversity action across the UK (see www.ukgap.org.uk) and, at a country level, the establishment of Scottish and English Geodiversity Forums[2] and associated Geodiversity Charters, which set out a shared vision and ambition for geodiversity and geoconservation (Scottish Geodiversity Forum 2012; English Geodiversity Forum 2014).

The practice of geoconservation has similarly grown; as valued sites have been identified, how do we then conserve them? Early responses to threat have been highlighted already and have included preventing damaging activity such as quarrying, managing people to reduce their impact, and protecting from the elements by constructing enclosing buildings. The first comprehensive guide to geoconservation, published by the Nature Conservancy Council in 1990, was *Earth Science Conservation in Great Britain: A Strategy* (Nature Conservancy Council 1990). This sets out the *raison d'être* for earth science conservation (now referred to as geoconservation), a long term and themed strategy for its delivery, and ways of conserving sites after selection. Importantly, it included a handbook of earth science conservation techniques which brought together the practical experience accrued largely from the 1950s onwards.

The concept of a site-based approach to geoconservation was introduced, recognising that the approach to geoconservation would vary according to the extent of the resource and the nature of the site. For example, an extensive and eroding chalk coastline is best managed by ensuring that coastal erosion is maintained; in contrast, an inland disused quarry with a finite fossil resource is likely to require clearance of vegetation to maintain an exposure, and may require the careful management of fossil collecting. On the basis of this distinction, the Earth Science Conservation Classification (ESCC) was established (Nature Conservancy Council 1990). The code allows a classification of sites into 'Exposure' (extensive) or 'Integrity' (finite or restricted in extent) sites, subdivided into a number of categories such as active quarries, disused quarries, coastal cliffs, active and static geomorphological sites, caves, and fossil or mineral sites. The ESCC was modified in 2006 with a distinction made between integrity and finite sites (Prosser, Murphy and Larwood 2006), but otherwise it has remained unchanged as the basis for guiding geoconservation activity in the UK.

---

[1] The UKGAP was developed through cooperation between geological and geoconservation organisations, groups and individuals across the UK and was formally launched in 2011.

[2] The Scottish and English Geodiversity Forums have brought together organisations, groups and individuals in Scotland and England with a common interest in promoting geodiversity and geoconservation, and supporting the delivery of the UKGAP.

Further syntheses of the practice of geoconservation include the Open University textbook, *Earth Heritage Conservation* (Wilson 1994), the *RIGS Handbook* (Royal Society for Nature Conservation 1999), a guide to voluntary geoconservation and the conservation of Local Geological Sites, and *Geological Conservation: A Guide to Good Practice* (Prosser, Murphy and Larwood 2006), which includes a range of case studies. Much of the geoconservation activity in the UK is documented in *Earth Heritage* magazine, which originated in 1968 as an *Information Circular* of the Nature Conservancy Council, then became the *Earth Science Conservation* magazine and, since 1994, *Earth Heritage* (see www.earthheritage.co.uk).

Systematic geoconservation across Europe

The stages of developing geoconservation seen in the UK (valuing; *ad hoc* geoconservation action; development of geologically inclusive nature conservation legislation at a national level and the compiling of inventories; geoconservation at a local level; and the widening of geoconservation beyond designated sites) provide a useful model to examine the parallel development of geoconservation across Europe. This has been documented in some detail by Erikstad (2008; 2013), Gray (2013), Wimbledon and Smith Meyer (2012) and Larwood *et al* (2013), and is further considered here from the earliest development of legislation to the present.

There is a long history of geoconservation, which is intimately linked to the development of nature conservation; geoconservation measures often compete with the conservation of biotic elements for priority as nature conservation practices and policies develop. The earliest nature conservation and environmental legislation in Europe often relates to the establishment of laws and regulations relating to the management of forests and hunting rights (much the same as in the UK) and has no direct or obvious geoconservation benefit. Examples of early written statements dealing with nature protection (beyond forests and hunting) date back to the 14th century in Lithuania, whilst a later Lithuania Statute of 1588 specifically protected river and stream courses and large stones, used to indicated land ownership (Satkunas *et al* 2012); although not explicitly geoconservation, this protected geodiversity.

Not only does Germany have some of the earliest examples of conserving individual geological sites, it also has possibly the oldest nature conservation legislation in Europe, providing recognition to geodiversity. Through the protection of monuments of the Grand Duchy of Hesse in 1902, natural monuments – including rocks, watercourses and trees of historical, scientific or scenic importance – were protected by law (Wiedenbein 1994). The Nordic countries soon followed, with a strong interest in the relationship between nature and culture developing from the late 19th century. This relationship was strengthened by the work of Finnish painters, bringing scenic landscapes to a wider public, whilst the expansion of road and rail networks in Sweden brought more people into contact with nature. This led in 1885 to the founding of the Swedish Tourist Association and, in 1909, to the establishment of the Swedish Society for Nature Conservation. In the same year, Sweden established nature protection acts conserving natural monuments, such as 'crooked knotty trees or remarkable erratic boulders', and declaring areas as national parks, establishing nine in the northern mountain region – the first national parks in Europe (Lundqvist *et al* 2012). In Norway (Erikstad 2012) the first Nature Conservation Act was established in 1910, stating that 'protected areas can be established to protect wild plants and animals and geological and mineralogical sites', in much the same way that the UK 1949 Act identified geological and physiographical features, but almost 40 years earlier. This inclusiveness has been carried through to Norway's present Nature Diversity Act 2009, where 'nature diversity' is considered to be a combination of biodiversity, geodiversity and landscape diversity, strengthening the early sound basis for geoconservation in Norway.

In Croatia, Marjanac (2012) notes that nature conservation was first encompassed in hunting law in 1893, and more specifically a law on cave conservation was enacted in 1900. The importance of both aesthetic and scientific value is emphasised and, in the publication of the Rule Book on Croatian National Parks in 1938, a quarry is proposed as a geological monument; there is some discussion of the challenge of conserving both geological and palaeontological objects, and a concern is expressed over the general lack of awareness of the value and sensitivity of geodiversity. Conserving a quarry for its geodiversity importance, grappling with the challenge of geoconservation – and particularly the lack of wider understanding for the need for geoconservation – remain as relevant today.

As environmental legislation develops, so does the approach to selecting places to protect. Early lists and inventories largely focus on caves and glacial features, particularly erratics. The Scottish Boulder Committee appears to be the earliest, but other examples include 40 glacial boulders described as monuments of nature in Belarus (1935) and the compilation of systematic data on erratics in Estonia by 1940 (901 listed and 210 protected). Germany (in Prussia) provides perhaps the earliest example of cross-cutting inventory, with the establishment of the first official department for nature conservation in Prussia in 1906 and the initiation of the task to undertake a systematic registration of sites by inventory, together with research and monitoring (Wiedenbein 1994). It is not, however, until the latter part of the 20th century that a 'modern' approach to inventories commenced; a Dutch working group started an inventory of earth science sites important for science and education in 1969 (Erikstad 2008), and in 1977 the previously described GCR was initiated in the UK. In 2015, at least 15 European countries have an active geodiversity audit and inventory (source International Union of Geological Sciences Geoheritage Task Group: www.geoheritage-iugs.mnhn.fr).

It is worth noting two geological locality terms that have emerged and are in common use across Europe – geosite and geotope. A geosite is a 'locality or area showing geological features of intrinsic scientific interest, features that allow us to understand the key stages in the evolution of the Earth' (Wimbledon and Smith-Meyer 2012, 19). It is a broad and catch-all label that includes both small sites and large areas, and is a common term used in site inventories. Some geomorphologists have employed the term 'geomorphosite' as an analogue of the geosite for their discipline. The Geosites Project was initiated by the International Union of Geological Sciences (IUGS) in 1994, with subsequent collaboration with ProGEO. Although there is currently no activity on this initiative, its intention was to provide a global inventory of key geosites (source IUGS Geoheritage Task Group: www.geoheritage-iugs.mnhn.fr).

The term 'geotope' tends to be used in German-speaking countries (Wiedenbein 1994a, 1994b). Stürm (1994, 28) defines a geotope as a 'distinct part of the geosphere of outstanding geological and geomorphological interest. They have to be protected against influences that could damage their substance, form or natural development.' They are distinct from geosites in that the definition highlights their 'outstanding' qualities and, as a consequence, the need for geotope conservation. Gray (2013) reviews the present legislative situation in Europe and the associated development of inventories, as do the contributors to Wimbledon and Smith-Meyer (2012), providing detailed accounts for each European country and outlining the origin of geoconservation through to the present situation.

GEOCONSERVATION: THE DUTCH APPROACH

The Netherlands provide an interesting comparison to the UK. Geologically, the Netherlands are much younger. The oldest rocks (Carboniferous rocks unconformably[3] overlain by Cretaceous chalk and sandstone) are confined to the south, in a country that is otherwise dominated by Pleistocene glacial and interglacial sediments and landforms, and the post-glacial development of today's low-lying landscape and coastline. It is a strongly altered landscape, artificially drained and in many areas intensively cultivated (over 80% of the land surface) with clay, sand, gravel, peat and chalk all exploited.

In 1908 the geologist J van Baren spent six years travelling and describing the country's geology. He noted the rapidly degrading landscape and warned of the impact of further destruction, but with little result (Gonggrijp 1994). Despite a number of attempts (with only two Dutch geological monuments being established in the 1920s), nothing succeeded until 1969 when the Gea Project was established to provide an inventory of the most important geological, geomorphological and pedological sites as a source of information for nature conservation and planning authorities – the first modern inventory for geoconservation purposes (Gonggrijp 1994; Erikstad 2008). A systematic inventory, it included a literature survey followed by field evaluation of sites on the basis of five criteria: rarity, condition, representativity [sic], scientific value and educational value.

A management approach was also developed (with similarities to the UK ESCC), based on management requirements, with three broad categories of site identified: geomorphologically important sites valued for their landforms and associated natural processes; buried features important for research; and exposures, largely man-made, such as pits and quarries. Despite this early impetus, with practical success such as the establishment of the glacial erratic Schokland Geological Park and interpretation of push moraines in the Dutch landscape (Gonggrijp 1994; Robinson 1989), there is currently no national organisation responsible for registering and monitoring geosites, and most activity is undertaken at a provincial level. In the mid-1990s, 12 geological monuments were selected in Utrecht province; 40 quarries were cleared in Limburg province, with the provision of interpretation, and more than 60 geological monuments were established in other provinces (Ancker and Jungerius 2012).

Ancker and Jungerius (2012) provide a detailed summary of the development of geoconservation legislative and protective measures, concluding that there remains a lack of effective legal protection for geodiversity at a national level in the Netherlands. Discussion with government and nature organisations is, however, once more raising an understanding of the value of geodiversity as part of the Dutch landscape and, importantly, the role that geodiversity (particularly the re-establishment and maintenance of natural processes) will have in the future management of the Dutch natural environment and landscape. With value established, threats will be understood and actions – geoconservation – will be supported by people; this is seen as a critical driver by the Dutch government.

---

[3] An unconformity occurs where directly overlying rocks (here the Cretaceous) are significantly younger than the underlying rocks (Carboniferous in this case). The difference in age between the Cretaceous and Carboniferous is approximately 200 million years. An unconformity is usually the result of an extended period of non-deposition or erosion.

EUROPEAN COOPERATION AND THE DEVELOPMENT OF INTERNATIONAL GEOCONSERVATION EFFORT

There is little early evidence of transnational cooperation and sharing of ideas in geoconservation, or at least it is not clearly documented. The development and inclusion of geology in the development of nature conservation legislation in the UK seems influenced by the inclusion of geological monuments in the development of the national park system in the USA (Prosser 2008; Thomas and Warren 2008), where the first national park, Yellowstone, was established in 1872. It seems reasonable that this had a similar influence on the development and adoption of national parks in Europe, which represent the earliest protection of wider landscapes, and that similar principles were translated into nature conservation legislation. The geological science community was certainly global. The first International Geological Congress (IGC) meeting in Paris in 1878 brought together geologists from North America and Europe, to encourage the advancement of earth science worldwide.

In 1948, when the IGC met in London, it recommended to 'all Governments represented at the Congress that, in any country where suitable action has not been already taken, lists of geological sites and districts of outstanding scientific importance shall be compiled (as has been done in England and Wales at the instigation of the Nature Reserves Investigation Committee), and that legislative measures shall be taken for safeguarding these sites and securing access to them' (Butler 1948, 363). The same Congress also agreed the formation of the International Union of Geology (later the International Union of Geological Sciences, or IUGS) to advise UNESCO on geological matters. The overall influence of this recommendation is not clear. It did, however, encourage Poland (Alexandrowicz 2012) to more actively and systematically consider geoconservation; it is good evidence of possibly the first international cooperative encouragement of geoconservation.

In 1988, for the first time, seven European countries (Austria, Denmark, Finland, Norway, the Netherlands, the Republic of Ireland and the UK) were invited to a meeting by the Dutch Research Institute for Nature Management to exchange geoconservation ideas (Robinson 1989). This reflected the then forward-thinking approach in the Netherlands, and the meeting established the European Working Group on Earth Science Conservation which later became ProGEO, the European Association for the Conservation of Geological Heritage (ProGEO 2014). The first international symposium (in Digne, France, in 1991) agreed the Digne Declaration, which sets out the shared value placed on geological heritage and the importance of its protection: 'the Earth retains the "memories" of the past, inscribed both in its depths and on its surface … the slightest damage could lead to irreversible losses for the future. In undertaking any form of development we should respect the singularity of this heritage' (Martini and Pagès 1994, 272). This was followed in 1993 by the UK Malvern International Conference on Geological and Landscape Conservation (O'Halloran et al 1994), which recognised the need to explore the establishment of an international earth science convention.

Since then ProGEO has held regular meetings and conferences, hosted regional European working groups on geoconservation, and adopted a protocol in 2012: 'Conserving our shared geoheritage – a protocol on geoconservation principles, sustainable site use, management, fieldwork, fossil and mineral collecting' (Wimbledon and Smith-Meyer 2012). The consequent sharing of knowledge and practical experience has led to a surge in geoconservation activity and cooperation at local, national and international levels (Erikstad 2008, 2013; Larwood et al 2013). In 2004 the Council of Europe adopted *Recommendation Rec(2004)3 on conservation of geological heritage and areas of special geological interest* (Council of Europe 2004). This emphasised the importance of member states identifying areas of special geological interest, developing protection and management measures and legal instruments, widening awareness, and encouraging geoconservation and international cooperation. Subsequently, the International Union for the Conservation of Nature (IUCN) World Conservation Congress meeting in Barcelona in 2008 adopted *Resolution 4.040: Conservation of geodiversity and geological heritage* (IUCN 2008), encouraging the involvement of government, independent sector groups and international organisations around the world.

In 2000, the European Geopark Network (EGN) was formed (Zouros and Martini 2003; http://www.europeangeoparks.org), with the establishment of four European Geoparks: Réserve Géologique de Haute-Provence (France); Natural History Museum of Lesvos Petrified Forest (Island of Lesvos, Greece); Geopark Gerolstein/Vulkaneifel (Germany); and Maestrazgo Cultural Park (Spain). In 2004, cooperation with UNESCO led to the Global UNESCO Network of Geoparks – the Global Geoparks Network (GGN) (Larwood *et al* 2013) – in which the EGN is now a regional network. The motto of the GGN is 'Celebrating Earth Heritage, Sustaining Local Communities'. Global Geoparks since 2015 have had UNESCO programme status. Global Geoparks are fundamentally about people, exploring and celebrating the links between communities and the Earth, with a strong emphasis on the contribution that Global Geoparks make to sustainable development. Reflecting the European origin of the geopark concept, currently 52 of the 90 Global Geoparks are in 19 European countries (Larwood *et al* 2013).

The global value of Europe's geodiversity is further emphasised through UNESCO's World Heritage Convention, which was agreed in 1972 (Larwood *et al* 2013). To be added to the World Heritage List, a site must fulfil at least one of ten selection criteria for outstanding universal value – six reflecting cultural qualities and four reflecting natural qualities. Specifically, Criterion viii ('outstanding examples representing major stages of Earth's history, including record of life, significant ongoing geological processes in the development of landforms, or significant geomorphic or physiographic features') must be fulfilled for a site to be inscribed as a geological World Heritage Site (see Figure 7.1B of the Jurassic Coast World Heritage Site). The first geological World Heritage Sites were the Galapagos Islands volcanic archipelago (Ecuador), the Nahanni National Park mountain ranges (Canada) and the caldera basin of the Yellowstone National Park (USA), which were inscribed in 1978.

In 1979 the first European geological World Heritage Site was inscribed in Croatia: the karstic landscape of the Plitvice Lakes National Park. Today, in the European region, there are 18 World Heritage Sites inscribed on the basis of their geodiversity (Criterion viii) out of a total of 39 (46%) natural World Heritage Sites inscribed in this region; this demonstrates the high and global value placed on Europe's geodiversity (source www.whc.unesco.org/en/list). The most recent additions in 2014 were Stevns Klint (Denmark), where chalk cliffs expose evidence of the meteorite impact and associated mass extinction at the end of the Cretaceous (65 million years ago), and the Wadden Sea (Netherlands, Germany and Denmark), the largest unbroken system of intertidal sand and mud flats in the world. The Wadden Sea supports a complex of freshwater and marine ecosystems with more than 5,000 recorded species, demonstrating the intimate link between geodiversity and biodiversity.

Geodiversity also has a close, and often defining, relationship with cultural heritage, which can be demonstrated through World Heritage Site designation. Firstly, the Giant's Causeway and Antrim Coast (Figure 7.1C) in Northern Ireland is a World Heritage Site on the basis of its lava plateau and exceptional natural beauty. It is so called due to the legend of two giants, the Irish Finn MacCool and his Scottish rival, Benandonner. According to legend, to settle a dispute, the giants built the Causeway between Scotland (where the basaltic lava flow is also visible on the Isle of Staffa) and Ireland – geodiversity explained through story-telling and myth. Secondly, the Frontiers of the Roman Empire World Heritage Site (northern England, Scotland and Germany), a cultural World Heritage Site, demonstrates the influence of geodiversity on the margins of the Roman Empire through the choice of the landforms that were selected to erect protective barriers.

The frontier in Germany (known as the Rhaetian Limes) stretches from the Rhine to the Danube, follows contours, skirts mountains, and connects and uses rivers. In Scotland, the Antonine Wall connects the narrowest point between the Firth of Forth and the Firth of Clyde. Most spectacularly, however, Hadrian's Wall (Figure 7.1A) is perched on the high escarpment of the Whin Sill (intruded approximately 300 million years ago); here, geodiversity has defined (and limited) the frontier of the Roman Empire, partly through the influence of the under-lying geology. Lastly, Studley Royal Park, including the ruins of Fountains Abbey (Figure 7.1D), in North Yorkshire in northern England, is a cultural World Heritage Site on the basis of its designed landscape that also demonstrates the influence of geodiversity. A deep valley, cut in the last Ice Age, provided the ideal protected location for the founding of the Cistercian Abbey in 1132; the surrounding geology (sandstone and limestone) provided the resources to construct the Abbey, and the meandering course of the river provided the inspiration and form for the designed 18th century landscape and complex of water features.

This connection between geodiversity and the role it plays in underpinning World Heritage properties inscribed for biological and cultural values is increasingly recognised, but not fully developed. Emphasising a cooperative approach to heritage – cultural and natural, and in particular the supporting value of geodiversity – will not only enable a representative, balanced and credible World Heritage List (Larwood *et al* 2013; Larwood and Parkinson in press), but can widen people's engagement, understanding and valuing of geodiversity at all scales – the starting point of geoconservation.

This cooperation and embedding of geoheritage geoconservation has led to an upsurge in research and publication, in particular within peer-reviewed journals. Two are worth noting. The *Proceedings of the Geologists' Association* (PGA) published a special issue on 'Geoconservation for science and society' in 2013 (Prosser *et al* 2013a), with 16 papers reflecting on the successes, opportunities and challenges for geoconservation from a conference convened by the Geologists' Association. Since this publication, the PGA has regularly published geoconservation papers and includes geoconservation within its publishing scope. In 2009, the first international journal *Geoheritage*, specifically dedicated to covering all aspects of global geoheritage (including geocon-servation), was launched. Published by Springer, it is a ProGEO-IUGS collaborative initiative with its origin in European geoconservation. Lastly, and demonstrating how far geoconserva-tion has advanced, a *History of Geoconservation* (Burek and Prosser 2008) collected 22 papers reviewing the progress to date in geoconservation; there is a history to review and reflect on, and a future to consider.

CONCLUDING THOUGHTS AND REFLECTIONS

Europe has a long history of geoconservation. This partly reflects the origin and growth of geological science across the continent, a striving to understand our geological origins, how we can harness the resources that our geoheritage offers, and a simple fascination and wonder at the diversity – geodiversity – that our geological past has created. The value that this has established for geodiversity has meant that, when it is threatened, people have taken action for geodiversity and have undertaken geoconservation. The examples cited in this chapter (and elsewhere in this volume) are illustrative and far from exhaustive; wherever one looks, as Wimbledon and Smith-Meyer (2012) amply illustrate, geoconservation is happening.

In some countries, such as the UK, geoconservation is advanced and mature; in others it is in its infancy, but everywhere advances are being made. Prosser *et al* (2013b) develop six future priorities for geoconservation, which build on our existing approach and seek to widen the support for geoconservation:

1. Continue to conserve geodiversity, building on existing strengths and successes.
2. Strengthen geoconservation as a discipline.
3. Make geoconservation relevant to society, especially to the management of the natural environment.
4. Inspire people to value and care for geodiversity.
5. Influence legislation, policy and practice.
6. Sustain and grow the resources needed for geoconservation and the numbers of people involved.

A common theme is the need to continue to widen and strengthen an awareness of geodiversity and an understanding of its value and importance to people and the environment. This has always been critical for successful geoconservation, and remains more so than ever where actions for geodiversity need to be justified and understood as a part of the conservation of the natural and cultural heritage, and the benefits brought to society – now and in the future.

REFERENCES

Alexandrowicz, Z, 2012 Poland, in *Geoheritage in Europe and its Conservation* (eds W A P Wimbledon and S Smith-Meyer), ProGEO, Oslo, 254–63

Ancker, J A M van den, and Jungerius, P D, 2012 The Netherlands, in *Geoheritage in Europe and its Conservation* (eds W A P Wimbledon and S Smith-Meyer), ProGEO, Oslo, 232–45

Bolner-Takács, K, Cserny, T, and Horváth, G, 2012 Hungary, in *Geoheritage in Europe and its Conservation* (eds W A P Wimbledon and S Smith-Meyer), ProGEO, Oslo, 158–69

Burek, C V, and Prosser, C D, 2008 The history of geoconservation: an introduction, in *The History of Geoconservation, Special Publication No 300* (eds C V Burek and C D Prosser), The Geological Society, London, 1–5

Butler, A J, 1948 The eighteenth session of the International Geological Congress, *Geological Magazine* 85 (6), 361–5

Cervato, C, and Frodeman, R, 2012 The significance of geologic time: cultural, educational and economic frameworks, in *Earth and Mind II: A Synthesis of Research on Thinking and Learning in Geosciences, Special Paper No 486* (eds K A Kastens and C A Manduca), Geological Society of America, Washington, 19–27

Chubb, L, 1945 *National Geological Reserves in England and Wales: Report by the Geological Reserves Sub-Committee of the Nature Reserves Investigation Committee. Conference on Nature Preservation in Post-War*

Reconstruction, Natural History Survey of Great Britain (2 edn), The Society for the Protection of Nature Reserves/British Museum (Natural History), London, 1–41

Council of Europe, 2004 *Recommendation Rec(2004)3 on conservation of geological heritage and areas of special geological interest*, available from: https://wcd.coe.int/ViewDoc.jsp?id=740629 [accessed 15 January 2015]

Doughty, P, 2008 How things began: the origins of geological conservation, in *The History of Geoconservation, Special Publication No 300* (eds C V Burek and C D Prosser), The Geological Society, London, 7–16

Doyle, P, 2008 A vision of 'deep time': the 'Geological Illustrations' of Crystal Palace Park, London, in *The History of Geoconservation, Special Publication No 300* (eds C V Burek and C D Prosser), The Geological Society, London, 197–205

Doyle, P, Bennett, M R, and Robinson, E, 1996 Creating urban geology: a record of Victorian innovation in park design, in *Geology on your Doorstep: The Role of Urban Geology in Earth Heritage Conservation* (eds M R Bennett, P Doyle, J G Larwood and C D Prosser), The Geological Society, London, 74–84

Duckeck, J, 2014 *Showcaves of the World*, available from: http://www.showcaves.com [accessed 19 August 2014]

Ellis, N, 2008 A history of the Geological Conservation Review, in *The History of Geoconservation, Special Publication No 300* (eds C V Burek and C D Prosser), The Geological Society, London, 123–35

— 2011 The Geological Conservation Review (GCR) in Great Britain – rationale and methods, *Proceedings of the Geologists' Association*, 122 (3), 353–62

Ellis, N V, Bowen, D Q, and Campbell, S, 1996 *An Introduction to the Geological Conservation Review, GCR Series No 1*, Joint Nature Conservation Committee, Peterborough, 131

English Geodiversity Forum, 2014 *Geodiversity Charter for England*, available from: www.englishgeodiversityforum.org.uk [accessed 10 January 2014]

Erikstad, L, 2008 History of geoconservation in Europe, in *The History of Geoconservation, Special Publication No 300* (eds C V Burek and C D Prosser), The Geological Society, London, 249–56

— 2012 Norway, in *Geoheritage in Europe and its Conservation* (eds W A P Wimbledon and S Smith-Meyer), Oslo ProGEO, 246–53

— 2013 Geoheritage and geodiversity management – the questions for tomorrow, *Proceedings of the Geologists' Association* 124 (3), 713–19

Evans, D, 1992 *A History of Nature Conservation in Britain*, Routledge, London

Foley, S, 1694 An account of the Giant's Causeway in the North of Ireland, *Philosophical Transactions of the Royal Society of London*, 18 (212), 170–5

Gonggrijp, G, 1994 Earth science conservation in the Netherlands, *Mémoires de la Société Géologique de France* 165, 139–47

Gordon, J E, 1992 Conservation of geomorphology and Quaternary sites in Great Britain: an overview of site assessment, in *Conserving our Landscape: Proceedings of the Conference, 'Conserving our Landscape: Evolving Landforms and Ice-Age Heritage', Crewe, May 1992* (eds C Stevens, J E Gordon, C P Green and M G Macklin), 11–21

Gordon, J E, and MacFadyen, C C J, 2001 Earth heritage conservation in Scotland: state, pressures and issues, in *Earth Science and Natural Heritage* (eds J E Gordon and K F Leys), Stationery Office, Edinburgh, 130–44

Gray, J M, 2013 *Geodiversity: Valuing and Conserving Abiotic Nature* (2 edn), Wiley, London

Gray, J M, Gordon, J E, and Brown, E J, 2013 Geodiversity and the ecosystem approach: the contribution of geoscience in delivering integrated environmental management, *Proceedings of the Geologists' Association* 124, 659–73

Hoffman, T, and Schönlaub, H P, 2012 Austria, in *Geoheritage in Europe and its Conservation* (eds W A P Wimbledon and S Smith-Meyer), ProGEO, Oslo, 30–9

Hose, T A, 2008 Towards a history of geotourism: definitions, antecedents and the future, in *The History of Geoconservation, Special Publication No 300* (eds C V Burek and C D Prosser), The Geological Society, London, 37–60

— in press (ed), *Appreciating Physical Landscapes: Three Hundred Years of Geotourism, Special Publication No 417*, The Geological Society, London

IUCN, 2008 *Resolution 4.040: Conservation of Geodiversity and Geological Heritage. Adopted by the 4th IUCN World Heritage Congress, Barcelona*, available from: http://intranet.iucn.org/webfiles/doc/IUCNPolicy/Resolutions/2008_WCC_4/English/RES/res_4_040_conservation_of_geodiversity_and_geological_heritage.pdf [accessed 15 January 2015]

Knell, S, 1996 Museums: a timeless urban resource for the geologist, in *Geology on your Doorstep: The Role of Urban Geology in Earth Heritage Conservation* (eds M R Bennett, P Doyle, J G Larwood and C D Prosser), The Geological Society, London, 104–15

Kriz, J, 1994 Conservation of geological sites, fossils and rock environments in Czechoslovakia, *Mémoires de la Société Géologique de France* 165, 101–2

Larwood, J G, Badman, T, and McKeever, P, 2013 The progress and future of geoconservation at a global level, *Proceedings of the Geologists' Association* 124, 720–30

Larwood, J G, and Parkinson, S, in press *Culturally Natural or Naturally Cultural?* Natural England–National Trust Report, Peterborough

Lundqvist, S, Fredén, C, Johansson, C E, Karis, L, and Ransed, G, 2012 Sweden, in *Geoheritage in Europe and its Conservation* (eds W A P Wimbledon and S Smith-Meyer), ProGEO, Oslo, 345–57

Marjanac, L, 2012 Croatia, in *Geoheritage in Europe and its Conservation* (eds W A P Wimbledon and S Smith-Meyer), ProGEO, Oslo, 81–91

Martini, G, and Pagès, J (eds), 1994 Actes du première symposium international sur le protection du patrimoine géologique, *Mémoires de la Société Géologique de France* 165

McMillan, A A, 2008 The role of the British Geological Survey in the history of geoconservation, in *The History of Geoconservation, Special Publication No 300* (eds C V Burek and C D Prosser), The Geological Society, London, 103–12

Milne Home, D, 1872 Scheme for the conservation of remarkable boulders in Scotland, and for the indication of their position on maps, *Proceedings of the Royal Society of Edinburgh* 7, 475–88

— 1884 Tenth and final report of the Boulder Committee; with Appendix, containing an abstract of the information in the nine annual reports of the committee: and a summary of the principal points apparently established by the information so received, *Proceedings of the Royal Society of Edinburgh* 12, 765–926

Natural England, 2010 *Valuing our Landscapes. The European Landscape Convention in Action: Making a Difference in Places that Matter*, Natural England, Peterborough

Nature Conservancy Council, 1990 *Earth Science Conservation in Great Britain – A Strategy* (and *Appendices – A Handbook of Earth Science Conservation Techniques*), Nature Conservancy Council, Peterborough

O'Halloran, D, Green, C, Harley, M, Stanley, M, and Knill, J (eds), 1994 *Geological and Landscape Conservation*, The Geological Society, London

ProGEO, 2014 www.progeo.se [accessed 10 January 2015]

Prosser, C D, 2008 The history of geoconservation in England: legislative and policy milestones, in *The History of Geoconservation, Special Publication No 300* (eds C V Burek and C D Prosser), The Geological Society, London, 113–22

— 2012 William Archibald Macfadyen (1893–1985): The 'father of geoconservation'?, *Proceedings of the Geologists' Association* 123, 182–8

— 2013a Our rich and varied geoconservation portfolio: the foundation of the future, *Proceedings of the Geologists' Association* 124 (4), 568–80

— 2013b Planning for geoconservation in the 1940s: an exploration of the aspirations that shaped the first national geoconservation legislation, *Proceedings of the Geologists' Association* 124 (3), 536–46

Prosser, C D, Brown, E J, Larwood, J G, and Bridgland, D R, 2013a Geoconservation for science and society, *Proceedings of the Geologists' Association* 124, 559–730

— 2013b Geoconservation for science and society – an agenda for the future, *Proceedings of the Geologists' Association* 124, 561–7

Prosser, C D, and King, A H, 1998 Regionally Important Geological and Geomorphological Sites: the origin and a forward view, in *Proceedings of the First UK RIGS conference* (ed P G Oliver), Hereford and Worcestershire RIGS Group, Worcester, 1–8

Prosser, C D, Murphy, M, and Larwood, J G, 2006 *Geological Conservation: A Guide to Good Practice*, English Nature, Peterborough

Reynard, E, 2012 Geoheritage protection and promotion in Switzerland, *European Geologist* 34, 44–7

Robinson, E, 1989 European geological conservation, *Terra Nova* 1, 113–18

Röhling, H, Schmidt-Thomé, M, and Goth, K, 2012 Germany, in *Geoheritage in Europe and its Conservation* (eds W A P Wimbledon and S Smith-Meyer), Oslo, ProGEO, 132–43

Royal Society for Nature Conservation, 1999 *The RIGS Handbook*, Royal Society for Nature Conservation, Newark, England [available at: www.geoconservationuk.org.uk]

Satkunas, J, Lincius, A, and Mikulenas, V, 2012 Lithuania, in *Geoheritage in Europe and its Conservation* (eds W A P Wimbledon and S Smith-Meyer), Oslo, ProGEO, 217–23

Scottish Geodiversity Forum, 2012 *Scotland's Geodiversity Charter*, available from: www.scottishgeodiversityforum.org.uk [accessed 10 January 2015]

Sheail, J, 1976 *Nature in Trust: The History of Nature Conservation in Britain*, Blackie, Glasgow

Sorby, H C, 1875 On the remains of fossil forest in the coal-measures at Wadsley, near Sheffield, *Quarterly Journal of the Geological Society of London* 31, 458–500

Stace, H, and Larwood, J G, 2006 *Natural Foundations: Geodiversity for People, Places and Nature*, English Nature, Peterborough

Stürm, B, 1994 The geotope concept: geological nature conservation by town and country planning, in *Geological and Landscape Conservation* (eds D O'Halloran, C Green, M Harley, M Stanley and J Knill), The Geological Society, London, 27–31

— 2012 Switzerland, in *Geoheritage in Europe and its Conservation* (eds W A P Wimbledon and S Smith-Meyer), ProGEO, Oslo, 358–65

Taylor, H A, 1995 Urban public parks, 1840–1900: design and meaning, *Garden History* 23, 204–8

Thomas, B A, 2005 The palaeontological beginnings of geological conservation: with case studies from the USA, Canada and Great Britain, in *History of Palaeobotany: Selected Essays, Special Publication No 241* (eds A J Bowden, C V Burek and R Wilding), The Geological Society, 95–110

Thomas, B A, and Cleal, C J, 2005 Geological conservation in the United Kingdom, *Law, Science and Policy* 2, 269–84

Thomas, B A, and Warren, L M, 2008 Geological conservation in the nineteenth and early twentieth centuries, in *The History of Geoconservation, Special Publication No 300* (eds C V Burek and C D Prosser), The Geological Society, London, 17–30

Wiedenbein, F W, 1994a German developments in earth science conservation, *Mémoires de la Société Géologique de France* 165, 119–27

— 1994b Origin and use of the term 'geotope' in German-speaking countries, in *Geological and Landscape Conservation* (eds D O'Halloran, C Green, M Harley, M Stanley and J Knill), The Geological Society, London, 117–20

Wilson, C, 1994 *Earth Heritage Conservation*, Milton Keynes, Open University

Wilson, E O (ed), 1988 *Biodiversity*, National Academy Press, Washington

Wimbledon, W A P, and Smith-Meyer, S (eds), 2012 *Geoheritage in Europe and its Conservation*, ProGEO, Oslo

Worton, G J, 2008 A historical perspective on local communities and geological conservation, in *The History of Geoconservation, Special Publication No 300* (eds CV Burek and C D Prosser), The Geological Society, London, 137–46

Zouros, N, and Martini, G, 2003 Introduction to the European Geoparks Network, in *Proceedings of the 2nd European Geoparks Network Meeting, Lesvos, Greece, 3–7 October 2001* (eds N Zouros, G Martini and M-L Frey), Natural History Museum of Lesvos Petrified Forest, 17–21

8

Geotourism in Britain and Europe: Historical and Modern Perspectives

Thomas A Hose

Early geotourism

This chapter provides an overview of the development of geotourism in some key areas of Britain, with some mention of key European examples. It examines how geotourism developed over several hundred years, and it contextualises geotourism within broader landscape tourism developments. Geotourism's antecedents include the late 18th and 19th century European aesthetic and nature conservation movements (see Figure 8.1) that contributed to modern sustainable tourism. The 'Romantic' movement in particular was an essential prelude to nature conservation and sustainable development, because it 'valued the spiritual over the material, and humans came to be seen as part of nature, not superior to it' (Hardy *et al* 2002, 476). It coincided with the first formal UK attempts, especially in the Lake District (Ritvo 2009), to promote and then preserve wildlife, landscapes and geosites. It was the Lake District (Hose 2008a) and the Scottish Highlands (Hose 2010a) that drew writers and artists from the 19th century's first quarter to record the splendour of, and acknowledge the geological influences on, their landscapes. They interpreted what lay before them through the cultural filter (Hose 2010b) of the 'Romantic' aesthetic movement (Hebron 2006), producing visual and literary images of landscape.

Aesthetic landscapes

Landscapes are socio-cultural constructs as much as they are physical entities. Since the Renaissance, European leisure travellers have selected landscapes that matched their quest for the novel, exotic and authentic. Specific expectations about the places and landscapes they planned to visit were created and dominated by images found in art galleries and guidebooks. Such landscapes are idealised views held in the mind, composed of key reference points mapped together by imagination, education and recollections. The perceptions of landscapes and the values ascribed to them by travellers are a combination of their processing of direct observation and subsequent interpretation, influenced by their education, knowledge, interests and expectations. Hence travellers, tourists, artists, writers, geographers, geologists and geomorphologists define, delineate, describe and depict landscapes from different mind-sets.

In mid-18th century Britain, Edmund Burke's (1729–97) *A Philosophical Enquiry into the Origin of Our Ideas of the Sublime and Beautiful* (1757) equated the sublime with astonishment, fear, pain, roughness and obscurity. However, the Romantics associated the sublime with the tumultuous chaos of mountains lying beyond rolling foothills, deep valleys and dangerous rocky precipices. William

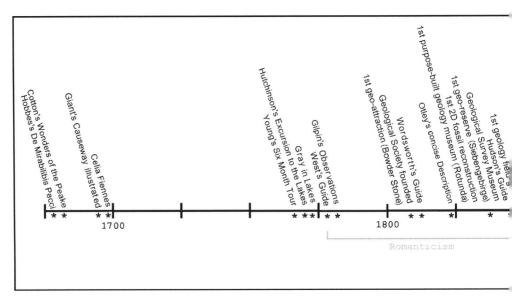

FIGURE 8.1 TIMELINE OF INFLUENTIAL ARTISTIC MOVEMENTS

Wordsworth (1770–1850) suggested, in the fifth and final edition of his *A Guide Through the District of the Lakes* (Figure 8.2D), that the landscape was 'the result of Nature's first great dealings with the superficies of the earth' (Wordsworth 1835, 35). The wild ruggedness of the Lake District's great rock masses, fells and lakes solicited awe and wonder from travellers previously accustomed to appreciating lowland southern Britain's agricultural and parkland landscapes as evidence of man's control of nature. Within the Picturesque movement, the softer effects stemming from nature's subsequent impacts – producing the variegation and harmony expressed by a river meander's curve or a lake shore, the grouping of their flanking trees, the interplay of light and shade, and the subtle colour gradations that framed the scene – were of interest to the traveller. From the late 17th century, this topographical approach was adopted by pioneering travellers and artists, who picture-framed landscapes from scenic viewpoints or 'stations', such as those named for the Lake District by Thomas West (1720–79) in his *A Guide to the Lakes* of 1778. The Romantic movement, adopted by artists, poets and travellers from around 1780 to 1850, was all-embracing, involving travellers' emotional reflections on landscapes and their evocation in visual art and literature. These aesthetic landscape movements (see Figure 8.1) reflected three interrelated elements: the travellers' nature and purpose; the meanings ascribed to, and understandings of, natural phenomena; and the shift from rural to industrial society, with the concomitant rise of the middle classes in numbers and influence. Late 20th century 'ruralism' (Hose 2012, 14) visualised landscapes employed for leisured 'aristocratic' pursuits that were being undertaken by the aspirant middle classes.

GEOLOGY AND LANDSCAPES

After the mid-18th century, scientific geology's field study was a mainly British occupation, but one with significant French and German contributions. It was in early 19th century Germany that the need to protect key geosites, especially from quarrying, was recognised. Its first nature reserve

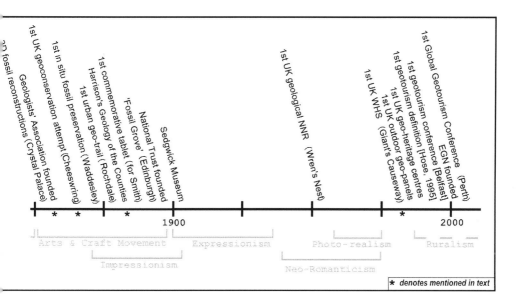

of 1836, the Drachenfels in the Siebengebirge uplands, was of some geological significance, as was the Teufelsmauer, protected in 1852 (Grube and Albrechts 1992, 16). The Drachenfels is formed of the remnants of a long-extinct volcano; its rocks have been quarried since Roman times, and were used to build Cologne Cathedral. It is the closest of all the hills in the Siebengebirge to the River Rhine, which facilitated the use of barges to transport its quarried rock. Quarrying ended when the Prussian government bought the quarry in 1836. In 1922 the first protection measures were put in place, and in 1956 it was declared a national park. The Drachenfels and its castle ruins were popularised by the Romantic movement, especially in poems by Lord Byron (1788–1824), Edward Bulwer-Lytton (1803–73) and Heinrich Heine (1797–1856) that ensured its international attention; most noteworthy was its mention in Byron's *Childe Harold's Pilgrimage* (published in four parts between 1812 and 1818), which established the poet's career. The hill's popularity was further aided when, in 1883, it became readily accessible by railway and became an established tourist attraction, which continues to the present day. The Teufelsmauer (or 'Devil's Wall') is a rock formation of hard Cretaceous age sandstones in the Harz Mountains. To prevent its total destruction by quarrying, it was given limited protection in 1832 and in 1852, by the head of the district authority. It has been protected as a nature reserve since 1935.

Meanwhile, in Britain, the Cheesewring, a granite tor above a quarry in Cornwall, was saved by restrictive clauses in quarrying leases of 1845 and 1865. Geological field study was principally undertaken by men of considerable social and economic standing, sometimes members of the established Church, suggesting a shift in the social climate that permitted ventures into 'wild' landscapes. This interest in geological field inquiry was part of a wider fascination with natural history, fostered by the realisation that 'even the amateur could hope to make significant contributions and participate in important national, even international scientific endeavours' (Bedell 2000, 4–5). They generally trod in the footsteps, and were guided by the publications, of earlier travellers engaged in agricultural, industrial and socio-economic reportage. Up to the mid-18th

Figure 8.2 The title pages of some significant historic geotourism texts

A. *De Mirabilibus: Being the Wonders of the Peak in Darby-shire, Commonly called The Devil's Arfe of Peak. In English and Latine. The Latine Written by Thomas Hobbes of Malmsbury. The English by a Person of Quality* (Thomas Hobbes 1678).

B. *The Wonders of the Peake* (Charles Cotton 1681), with frontispiece of Peak Cavern in Castleton.

C. *A Guide to the Lakes: Dedicated to the Lovers of Landscape Studies, and to All Who Have Visited, or Intend to Visit the Lakes in Cumberland, Westmorland and Lancashire* [3 edn] (Thomas West 1784), with frontispiece of Grasmere.

D. *A Guide Through the District of the Lakes in The North of England, with A Description of the Scenery etc. For the Use of Tourists and Residents* [5 edn] (William Wordsworth 1835), with fold-out topographical map.

Note that the volumes have not been reproduced to the same scale.

century, the 'preferred rural landscape was generally a humanised scene of cultivation, evidence of the successful control of nature' (Towner 1996, 138). The 'wild' areas, like those geologically mapped in the Lake District by Jonathan Otley (1766–1856) and Adam Sedgwick (1785–1873), were seen by most people at the time as waste places of no economic worth.

By the late 18th century, interest and pleasure in wild areas is found in travellers' writings. Early leisure travellers venturing into them were directed by the selections of authors, diarists, poets and artists promoting the landscape aesthetic movements. Initially, the Peak District, considered the birthplace of British geotourism (Hose 2008a), was favoured by Britons. England's first recorded geotourist was Celia Fiennes (Morris 1949), a privileged late-17th century horseback traveller who went to the Peak District's Ashbourne copper mines in 1698. The area is also where the first tourist guides were published; its major attractions were organised, described and promoted by the poem and guide of Thomas Hobbes (1588–1679) and Charles Cotton (1630–87), respectively *De Mirabilibus Pecci: Being the Wonders of the Peak in Darby-shire* (Figure 8.2A), published in English in 1678, but first published in Latin in 1636, and *The Wonders of the Peake* of 1681 (Figure 8.2B); both works recognised seven 'wonders', of which five had a geological basis.

THE LAKE DISTRICT, ROMANTICISM AND GEOTOURISM

A literary geotourism legacy

The Lake District is significant to geotourism's development because its exploration opened up what was previously considered to be a wild and remote region to leisure travellers. Historians, writers of literary criticisms and poets have charted its rise over some 200 years from comparative obscurity to one of the most visited parts of Britain. From the 1750s, travellers (popularly called 'Lakers', as the area is also known as 'The Lakes') visited the region, mainly because of its landscapes and antiquities. The poet Thomas Gray (1716–71) toured in 1767 and 1769; the letters describing his second tour were published in William Mason's posthumous edition of his work in 1775 and helped to establish the region's main stopping-points – later to become tourist 'stations'. Gray visited Ullswater from Penrith, used Keswick as a base to explore Derwentwater, Borrowdale, Bassenthwaite and Castlerigg stone circle, and then passed the foot of Helvellyn to Grasmere, Ambleside and Windermere, finishing at Kendal. Reflecting the travellers' mind-set of his day, he likened his Derwentwater to Borrowdale journey to crossing Alpine passes where travellers were threatened by avalanches. Celia Fiennes had ridden through the Lake District in 1698 and, viewing it as an unprofitable wilderness, merely recorded Windermere's potted char and some bread recipes. Daniel Defoe in the 1720s – with a like mind-set – recorded the Lake District as barren and frightful. The artist William Gilpin (1724–1804) toured in 1772 and the agrarian writer Arthur Young (1741–1820) in 1768. Gilpin, in *Observations on the River Wye and several parts of South Wales, etc. relative chiefly to Picturesque Beauty; made in the summer of the year 1770* of 1782, remarked much as Gray had done on the road into Borrowdale: 'grew wilder, and more romantic … riding along the edge of a precipice, unguarded by any parapet, under impending rocks, which threaten above' (Gilpin 1786, Volume 1, 187); such an actual alpine scene is captured in J M W Turner's 1804 painting, *The Passage of the St Gothard*. Young, focused on agricultural improvement, described Derwentwater as elegant but the surrounding mountains as wild with dreadful chasms; he likewise observed, in stark contrast to his praise of the verdant well-tended fields around Kendal, that: 'Twelve of the fifteen miles from Shapp to

Kendal are a continued chain of mountainous moors, totally uncultivated, one dreary prospect'
(Young 1770, 169).

By the 18th century, illustrated guidebooks on the Lake District's antiquities, such as William
Hutchinson's (1732–1814) two-volume *The History of the County of Cumberland* (1794), had
been published. Twenty years earlier, Hutchison justified publication of *An Excursion to the
Lakes, in Westmoreland and Cumberland, August 1773* (Hutchinson 1774) on the grounds that:
'When ever I have read the descriptions given by travellers of foreign countries, in which their
beauties and antiquities were lavishly praised, I have always regretted a neglect which has long
attended the delightful scenes at home' (Hutchinson 1774, 1). In 1778 Thomas West (1720–79)
compiled the Lake District's principal locations and sights, with recommended 'stations', into
its first tourist guidebook, *A Guide to the Lakes: Dedicated to the Lovers of Landscape Studies, and
to All Who Have Visited, or Intend to Visit the Lakes in Cumberland, Westmorland and Lancashire*
(Figure 8.2C); he thus established the pattern for later tours and guidebooks. The *Guide* provides
descriptions of the precision necessary to find his numbered stations before the advent of detailed
maps; Coniston's first station above Nibthwaite could be found as the lake came into view, 'by
observing an ash tree on the west side of the road, and passing that till you are in a line with
the peninsula' (West 1778, 50–1). It was one of the first to promote the 'Picturesque' and was
a major commercial success. West had a museum near Coniston and he published its visitors'
names in the local weekly newspaper; its popularity can be gauged from the 1,540 persons listed
in 1793. West was amongst the first writers to challenge the perception of the north of England
as a wild and savage place.

Following his 1786 Lake District tour, Arthur Young commented on 'a vast many edges of
precipices, bold projections of rock, pendent cliffs, and wild romantic spots, which command the
most delicious scenes, but which cannot be reached without the most perilous difficulty: To such
points of view, winding paths should be cut in the rock, and resting-places made for the weary
traveler' (Young 1770, 155); he also suggested that trees obscuring views should be pruned or
felled. From 1783, Peter Crosthwaite (1735–1808) published maps showing both West's and his
own stations. Crosthwaite, after retiring from sea duties with the East India Company, initially
worked as a Lake District guide before establishing, in 1781, the museum in Keswick from which
he sold his maps and other aids to the region's Romantic travellers. He improved physical access
to some of his stations; for example, for one of his two mapped stations above Derwentwater
he had steps cut and a cross marked on the ground. The popular stations often had structures provided for travellers. These were not universally welcomed; for example, in 1799, James
Plumtree considered that Windermere's first station somewhat lacked rustic charm. Plumtree, of
whom little seems to be known, completed and wrote up, but unfortunately never published,
two pedestrian tours (Plumtree 1798; 1799) at the end of the 18th century that included Lake
District elements in some detail.

Artists recorded the region's lakes, mountains and curiosities; these were occasionally published,
sometimes with accompanying prose or poetry, as sets of engravings. Artworks in their original
forms were primarily viewed by the social elite in private and commercial galleries; from the
1840s, some were viewed by the middle classes in the newly opened public art galleries. For
those unable to view the originals, due to restrictions of either social class or geography, the
emergence of commercial lithography facilitated the production and distribution of fair copies.
Photography was increasingly employed, mainly as a monochrome recording medium, from the
second half of the 19th century. Thus, for anyone unfamiliar with the Lake District's landscapes,

their expectations and impressions were influenced by various artistic visualisations, some of which appeared as engravings in guidebooks. John Constable (1776–1837), on his only visit in the autumn of 1806, sketched and painted around Kendal, Brathay, Skelwith, Thirlmere and Windermere; his watercolours *Windermere* and *The Castle Rock, Borrowdale* reveal his familiarity with geology. Joseph William Mallord Turner (1775–1851) made a living as a young artist producing topographical works; he exhibited two Lake District paintings at the Royal Academy in 1798, with *Morning Amongst the Coniston Fells, Cumberland* being noteworthy. His artistic rival, John Glover (1767–1849), exhibited Lake District paintings from 1795 onwards. Glover lived in the Lake District from around 1818 to 1820. He had a considerable reputation, with his own gallery in London's fashionable Old Bond Street; his c.1820 *Thirlmere* painting and c.1831 *Derwentwater* pen and wash drawing are typical of his faithful, detailed depictions.

Derwentwater was very popular with early tourists. Thomas West's late 18th century guidebook noted that its view from the Cockshott Hill 'station' was close to the ideal requirements of the 'picturesque' because, in 'a spacious amphitheatre, of the most picturesque mountains imaginable, an elegant sheet of water is spread out before you, shining like a mirror, and transparent as chrystal; variegated with islands … clothed with forest verdure' (West 1778, 89–90). The architect Henry Holland (1745–1806) disparagingly commented in 1811 that 'nobody you know, travels now a days without writing a quarto to tell the world where he has been, etcetera, what he has beheld' (Barton 1998, 3), and clearly would have welcomed the well-researched and more focused accounts that began to appear within a decade.

The region's industrial and mining sites were ignored when it was promoted by the 'Lake School' of Romantic poets (Samuel Taylor Coleridge, Robert Southey and William Wordsworth). It was promoted to discerning elite tourists by William Green's (1760–1823) *The Tourist's New Guide* of 1819, a detailed volume well known to, and fulsomely praised by, Wordsworth (1770–1850) in his own guidebook. Starting as a surveyor in 1778, Green produced a significant body of detailed and scientifically observed aquatints, etchings and watercolours. From about 1800 he lived in Ambleside, with a Keswick studio, as a successful topographic artist. Whilst Green popularised the region with pictures, Wordsworth did so particularly poetically. Wordsworth's guidebook was innovative in that, alongside descriptions of what could be seen, it linked landscape features to natural history (including geology), local history and people; it was an 'alternative to what its author considered the superficial and exploitive approach to the Lakes taken by many other works, which treated landscape and village as aesthetic spectacles' (Buzzard 1993, 29–30). It was developed from his anonymous text accompanying the Reverend Joseph Wilkinson's (1764–1831) volume of engravings, *Select Views in Cumberland, Westmoreland and Lancashire* (1810), and in 1820 it became an appendix to his River Duddon sonnets. By 1822 it was a guidebook in its own right and went to several editions up to the 1840s. Later editions included three geology letters commissioned from Adam Sedgwick; these were neither a novel nor a singular natural history inclusion, because the 1842 edition had also included botanical notes.

The Lake District's later guidebooks were much more than mere station descriptions; they were holistic accounts for landscape students. Guidebooks such as Charles Mackay's (1814–89) illustrated *The Scenery and Poetry of the English Lakes: A Summer Ramble* (1846) were particularly noteworthy for their text, focusing travellers' attentions upon site-centred poetry and other literary allusions, even if their literary allusions were sometimes geographically inappropriate; for example, Mackay used his account of Stock Ghyll Force near Ambleside as an excuse to quote the waterfall description in Shelley's *Alastor, or The Spirit of Solitude* of 1816, in which the narrator recounts the life of a poet zealously pursuing the most obscure part of nature whilst seeking out the peculiar truths in obscure places (which, in that poem's case, were those of the Caucasus Mountains) – perhaps not unlike the field geologist of the time. The invention of such literary landscapes is the enduring legacy of these early travel writers, and it pervades many modern geopark and protected landscape publications, reaching out to a wider audience rather than dedicated geotourists. The impetus for much of the original landscape discovery and literary publication was the closure of continental Europe to the British from 1789 to 1815, because of the French Revolution and the Napoleonic Wars; its landscapes were then promoted, surprisingly to the modern mind, as lookalikes for Europe's alpine regions, with which Wordsworth and his literary and artistic contemporaries were themselves familiar from their Grand Tours.

Few 19th century geology guidebooks were published for the Lake District. In 1823 Jonathan Otley, its earliest geologist, published the first populist account, *A Concise Description of the English Lakes … With Remarks on the Mineralogy and Geology of the District*, which went (with various titles) to a final seventh edition in 1843. Otley was the pioneer who unravelled the complex stratigraphical and structural geology of the Lake District. Until he was 25 he worked, making baskets, with his father at Loughrigg near Grasmere. Having by then developed the skills of watch and clock-repairing, he moved in 1791 to Keswick where he established himself in business with a clock and watch-repairing business; its success allowed him to develop his interests in map-making, geology, meteorological observations and natural history. He published the first map of the Lake District's geology in 1813. Its accompanying guidebook of 1823 eventually ran to eight editions by 1849 and sold more than 8,000 copies (Smith 2000, 32). His earliest recorded geological work is a paper on the plumbago Black Lead Mine in Borrowdale, read before the Manchester Philosophical Society on 27 December 1816. He published an essay on the geological structure of the Lake District in the *Philosophical Magazine* in 1820, and also in the first volume of *The Lonsdale Magazine*, a short-lived local journal, in the same year. However, because he willingly shared his knowledge with the likes of Adam Sedgwick (whom, from the summer of 1823, he guided on several excursions that began a long association between them) and John Phillips (1800–74), who also published on the region, his contribution is today generally not appropriately acknowledged, having been overshadowed by such better known and connected academic geologists. John Hudson's *Complete Guide to the Lakes* (1842) was supplemented by geological notes, again written by Adam Sedgwick. Towards the end of the 19th century, quite detailed English county-based geological accounts appeared in trade directories; William Jerome Harrison's (1845–1908) *Geology of the Counties of England and Wales* (1882) was a bound volume of such accounts.

Rock features as geo-attractions were an early Lake District commercial geotourism enterprise; in 1798 Joseph Pocklington (1736–1817) made the Bowder Stone (a 1,250-tonne perilously balanced boulder) in Borrowdale the first such attraction. Pocklington, who had been born in Newark into a then successful Nottinghamshire banking family (although the family bank eventually failed in 1809), went up to Jesus College, Cambridge, but he never graduated. He then seemingly toured Britain and produced a sketchbook of his travels. He inherited a legacy when only 26 that provided him with the means for a life of leisure, and also eventually funded his 1778 purchase of one of the islands on Derwentwater, upon which he built a house and numerous follies. By 1807, along with erecting a cottage for the resident guides to the Bowder Stone, Pocklington had also set up a druid stone, erected a small chapel, and affixed a ladder for visitors to clamber atop the Bowder Stone itself. Rock fragments around its base were also cleared away and a hole dug through the base so visitors could shake hands from either side. Green's 1819 illustrated guidebook noted a further decline in its natural and contemplative appeal, due to the guardian presenting visitors with an 'exordium preparatory to the presentation of a written paper, specifying the weight and dimensions of the stone' (Green 1819, Volume 2, 134). West had described its location 40 years earlier: 'Bowdar-stone … a mountain in itself, the road winds round its base. Here rock riots over rock, and mountain intersecting mountain, form one grand sweep of broken pointed crags' (West 1778, 100). Until the last quarter of the 20th century, there were no other commercial Lake District geotourism attractions.

Some present-day Lake District geotourism sites

Modern geotourists have fairly free access to one of the UK's most visited national parks. There is an ever-expanding range of populist leaflets and books on its geology. The fortunes of its geo-attractions have been varied. For such a geologically rich area, there are surprisingly few dedicated visitor sites and most are mining-related.

Keswick Pencil Museum

Arguably the most unusual geological attraction is the Keswick Pencil Museum. Pencils were originally manufactured from locally sourced graphite. Graphite was first found in Borrowdale in the early 1500s, but the first record of a Keswick factory making pencils is in 1832; by 1854 there were four such enterprises. The last local graphite mine closed in 1890 and modern pencils are made from an artificial material at a new factory opened in 2008. The original Cumberland Pencil Museum opened, with a reported 80,000 annual visitors, in 1981 in a 1950s concrete prefabricated building (formerly the works canteen) at the Southey factory site. The museum displays barely mention local geology, but at least visitors enter them through a replica of an early graphite mine – the original Seathwaite Mine.

Threlkeld Quarry and Mining Museum

The Threlkeld Quarry and Mining Museum near Keswick is within a former granite quarry. Opened in the 1870s, it supplied railway ballast to the Penrith-Keswick line and then the Crewe-Carlisle railway; it also supplied roadstone, kerbing and dressed stone for facing buildings. Threlkeld granite was also used in the Manchester Corporation Waterworks Thirlmere reservoir scheme. Unlike Shap granite, the stone is unsuitable for polishing and decorative uses. In 1892 a pre-cast concrete works was established to produce the Threlkeld concrete flagstones used in many towns in northern England; they were made in the flag sheds now refurbished as the Blencathra Business Centre near the museum's entrance. Eventually it was producing granite chippings, also sourced from two other nearby quarries, coated with bitumen at a plant within the site. The quarry closed in 1982, whereupon demolition contractors removed everything saleable, leaving a devastated derelict site.

In 1992, Lakeland Mines and Quarries Trust took a lease with the intention of developing a museum, but in the meantime the ravages of weather and vandals had caused serious deterioration of the remaining buildings. Fortunately, because the rock was initially blasted loose from the quarry face with black powder and then a mixture of fertiliser and diesel fuel – both gentle explosives that shift out blocks of rock without shattering them – the remaining quarry faces were relatively stable. Trust members repaired the buildings and landscaped the site. In 1995 the Trust was formally wound up, handing over the site to the Museum Company (the trading arm of the Trust) which now runs the site. In 1995 the Caldbeck Mining Museum joined the project, facilitating the display of both local quarrying and mining industrial history. The final phase of quarry activity was illustrated by the unique working collection of Ruston Bucyrus excavators; the museum is now the home of the Vintage Excavator Trust. In late 2004, the Caldbeck Mining Museum collection was removed. The resultant new space was used for a new mining section, opened in mid-2005, with a new lighting, drilling and explosives room.

The museum now houses the Mining Room, with artefacts, plans and photographic records of the explorations of many local metalliferous mines, where representative displays of local minerals can be seen, together with a display on mine lighting, drilling and explosives. The Quarry Room explains the relationship between the local geology and the quarrying of limestone, sandstone, granite and slate; photographs and rock samples from over 50 quarries are displayed to illustrate the special features of each. Samples of rocks from all the important local rock formations are also keyed to a large table map. The museum now has a new mining section, developed with the Cumbria Amenity Trust Mining History Society, and a guided tour through a reconstructed lead/copper mine is now possible, together with mineral panning. There is also a display of excavator machinery and an engine house with a steam engine. The quarry floor has a narrow-gauge railway, hauling passenger trains in the holiday season; it enables visitors to view the quarry faces in safety. It is still a private museum staffed by volunteers. The quarry has RIGS and Local Geology Site designation, because it shows the contact between the Skiddaw slates and the Threlkeld granite; the metamorphic aureole (the ring of slates around the granite that was altered by the latter's intense heat and its hot gases and fluids that escaped into the slates) contains the site's greatest mineral interest.

Force Crag Mine

Force Crag Mine lies 7km to the west of Keswick, at the head of the remote Coledale Valley. The mine was first worked in the 16th century, with a 1578 ore report showing the existence of a lead-silver vein. It was mined for lead between 1839 and 1865, and from 1867 for zinc and barytes. When it closed in 1991 it was the Lake District's last working metal mine. Built in 1908 to 1909 to separate the minerals from the rocks within which they are encased, its crushing mill survives, along with ore refining machinery from the 1980s and earlier; it is the only former metal mine in the UK that has kept its processing plant in anything like complete order. The mill was restored in 2004; visitors can see *in situ* what was mined and can follow the mineral processing flow through the old mill plant. Force Crag is a Scheduled Ancient Monument and a geological SSSI in the ownership of the National Trust. Admission to the mine building is only via booked tours; there is no permitted access to the mine itself.

Nenthead Mines, Alston

The site of Nenthead Mines, mainly owned by Cumbria County Council, lies in a remote valley near Alston. It was a major centre for zinc, lead and silver mining, with its first smelting mill built in 1737. The last mine closed in 1961 but mining exploration for zinc began in 2013. Today several mine entrances, an 80m engine shaft, spoil heaps and old buildings are linked by a network of paths and tracks. Its conserved buildings include a small museum and interpretation display. It is managed by the Nenthead Mines Conservation Society, working with the County Council, with the remit to advance public education about mining and smelting at Nenthead by researching, restoring, maintaining and exploring the historical and natural features relating to mining and smelting. In July 1996 the Nenthead Mines Heritage Centre was officially opened, with an adjacent shop and café, in buildings renovated by the North Pennines Heritage Trust. Its displays introduced visitors to the site, providing Victorian mining artefacts and an interactive geology display. The site also had other buildings with interpretive material, such as 'The Power of Water', a full-scale interactive display of working waterwheels, and the 100m deep Brewery Shaft that visitors could look down into. The centre also provided a panning activity for children to find minerals of their own, and a number of guided walks around the site. However, the main attraction for visitors was the opportunity to take a guided tour into Carr's Mine, a 300-year-old lead, zinc and silver mine. Unfortunately, in January 2013 the Trust went into administration and the centre was closed. Then, in November 2014, Cumbria County Council announced plans to auction some buildings and associated yards on the site that were formerly the Nenthead Mines Heritage Centre, and by early 2015 the sale had been completed to an undisclosed buyer.

Haig Colliery Museum, Whitehaven

Coal mining in the Whitehaven area began in the 13th century and ceased in the 1980s. Haig Colliery, with its extensive undersea workings sunk to a depth of 400m, was opened by the Whitehaven Colliery Company in 1916 and closed in 1986. After closure, many of its buildings were demolished as the site was landscaped to form the Haig Enterprise Park. However, the winding house, headgear, engines and overhead cranes had Grade II listed building and Scheduled Ancient Monument status; they were restored from 1994 onwards as part of the original on-site museum. A new visitor centre building was constructed in 2014 as part of a £2.3 million project, housing a reception area, an exhibition, an education suite, café and toilets. The new displays will cover the heritage coastline and, working with the Colourful Coast partners, reflect the area's flora and fauna, showcasing environmental changes from the industrial to the developing semi-natural landscape, following the closure and demolition of nearby heavy-chemical works; however, the geology of the superb sandstone cliffs is not discussed in the displays. As part of the project, the headgear was refurbished to reflect its appearance when originally constructed. A new ground floor entrance into the winding house was created, leading into a new mining exhibition spread across two floors. The building's huge middle room, the Powerhouse, was restored with a new floor to become a multi-function facility for functions and exhibitions. One of the engine halls and its engine was mothballed awaiting new funding. The bulk of the project's building works was completed and the museum reopened to the public in late 2014.

Saltcom Pit heritage site

Saltcom Pit was sunk between 1729 and 1731, below the sandstone cliffs of what is now Haig Colliery, to reduce the depth of the shaft needed to develop undersea coal workings some 2km under the Irish Sea. The pit, owned by Sir James Lowther, was the UK's first undersea coalmine and was worked until 1848. It is now a Scheduled Ancient Monument, as one of the finest examples of an 18th century nucleated coalmine site. Its sandstone-built engine and winding house (that housed a Newcomen-type atmospheric engine to drain the mine) is somewhat dilapidated but mainly intact. The pit was interpreted by several panels, of which only one survives, and made an interesting use of 'railway track' embedded in the ground and stamped with quotes from the mine's workers; the panel has an excellent empathetic social history storyline, but no mention of the geology. It was clearly a well thought out and innovative interpretative project by the Colourful Coast team, but has been somewhat neglected.

Honister Slate Mine

Possibly the most commercially successful geotourism venture in the Lake District is the Honister Slate Mine at the top of the Honister Pass in Borrowdale. It is some 300 years old and shows a variety of traditional and modern working methods; relatively small-scale extraction of the distinctive green roofing slate continues today. This was once exported widely, at first from Peel Harbour in 1658 when Sir Christopher Wren used it for the Chelsea Royal Hospital and Kensington Palace, and then by the Furness Railway from 1846. Today the slate is also used to manufacture souvenirs and garden ornaments. The visitor centre has several information panels dealing with the site's history, but with little mention of geology. The old quarry houses a *via ferrata* route that has proven to be very popular with visitors.

The sites summarised

These geotourism sites are either in commercial ownership (Keswick Pencil Museum and Honister Slate Mine), private trust ownership (Threlkeld Mine and Force Crag Mine), or mixed management and ownership with some local authority involvement (Haig Colliery, Nenthead Mines and Saltcom Pit). However, the removal of a major collection from one (Threlkeld Mine) and the sale of a visitor centre building at another (Nenthead Mines) indicate the fragility and short-term nature of some trust geotourism projects, despite the considerable and unrecognised input of volunteers.

Protecting the Lake District from over-commercialisation

By the mid-19th century the Lake District was readily accessible to tourists, but over-commercialisation and large numbers of tourists were threatening its popular landscapes. Wordsworth, who had done so much to popularise it to discerning elite tourists, did his best to stop early mass tourism by the lower middle and working classes who were travelling from the nearby industrial and mill towns. He attempted, in two 1844 public letters to the *Morning Post* (see Bicknell 1984, 186–98), and in his sonnet *On the Proposed Kendal and Windermere Railway*, also of 1844 and published in the *Morning Post* on 16 October 1844, to prevent its construction. He argued that it would disgorge tourists at Bowness above Lake Windermere, with 'cheap trains pouring out their hundreds at a time along the margin of Windermere', and drinking establishments run by 'the lower class of innkeepers', who were keeping pace with these 'excitements and recreations'.

John Ruskin (1819–1900), Wordsworth's spiritual successor, echoed his disapproval 30 years later when an extension to the railway was proposed, as it had by then created (see Ousby 1990, 192) a 'steam merry-go-round' carrying 'stupid herds of modern tourists' who were dumped 'like coals from a sack', and he certainly did not 'want to let them see Helvellyn while they are drunk'. However, condemnation of the railway and of mass tourism was not universal. One of the original railway's contracting engineers, George Heald (1816–58), penned a poem, which partly paraphrased Wordworth's *Morning Post* poem, as a riposte on 15 April 1847 to Wordsworth. Heald argued in favour of the railway's democratising influence and its cultural and social benefits; he also accused Wordsworth of obstructing the opportunities the railway could bring. Largely forgotten today, Heald was one of the key engineers in the early development of railways – amongst the company of George Stephenson (1781–1848), Robert Stephenson (1803–59) and Isambard Kingdom Brunel (1806–59) – and was notable for his role in building the routes that formed part of the Caledonian Railway, the Grand Junction Railway, the Lancaster and Carlisle Railway, and the North Midland Railway. Harriet Martineau (1802–76), in *A Complete Guide to the English Lakes* (1855), encouraged mass tourists to spend a day in the mountains, possibly with a guide. She would almost certainly have welcomed the *via ferrata* at Honister. Egalitarian versus elite access to treasured landscapes such as the Lake District has been a particular issue since the mid-19th century, partly resolved today by the (over)pricing of many activities and local accommodation.

SCENIC LANDSCAPES, GEOPARKS AND GEOTOURISM

Having considered geotourism's background and its relationship with geoconservation, history and development by using the example of the Lake District, their relevance to present-day geotourism should be noted. World heritage designation has been achieved by only two geotourism areas in the UK – the Dorset and East Devon (the 'Jurassic') Coast, and the Giant's Causeway – neither of which, unlike some UK national parks, has ever sought geopark status. Geoparks, as noted in Chapters 1 and 7, are *the* major 21st century geotourism development in the UK and Europe. They are perhaps a natural development of the 19th century Romantic approach to landscape and, as in the Lake District's Kendal to Windermere railway scheme, were developed to regenerate local economies and indirectly open up the landscape to mass tourism. Hence, they are likely to reignite the debate that Wordsworth's elite approach to access initiated. The key geoconservation issue that should have been learnt from the Lake District is the impact of increasing tourists' access by creating new paths, trails and structures, with the increasing pressures that higher numbers of tourists inflict on the landscape. Perhaps the greatest impact on the landscape, and one that would have been much lessened by railways, has been the post-1960s development of dual-carriageway roads (particularly the A591 between the foot of Helvellyn and Thirlmere) and motorways. Further, there is the consideration that, in making the landscape readily accessible to too many tourists, whether by railway or road, something of its inherently spiritual nature is lost. Perhaps the greatest irony in this context is that it is the houses of Wordsworth and Ruskin (and of Beatrix Potter) that now attract the highest numbers of visitors. However, the argument could be advanced that these are sacrificial 'honey spots' and that at least the remote locations are only accessible on foot to the dedicated lover of the Lake District landscape.

From the middle of the 20th century, a new UK national statutory impetus to landscape tourism was provided, as noted in Chapter 7, by the 1949 National Parks and Access to the Countryside Act; this permitted the recognition, designation and protection of areas of outstanding scenery as National Parks and Areas of Outstanding Natural Beauty (AONBs). However, only the National Parks were statutorily provided with the measures and funding to facilitate and promote access for tourists, and progress was exceptionally slow (Green 1996, 102–5); this was partly due to resistance, with undue fears over the damage caused by visitors, from established interests such as major landowners and the farming communities. There was some recognition that the National Parks and AONBs were facing pressure from unsuitable activities, and means were sought to provide countryside access and facilities closer to the main urban areas; the 1968 Countryside Act therefore promoted urban fringe scenic tourism through the establishment of Country Parks for informal recreation. Some of these Country Parks, such as Park Hall in Staffordshire, were established on redundant extractive industry sites, but this geological interest was usually ignored in their interpretative provision.

On an international level, UNESCO recognition has been available since 1972 for a variety of landscapes and sites. The most significant is citation within the World Heritage List; this has around 630 sites, 7% of which are inscribed primarily for their geological interest. However, not all significant geosites and geomorphosites meet the 'outstanding universal value' criterion required by the World Heritage Convention. Hence, an alternative recognition was required, and geoparks were proposed to promote landscapes on a holistic, rather than a purely geological, basis. In 1999 the geoparks programme, in which the original geotourism approach (Hose 1995) was incorporated within the proposal documentation (UNESCO 2000), envisaged that they would recognise the relationships between people and geology and the potential for economic development. The programme's major benefit was to focus attention directly on geosite and geomorphosite conservation, and the related issue of sustainable development. Ideally, the geological interest should be allied to some archaeological, cultural, historical or ecological interest. Sustainable geotourism within geoparks is likely to be promoted by the sale of local products and souvenirs; creation of new products with geological connotations; reinforcement of local hotel and restaurant business through tourists and visitors; creation of new jobs linked to geology, guides, technicians and cottage industry; and support to local transport.

Geoparks are required to provide adequate educational provision in order to maintain membership. Significantly for geoconservation, the sale of geological material, whether local or imported, is prohibited within their boundaries – although nobody seems to have considered the impact on established jewellery businesses. This ban has created issues at classic localities (such as the 'Jurassic Coast') with a long history of collecting, preparing and retailing fossils and minerals to their visitors. UNESCO supported the establishment of European Geoparks as a small network of sites with significant geoheritage and sustainable development strategies. In June 2000 the European Geoparks Network (EGN) was established, as noted in Chapter 7, by four founding geoparks that had signed a convention to share information and expertise.

One of these four initial geoparks was Aliaga (located within the Cultural Park of Maestrazgo) in Spain. The country has three other geoparks: Cabo de Gata in Almería; Sobrarbe in the central Pyrenees; and the Subbetic mountains in Córdoba. Spain also has a similar category of geotourism promotion, with 'Geological Parks' such as the Chera Geological Park (in Valencia) and the Isona Cretaceous Park (in Lérida). In April 2001 the EGN signed a collaborative agreement with, and placing it under the auspices of, UNESCO's Division of Earth Sciences. In October 2004 the EGN signed the Madonie declaration, recognising the EGN as the official branch of the UNESCO Global Geoparks Network for Europe. Today, the EGN is represented in ten countries with some 30 geoparks. However, not all European or UK geoparks are members of the EGN. Some geoparks begin as national geoparks and then become candidates for EGN membership. For example, in Germany the Swabian Alb Geopark was established in 2002 as a national geopark and as an EGN in 2005. Some geoparks have chosen, sometimes due to funding issues, to leave either the EGN and/or the UNESCO Global Geoparks Network; for example, the Abberley and Malvern Hills Geopark withdrew from both in 2008.

Geoparks have no national statutory designation or protection in their own right; whatever protection they do have comes via other designations, such as (in the UK) National Park or AONB status. The first UK geopark was the Marble Arch Caves in Northern Ireland in 2001. The first in England was the North Pennines in 2003. The recognition of the Fforest Fawr Geopark in 2006 (within the Brecon Beacons National Park) suggests that the two regions that were most significant in geotourism's early development – now the most popular national parks in the UK – could be similarly recognised, and the relationship between their geology and scenic beauty could then be better promoted. Focused on economic and social regeneration, geoparks continue to develop the trends in interpretative provision resulting from the 1970s industrial heritage boom that had similar desired outcomes.

Summary

The development of geotourism required a fundamental shift in the way in which landscapes were perceived and then exploited for tourism. The recognition that wild landscapes were worthy places to visit was essential. The motivations of curiosity and aesthetic value appeared before scientific value for travellers. The change in artists' visualisations and travellers' writings, and their impact on landscape recognition and promotion, still influences modern geotourism provision. The 'Romantic' movement's legacy to modern travellers and tourists, and for modern populist geotourism provision, is the established preference for participants to visit attractive 'wild' or 'natural' landscapes, rather than the 'controlled' agricultural landscapes and the 'brutal' locations of heavy industry and mining. Modern geotourism partly seeks to make the latter more acceptable places for geotourists. Improvements during the 18th and 19th centuries in physical and intellectual access, to encourage elite tourists to visit previously virtually inaccessible locations, as exemplified by the Peak District and the Lake District, have modern parallels in the niche tourism developments such as geotourism. They also indicate how readily these can become the precursors of mass tourism. The case studies suggest that challenges related to increased public awareness of and access to wild areas, commonly considered as new issues by today's geotourism stakeholders, are actually akin to those that emerged almost 200 years ago. As noted in several of the succeeding 'case study' chapters, the past is really the key to developing geotourism today.

References

Acott, T G, and La Trobe, H L, 1998 An evaluation of deep and shallow ecotourism, *Journal of Sustainable Tourism* 6 (3), 238–53

Baird, J C, 1994 Naked rock and the fear of exposure, in *Geological and Landscape Conservation* (eds D O'Halloran, C Green, M Harley, M Stanley and J Knill), The Geological Society, London, 335–6

Barton, H A, 1998 *Northern Arcadia: Foreign Travellers in Scandinavia, 1765–1815*, Southern Illinois University Press, Carbondale and Edwardsville

Bedell, R, 2000 *The Anatomy of Nature: Geology and American Landscape Painting, 1825–1875*, Princeton University Press, Princeton

Bicknell, P (ed), 1984 *The Illustrated Wordsworth's Guide to the Lakes*, Congdon and Weed, New York

Buckley, R, 2003 Environmental inputs and outputs in ecotourism: geotourism with a positive triple bottom line? *Journal of Ecotourism* 2 (1), 76–82

Burek, C V, 2012 The role of LGAPs (Local Geodiversity Action Plans) and Welsh RIGS as local drivers for geoconservation within geotourism in Wales, *Geoheritage* 4 (1–2), 45–63

Burke, E, 1757 *A Philosophical Enquiry into the Origin of Our Ideas of the Sublime and Beautiful*, R and J Dodsley, London

Burton, A, and Burton, P, 1978 *The Green Bag Travellers: Britain's First Tourists*, Andre Deutsch, London

Buzzard, J, 1993 *The Beaten Track: European Tourism, Literature, and the Ways to Culture 1800–1918*, Oxford University Press, Oxford

Conybeare, W D, and Phillips, W, 1822 *Outline of the Geology of England and Wales*, William Phillips, London

Cotton, C, 1681 *The Wonders of the Peake*, Joanna Brome, London

Crosthwaite, P, 1783 *An Accurate Map of the matchlefs Lake of Derwent (fituate in the moft delightful Vale which perhaps ever Human Eye beheld) near Keswick, Cumberland; with Weft's eight stations*, scale 3 inches to 1 mile, engraved by S Neele (variously corrected and published 1788, 1794, 1809, 1819; reprinted 1863), Peter Crosthwaite, Keswick

Defoe, D, 1724–27 *A Tour thro' the Whole Island of Great Britain, Divided into Circuits or Journies* (in 3 vols), J M Dent, London

Doughty, P, 2008 How things began: the origins of geological conservation, in *The History of Geoconservation: Special Publication No 300* (eds C V Burek and C D Prosser), The Geological Society, London, 7–16

Dowling, R K, and Newsome, D (eds), 2008a *Inaugural Global Geotourism Conference (Australia 2008) Conference Proceedings*, Edith Cowan University, Perth

— (eds) 2008b *Geotourism*, Elsevier, London

Dowling, R K, and Newsome, D, 2010 The future of geotourism: where to from here?, in *Geotourism: The Tourism of Geology and Landscapes* (eds D Newsome and R K Dowling), Goodfellow, Oxford, 231–44

Frey, M-L, 2008 Geoparks – a regional European and global policy, in *Geotourism* (eds R K Dowling and D Newsome), Elsevier, London, 95–117

Gilpin, W, 1786 *Observations, Relative Chiefly to Picturesque Beauty, Made in the Year 1772, on Several Parts of England: Particularly the Mountains and Lakes of Cumberland and Westmorland* (in 2 vols), R Blamire, London

— 1789 *Observations, Relative Chiefly to Picturesque Beauty, Made in the Year 1776, On Several Parts of Great Britain; Particularly the High-Lands of Scotland* (in 2 vols), R Blamire, London

Green, B, 1996 *Countryside Conservation: Landscape Ecology, Planning and Management* (3 edn), Spon, London

Green, W, 1819 *The Tourist's New Guide, Containing a Description of the Lakes, Mountains, and Scenery in Cumberland, Westmorland, and Lancashire, with Some Account of Their Bordering Towns and Villages. Being the Result of Observations Made During a Residence of Eighteen Years in Ambleside and Keswick* (in 2 vols), Lough and Company, Kendal

Grube, A, and Albrechts, C, 1992 Earth science conservation in Germany, *Earth Science Conservation* 31,16–19

Hardy, A, Beeton, J S, and Pearson, L, 2002 Sustainable tourism: an overview of the concept and its position in relation to conceptualisations of tourism, *Journal of Sustainable Tourism* 10 (6), 475–96

Harrison, W J, 1882 *Geology of the Counties of England and Wales*, Kelly and Company/Simpkin, Marshall and Company, London

Hebron, S, 2006 *The Romantics and the British Landscape*, British Library, London

Hobbes, T, 1678 *De Mirabilibus: Being the Wonders of the Peak in Darby-shire, Commonly called The Devil's Arfe of Peak. In English and Latine. The Latine Written by Thomas Hobbes of Malmsbury. The English by a Person of Quality*, William Crook, London

Hose, T A, 1995 Selling the story of Britain's stone, *Environmental Interpretation* 10 (2), 16–17

— 1997 Geotourism – selling the earth to Europe, in *Engineering Geology and the Environment* (eds P G Marinos, G C Koukis, G C Tsiamaos and G C Stournass), Rotterdam, A A Balkema, 2955–60

— 2000 European geotourism – geological interpretation and geoconservation promotion for tourists, in *Geological Heritage: Its Conservation and Management* (eds D Barretino, W A P Wimbledon and E Gallego), Instituto Tecnológico Geominero de España, Madrid, 127–46

— 2008a Towards a history of geotourism: definitions, antecedents and the future, in *The History of Geoconservation, Special Publication No 300* (eds C V Burek and C D Prosser), The Geological Society, London, 37–60

— 2008b The genesis of geotourism and its management implications, *Abstracts Volume, 4th International Conference, Geotour 2008, Geotourism and Mining Heritage*, AGH University of Science and Technology, Krakow, 24–5

— 2010a Volcanic geotourism in West Coast Scotland, in *Volcano and Geothermal Tourism: Sustainable Geo-Resources for Leisure and Recreation* (eds P Erfurt-Cooper and M Cooper), Earthscan, London, 259–71

— 2010b The significance of aesthetic landscape appreciation to modern geotourism provision, in *Geotourism: The Tourism of Geology and Landscapes* (eds D Newsome and R K Dowling), Goodfellow, Oxford, 13–25

— 2011 The English origins of geotourism (as a vehicle for geoconservation) and their relevance to current studies, *Acta geographica Slovenica* 51 (2), 343–60

— 2012 3G's for modern geotourism, *Geoheritage* 4 (1–2), 7–24

Hudson, J, 1842 *Complete Guide to the Lakes, in Cumberland and Westmorland, August 1793*, Hudson and Nicholson, Kendal

Hutchinson, W, 1774 *An Excursion to the Lakes, in Westmoreland and Cumberland, August 1773*, J Wilkie/ W Goldsmith, London

— 1794 *The History of the County of Cumberland, and some Places Adjacent, from the Earliest Accounts to the Present Time: Comprehending the Local History of the County, its Antiquities, the Origin, Genealogy and Present State of the Principal Families, with Biographical Notes; its Mines, Minerals and Plants, with Other Curiosities, either of Nature or of Art* (in 2 vols), F Jollie, London

Jenkins, J M, 1992 Fossickers and rockhounds in northern New South Wales, in *Special Interest Tourism* (eds B Weiler and C M Hall), Belhaven, London, 129–40

Leafe, R, 1998 Conserving our coastal heritage – a conflict resolved, in *Coastal Defence and Earth Science Conservation* (ed J Hooke), The Geological Society, London, 10–19

Mackay, C, 1846 *The Scenery and Poetry of the English Lakes: A Summer Ramble*, Longman and Company, London

Martineau, H, 1855 *A Complete Guide to the English Lakes*, J Garnett, Windermere

Mavor, W, 1798–1800 *The British Tourists; or Traveller's Pocket Companion, through England, Wales, Scotland, and Ireland. Comprehending the Most Celebrated Tours in the British Islands* (in six vols), E Newberry, London

Morris, C (ed), 1949 *The Journeys of Celia Fiennes*, Cresset Press, London

Murray, Hon Mrs M, 1799 *A Companion and Useful Guide to the Beauties of Scotland, to the Lakes of Westmoreland, Cumberland, and Lancashire, and to the Curiosities in the District of Craven, in the West Riding of Yorkshire. To which is Added, a More Particular Description of Scotland, Especially that Part of it Called the Highlands*, George Nicol, London

Nabholtz, J R, 1964 Wordsworth's 'Guide to the Lakes' and the Picturesque Tradition, *Modern Philology* 61 (4), 288–97

Otley, J, 1818 *The District of the Lakes: A Topographical Map* (at a scale of approximately one inch to four miles) [originally sold in linen-mounted format in a pouch designed for the pocket; then later mounted inside the front cover of the various editions of Otley's guidebook from 1823 onwards]

— 1823 *A Concise Description of the English Lakes, the Mountains in their Vicinity, and the Roads by which they may be Visited; With Remarks on the Mineralogy and Geology of the District*, published privately, Keswick, J Richardson, London, A Foster, Kirkby Lonsdale

Ousby, I, 1990 *The Englishman's England: Taste, Travel and the Rise of Tourism*, Cambridge University Press, Cambridge

Phillips, W, 1818 *A Selection of Facts from the Best Authorities, Arranged so as to Form an Outline of the Geology of England and Wales*, Phillips, London

Plog, S C, 1974 Why destination areas rise and fall in popularity, *The Cornell Hotel and Restaurant Administration Quarterly* 15, 55–8

Plumtree, J, 1798 A Journal of a Pedestrian Tour by the Caves in the West Riding of Yorkshire to the Lakes and Home thro Parts of North Wales in the Year 1797, unpublished manuscript, Cambridge University Library

— 1799 A Narrative of a Pedestrian Journey through some parts of Yorkshire, Durham and Northumberland to the Highlands of Scotland and Home by the Lakes and Some Parts of Wales in the Summer of the Year 1799, unpublished manuscript, Cambridge University Library

Read, S E, 1980 A prime force in the expansion of tourism in the next decade: special interest travel, in *Tourism Marketing and Management Issues* (eds D E Hawkins, E L Shafer and J M Rovelstad), George Washington University Press, Washington DC

Ritvo, H, 2009 *The Dawn of Green: Manchester, Thirlemere and Modern Environmentalism*, University of Chicago Press, Chicago

Robinson, E, 1998 Tourism in geological landscapes, *Geology Today* 14 (4), 151–3

Selincourt, E (ed), 1977 *Wordsworth's Guide to the Lakes, The Fifth Edition (1835) with an Introduction, Appendices and Notes Textual and Illustrative*, Oxford University Press, Oxford

Sillitoe, A, 1995 *Leading the Blind: A Century of Guidebook Travel 1815–1911*, Macmillan, London

Slomka, T, and Kicinska-Swiderska, A, 2004 Geotourism – the basic concepts, *Geoturystyka* 1, 5–7

Smith, R A, 2000 Jonathan Otley: A pioneer of Lakeland geology, *Geology Today* 16 (1), 31–4

Stueve, A M, Cock, S D, and Drew, D, 2002 *The Geotourism Study: Phase 1 Executive Summary* [www.tia. org/pubs/geotourismphasefinal.pdf]

Towner, J, 1996 *An Historical Geography of Recreation and Tourism in the Western World 1540–1940*, Wiley, London

UNESCO, 2000 *UNESCO Geoparks Programme Feasibility Study*, UNESCO, Paris

UNWTO, 1997 *What Tourism Managers Need to Know: A Practical Guide for the Development and Application of Indicators of Sustainable Tourism* [http://www.worldtourism.org/cgi-bin/infoshop.storefront/ EN/product/1020-1]

Weiler, B, and Hall, C M (eds), 1992 *Special Interest Tourism*, Belhaven, London

West, T, 1778 *A Guide to the Lakes: Dedicated to the Lovers of Landscape Studies, and to All Who Have Visited, or Intend to Visit the Lakes in Cumberland, Westmorland and Lancashire*, Richardson and Urquhart, London/ W Pennington, Kendal

Wilkinson, J, 1810 *Select Views in Cumberland, Westmoreland and Lancashire by the Rev. Joseph Wilkinson, Rector of East and West Wretham in the County of Norfolk, and Chaplain to the Marquis of Huntley*, R Ackerman, London

Wordsworth, W, 1820 *The River Duddon, A Series of Sonnets: Vaudracour and Julia: and Other Poems. To which is annexed, a Topographical Description of the Country of the Lakes, in the North of England*, Longman, Hurst, Rees, Orme and Brown, London

— 1835 *A Guide through the District of the Lakes in the North of England, with a Description of the Scenery etc. For the Use of Tourists and Residents* (5 edn), Hudson and Nicholson/Longman & Co., Moxon, and Whittaker & Co., Kendal/London

Young, A, 1770 *A Six Months Tour through the North of England* (Vol 3), W Strahan and W Nicoll, London

9

Protecting and Promoting the Geoheritage of South-Eastern Europe

THOMAS A HOSE AND DJORDJIJE A VASILJEVIĆ

THE IMPLICATIONS OF A COMPLEX POLITICAL HISTORY

The region examined within this chapter lies within the Italian and Balkan peninsulas (see Figure 9.1). Its countries exemplify, probably better than any others in Europe, the considerable variation in its geoheritage protection and geotourism (see Hose and Vasiljevic 2012). It has different legislative approaches and governmental and non-governmental organisations (NGOs), whose variations reflect the legacy of the region's complex political history. The region was especially important in the emergence of scientific geology in the 17th century and in continental plate studies in the 20th century. Its unifying politico-cultural contexts are the now long-defunct Ottoman and Hapsburg empires, which at various times held territories in the region. The former was particularly significant up to the close of the Great War (1914–18), encompassing at its greatest extent Europe east of the Adriatic Sea and around the Black Sea. The Habsburg Empire, particularly globally dominant in the 16th to the early 19th centuries, encompassed much of the Italian Peninsula, and stretched eastwards from the northern Adriatic Sea to enclose central Europe and even the Netherlands. It has been argued that the empire's impact on present-day cultural norms, especially belief and trust in governmental systems and their officers, can still be found – as in the former Habsburg Empire's eastern borderlands (Becker *et al* 2011).

Whilst Italy was *the* early centre of European geological inquiry (Vai and Caldwell 2006), the Balkan countries were peripheral in geological research until the mid-20th century; the stimulus was provided by the USSR's post-war expansion, when scientists contributed to a better understanding of early crustal evolution. In the early 1990s, with the demise of USSR-style Communism, many of south-eastern Europe's countries underwent dramatic political, economic and social changes; these eventually led them into greater integration with wider European political institutions and, as an accidental consequence, nature conservation measures. Nature conservation and allied geoconservation approaches differ considerably across the countries of the region; consideration is given here to geoheritage conservation and promotion in Italy and Albania as exemplars. Turkey is also examined in some detail, because it shows an interesting relationship between the region's complex politics and its own unique culture. Numerous travelogues, but few geological texts, provide an overview of southern Europe, whereas south-eastern Europe's geology and tourist attractions are both well delineated and described. Nevertheless, in synthesising this account, the often outdated, at best subjective – and at worst prejudiced, and even conflicting – nature of much of the readily available published information has been apparent.

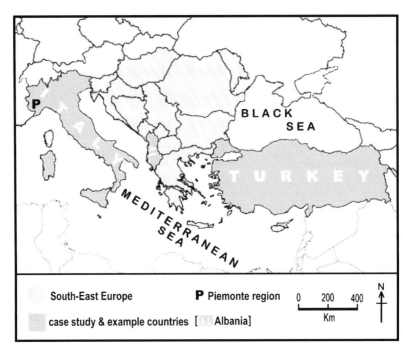

FIGURE 9.1 MAP SHOWING THE REGION, AND PARTICULAR COUNTRIES, CONSIDERED
IN CHAPTER 9.

EUROPEAN GEOCONSERVATION AND ProGEO

European geoconservation has been particularly well advanced in practice and theoretical underpinning (O'Halloran *et al* 1994), and its history is well documented (Burek and Prosser 2008).
Commencing in the 19th century, countries in northern and western Europe were particularly at
the forefront of modern geoconservation. Southern and south-eastern Europe are now beginning
to lead in some areas, such as the preservation and conservation of dinosaur footprint trackways,
as in Portugal (Santos, Silva and Rodrigues 2008), and fossils, as in Spain (Alcala *et al* 2012).
Successful geoconservation requires legislation, with its enforcement underpinned by reliable
geosite inventories. ProGEO, the European Association for the Conservation of the Geological
Heritage (see also Chapter 7), has been a major force in promoting pan-European geoconservation since the 1990s. The origin of this remarkable NGO was a geoconservation workshop held
in 1988 in Leersum, Holland. Its first development body, 'The European Working Group on
Earth-Science Conservation', was founded at that meeting. Following annual international meetings in northern and western Europe, ProGEO held its first meeting in south-eastern Europe
in Budapest in 1994. That followed one the previous year, held in Mitwitz-Cologne, Germany,
at which 'The ProGEO First General Assembly' was held – the formal start for ProGEO, at
which its Articles for constituent groups were formally adopted. In 1995 ProGEO held its first
sub-regional meeting in Bulgaria, with papers from the meeting published in two special issues
of *Geologica Balcania* in 1996.

ProGEO was officially registered (in Sweden) in 2000. Its stated aims were to promote the conservation of Europe's rich heritage of landscape, rock, fossil and mineral sites; to inform a wider public of the importance of this patrimony, and of its relevance to modern society; to advise, in member countries and in Europe as a whole, those responsible for protecting our Earth heritage; to organise and participate in research into all aspects of planning, science, management and interpretation that are relevant to geoconservation; to involve all countries in Europe, exchanging ideas and information in an open forum, and taking a full part in conservation in a global setting, including the formulation of conventions and legislation; to work towards an integrated European listing of outstanding geoscience sites, thus enabling full support to be given to the work of other international bodies, as well as to national initiatives towards site protection; and to achieve an integrated approach to nature conservation, promoting a holistic approach to the conservation of biological and physical phenomena.

Its 1999 Third International Symposium, 'Towards the Balanced Management and Conservation of the Geological Heritage in the New Millennium', led to the publication of two major texts (Barettino and Gallego 1999; Barrettino, Wimbledon and Gallego 2000) with summary accounts, including a whole section devoted to geoconservation, geoheritage management and the promotion of conservation in the region described in this chapter. The first meeting of ProGEO's 'Regional Working Group SW Europe' (covering France, Italy, Portugal and Spain) was held in 2010. A recent summary of European geoconservation history and measures, with some mention of geoheritage promotion (Wimbledon and Smith-Meyer 2012), has highlighted ProGEO's work and provides detailed individual country accounts.

Geoheritage in the Italian Peninsula

Geography and geology of the Italian Peninsula

The Italian, or Apennine, Peninsula is the central and smallest of the three major southern European peninsulas, and extends from the Po Valley in the north some 1,000km to the central Mediterranean Sea in the south. It is made up of the smaller Calabria, Salento and Gargano peninsulas. Sicily, Elba (and other smaller islands such as Palagruža, belonging to Croatia) are usually considered as islands off the peninsula and are grouped with it. It is bordered by the Tyrrhenian Sea on the west, the Ionian Sea on the south, and the Adriatic Sea on the east. Northwards it is confined by the Italian (Dolomite) Alps. Its backbone is formed by the Apennine Mountains (geologically, a folded mountain chain, the Appenides). It is today one of Europe's most tectonically, and certainly most volcanically, active areas. Consequently, its coast is mostly cliff lined and it has active volcanoes, some of which are used to define volcanic eruptive types. The peninsula, except for the microstates of San Marino and Vatican City, lies within the state of Italy. This, with an area of 301,217 km², is itself subdivided into 20 semi-autonomous political regions.

A lengthy account of Italian geology is provided by Doglioni and Flores (1997). Geologically, Italy consists of three main elements: the volcanoes of the Tyrrhenian Sea; the Appenides; and the Dolomite Alps. Separating and surrounding these are major river plains (such as that of the River Po). However, this apparent simplicity belies a complex geological structure that is a consequence of the interaction of past and present tectonic plate margins (of Africa and Europe) that have led to the much folded and welded nature of its continental crust. Given that a full consideration of the entire country's geology is beyond the scope of this chapter, but that some detailed knowledge is required to appreciate its geoheritage, the national overview of conservation legislation below is followed by a more detailed exploration of geoheritage conservation in the Piemonte region.

Italian legislation and nature conservation

A complex political history, with frequent changes of central government over the past 70 years, makes tracking the development and efficacy of the Italian legislation and government departments and agencies involved with nature conservation, let alone geoconservation, a real challenge. This is especially so when both central and regional governments may enact conservation and planning/development legislation, although the former is commonly focused on wildlife hunting and habitat protection. A useful summary of Italian geoconservation (Andrea and Zarienga 2000) was published as a consequence of ProGEO's 1999 'Third International Symposium' and its later collaborative project (Wimbledon and Smith-Meyer 2012). The work of the Italian Society for Environmental Geology (SIGEA), established in 1992, has been outlined by Gisotti and Burlando (1998). Its remit is to deliver public health and safety, conservation of natural and inhabited environments, and sustainable use of land and its resources.

The earliest modern nature conservation laws in Italy revolved around the establishment of National Parks. The creation of these in the Italian Alps was a major public issue in the early 20th century; two were eventually created by Mussolini's Fascist government – Gran Paradiso (in 1922) and Stelvio (in 1935). From the same period, the 1939 Protection of Natural Beauties Act led to Italy's national and regional legislation becoming increasingly focused on protecting natural heritage, but only latterly on recognising and conserving geoheritage. Late 20th century legislation in Italy began to focus on geosite and geomorphosite recognition and protection. Law No 657 (1974) established the Ministry of Environmental and Cultural Heritage; this was later, by Presidential Decree No 805 and Law No 431 (1985), enabled to engage in protecting some geosites as special areas. Law No 394 (1991) focused on creating protected areas and specifically defined the types of site to be protected, including 'one or more physical, geological, geomorphological formation of national or international significance due to their natural, scientific, aesthetic, cultural and recreational value'. Most regional legislation has since adopted a similar approach. Law No 394 (1991) also introduced the Public Inventory of the Protected Natural Areas, crucial to protected area planning. Coupled with regional vulnerability profiles, it was intended to present an up-to-date picture of all natural – including geological – assets in Italy, and introduced the first three-year study programme of the protected areas. It also envisaged a Charter for Nature, identifying 48 major natural areas termed 'landscape systems'; within these, the main habitat types were identified by the CORINE project as bio-types, with no references to either geosites or geomorphosites. The CORINE (Coordination of Information on the Environment) project was initiated in 1985 by the European Union, as a prototype project working on many different environmental issues. Its databases and several of its programmes, one of which is an inventory of land cover in 44 classes and presented as a 1:100,000 map, were taken over by the European Environment Agency. The Public Inventory of the Protected Natural Areas mandated those autonomous regions that so chose to create regional parks and natural reserves, and some 218 were eventually created. Law No 157 (1992) provided for the protection of wildlife, and indirectly (by placing restrictions on hunting) protected some geosites. A 1997 Presidential Decree introduced regulations on the performance of the relative directive 92/43/CEE for the conservation of natural habitats. Law No 490 (1999) was a 'Framework Law on Cultural and Environmental Heritage', Article 1 of which placed geosites and geomorphosites under the control of the Ministry of Culture and Environmental Heritage; further, its Article 140 included criteria for the creation of geosite inventories, but stated that any so-listed sites must also be recognised by the provincial authorities.

Twenty-first century legislation, especially at regional level, has accelerated the development of geosite inventories for incorporation within planning frameworks, to ensure their designated protection. Most notably, Law No 77 (2004) concerning 'The Italian Code of Cultural and Landscape Assets' (which stretches to some 153 pages) adopts a holistic standpoint to the country's landscapes, in that they are recognised to be the consequence of the interactions between anthropogenic and natural elements; its Article 136 states that geosites are specifically protected under its auspices. It requires local institutions to compile a list of geosites in their area, and these must be included within their town and country planning framework. As such, it is the only national protection afforded to Italian geosites. Law No 77 provided: 'Special measures for the protection and the fruition of Italian cultural, landscape and natural sites, inscribed on the "World Heritage List", under the protection of UNESCO' for those few exceptional geosites. Within Italy's regions, some have chosen to enact their own geoheritage protection measures. Emilia Romagna (in 2006), Liguria (in 2009) and Apulia (in 2009) have all employed the powers initially given to them in Law No 394 (1991) and used the 2004 Assets Law to create their own specifically geodiversity regional laws to promote geoheritage and geosite inventories, along with geoconservation. Most recently, 'Regulations for the identification, cataloguing and protection of geosites in Sicily' were issued in 2012; these established its regional geosites inventory. Italy's present Ministry of the Environment has now affirmed 18 National Parks and 147 Government Land Reserves, amounting to around 5% of Italy's land area; this work has benefited from the input of governmental and NGO nature conservation bodies, as well as that of some inspirational individuals.

Italian nature conservation and geoheritage promotion

Italian geoscientists, with the assistance of ProGEO members from across Europe, have developed national criteria for designating their country's geoheritage and geoconservation. These have been helpful to the regional authorities specifically responsible for founding new parks, reserves and natural monuments. Their efforts have been especially directed towards producing regional and national inventories of geosites, particularly by defining projects that enhance and promote geoheritage, together with geoconservation activities. Such initiatives have included, in the Marche region, a Regional Environment and Landscape Planning project undertaken in 1991, in Sardinia a census of 'natural monuments' undertaken from 1989 (but see Barca and Di Gregorio 1991 and also Di Gregorio and Ulzega 2002), and in Lazio a census of geosites around Rome. Similar projects were also developed in Abruzzo and Piemonte (Ghiraldi *et al* 2009). Hence, it can be argued that nature conservation legislation has been incrementally developed and its enforcement gradually improved.

Italian geologists, together with other technical and natural scientists, have had a central role in protected area planning; this has involved the recognition of candidate areas through the identification of scientific and environmentally significant assets. They have played a significant role in developing protected area management strategies to ensure the appropriate use of natural resources. These strategies must consider, especially with the emergence of geotourism, the environmental and socio-economic demands on protected areas. This connection between geology and protected areas has already produced commendable results. Particularly noteworthy are the themed trails (some of them geotrails) developed in national and regional parks, as well as nature reserves. Examples include a landform trail in the Marmitte dei Giganti Nature Reserve, Lombardia, a geotrail in the Conero Nature Park, Marche and a glacier trail in the Alta Val Sesia Nature Park, Piemonte.

Piemonte's geoheritage and its promotion

Piemonte is Italy's second largest region. It borders France and Switzerland and abuts the Italian regions of Lombardy, Liguria, the Aosta Valley and Emilia Romagna. It is a region of remarkably varied landscapes and habitats, with high mountains (43.3%), hills (30.3%) and large river plains (26.4%). It is surrounded on three sides by the Alps (including Monviso, where the Po rises, and the Monte Rosa massif). Research and conservation activities for recognising, describing and popularising geosites and geomorphosites, as in other Italian regions, are relatively recent activities. The first initiative was the publication of the *Carta Geomorfologica degli Elementi di Interesse Paesaggistico del Parco Nazionale del Gran Paradiso*, accompanied by cards describing selected geosites (Giardino and Mortara 2001). Further stimuli were provided by two populist publications about the exploitation of geomorphosites within Torino Province (Giardino and Mortara 2004). Following cooperation between the managing authority of Asti Province Natural Parks and the Department of Earth Sciences, University of Torino, an inventory of 219 geosites was compiled by the end of 2004; they were all in Asti Province and the Torino Hill area. Research was later undertaken in several other areas of Piemonte.

Following an agreement between the University of Torino's Earth Science Department and the Museo Regionale di Scienze Naturali (Natural Science Museum of Torino), a Piemonte region project was developed to collect and collate geoheritage information. This necessitated the production of an inventory and appraisal of its most important geosites, coupled with programmes to promote their usage and popularisation. Bibliographic research and fieldwork defined the state of geosite studies and identified geoconservation projects developed by public and private institutions; over 300 sites with geological interest were identified. Most geosite descriptions were cursory, but provided locality and basic geological/geomorphological information. The use of the Geographical Information System (GIS) and digital mapping resources made the data readily accessible and useful for land-use management; the development of an online GIS application and some attempt to make the information intellectually more accessible to a wider audience than geologists was noteworthy. This innovative geoheritage work has been described and partly evaluated (Ghiraldi *et al* 2009). In 2011 a development of the initial work, entitled the 'ProGEO-Piemonte' project, was proposed by the University of Torino.

This three-year multi-disciplinary project was devised to generate a new conceptual and operational approach to managing Piemonte's geoheritage, on the assumption that geosites can serve both public and private interests. From the already created geosite inventory, nine strategic geothematic areas representative of the region's geodiversity were selected for further investigation. Each was characterised by high scientific research potential, possibilities for the enhancement of the public understanding of science, recreational opportunities, and economic benefit to local communities. The scientific work generated knowledge and understanding of Piemonte's significant geological history, climate and environmental changes, natural hazards, soil processes and geo-resources; this information underpins an envisaged range of geoheritage interpretative and promotional resources based around geosites, museums, historic extractive industry sites and nature trails.

The work is managed and delivered by what are basically local geodiversity action plans (or LGAPs), an approach developed in the UK (Burek and Potter 2002; Burek 2012). These are developed with local partners to determine the management requirements for specific localities. It was anticipated that recognition of the economic value of the region's geodiversity would aid regional guidelines for geoconservation and geosite management, through an integrated quality management system appropriate for tourism and sustainable development strategies. The €400,000 project is evaluated annually by an international advisory board. Two interim accounts have been published, with one (Ferrero *et al* 2012) particularly noteworthy for its use of a Rubik's Cube analogy for Piemonte's geodiversity and the nine geothematic regions. The other (Zunino *et al* 2012) is an assessment of the efficacy of geosite inventories in relation to two quarries with noteworthy palaeontological resources; Verrura Savoia was opened in 2010 as a public geotrail with several interpretative panels, but the proposal to do likewise at Valle Ceppi has yet to be realised.

Piemonte exemplifies Italy's fostering of regional conservation measures legislation by enabling national legislation, and reflects the state's partly decentralised government. However, the inherent bureaucracy of successive changes of government potentially delays, if not actually hinders, effective national geoheritage protection. Italy shows the value of recognising and adopting geoheritage management approaches from elsewhere in Europe, together with the significant role played by universities and NGOs. Piemonte shows that Italy has, in recent years, been at the forefront of innovative geoheritage inventory methodology and geoheritage promotion.

THE BALKAN PENINSULA

Geography and geology of the Balkan Peninsula

The Balkan Peninsula takes its name from the Balkan Mountains; these stretch from the east of Bulgaria to easternmost Serbia. It covers 666,700km² of south-east Europe and is largely mountainous, with ranges trending north-west to south-east. The main ranges are the Balkan Mountains, running from Bulgaria's Black Sea coast to its border with Serbia; the Rhodope Mountains in southern Bulgaria and northern Greece; the Dinaric Alps in Bosnia and Herzegovina and Montenegro; the Šar Massif, running from Albania to Macedonia; and the Pindus range, running from southern Albania into central Greece and the Albanian Alps. The Balkans border the Adriatic Sea on the north-west, the Ionian Sea on the south-west, the Mediterranean and Aegean Seas on the south and south-east, and the Black Sea on the east and north-east. The highest mountain of the region is Musala (at 2,925m) in the Rila Mountains in Bulgaria. Mount Olympus in Greece is the second highest (at 2,917m). Seven countries (Albania, Bosnia and Herzegovina, Bulgaria, Greece, Kosovo, Macedonia and Montenegro) lie wholly and six (Serbia, Romania, Turkey, Italy, Croatia and Slovenia) partly within the peninsula. Balkan geology has been summarised by Zagorchev (1996a).

Balkan geoconservation measures

The ProGEO Working Group 1 (WG-1) was founded in 1995 with Albania, Bulgaria, Greece, Macedonia, Romania, Slovenia and Yugoslavia as its first members. The dissolution of Yugoslavia in the 1990s led to a complicated political situation in the Balkans and also impacted on geoconservation, whose measures in the Balkans, excepting Albania, are summarised in Table 9.1. The

Country	Lead Body	Major Statutory Protection
Bulgaria	Bulgarian National Service for the Protection of Nature (1994) Ministry of the Environment and Waters (1994) Bulgarian National Group for Geological Conservation (BNGGC)	First protected geosite, at Pobitite Kamani, recognised as geological monument (1937) Decree on Natural Protection (1960) - 200 protected nature sites. Further legislation (1 saw 1,500 more protected nature sites. Newer legislation (1998) - 6 preserved area types: reserves, national parks, natural bea spots, supported reserves, nature and prese parks, and the geosites within them.
Bosnia and Herzegovina		
Croatia	Geoconservation officially work of Institute of Quaternary Paleontology and Geology Croatian Academy of Sciences and Arts.	National Strategy for Environment Protectio (1996) distinguished geosites and geoconservation activities. Law of Nature Conservation (2005) recogni and protected minerals, resources and foss
Greece	Ministry of Culture Institute of Geology and Mineral Exploration of Greece (IGME)	
Macedonia	Institute for Protection of Natural Rarities	
Romania	National and Regional Commissions for the protection of natural monuments; National Protected Areas Network (NPAN)	First nature conservation law (1930) Nature protection laws (1950, 1973, 1995)
Serbia	Institute for Nature Protection of Serbia (established 1948)	Law for conservation of cultural properties a natural rarities (1946) Environmental Law (1991)
Slovenia	Environmental Agency of Slovenia.	Nature Conservation Law (1999)

main aims of WG-1 were to exchange information about geosites, to establish national nature conservation policies and strategies, to establish a list of geoheritage sites and to nominate sites for the World Heritage List (Zagorchev 1996b). After 1995, with national representatives implementing ProGEO objectives, geoconservation's activities increased. In late 1997, the Union of Albanian Geologists for the Geological Science Heritage (ProGEO Albania) was founded. Albania illuminates approaches to Balkan geoheritage protection and promotion issues and solutions, and is discussed below.

ojects of Note	Notes
ite designation piecemeal rather than systematic ner with an established methodology until first Regional EO meeting, 'Conservation of the geological heritage in n-East Europe' convened (1995).	only 50 nature sites protected by late 1950s.
GC initiated (1995) geodiversity inventory work; first ework list of Bulgaria published (1998).	First geoheritage list, of 55 geosites (1964). 224 protected sites - mostly geomorphosites (1974). > 360 protected geosites (2010).
Geological heritage of Bulgaria: sites of special scientific st' (1996 – 2000); database of > 200 geosites fully ribed and registered (2003).	By 2007 > 3,500 nature (few geological - natural landscape and rock formations) protected sites.
onmental Action Plan (2003) - geoheritage preservation crete element.	Several associations and non-governmental organizations engaged in geoheritage protection projects but mainly tourism focussed.
identified geosites (1982)	IGME, several universities and other institutions (e.g. Natural History Museum of the Lesvos Petrified Forest, Natural History Museum of Crete) engaged in geoconservation.
nstitute for Protection of Natural Rarities responsible for neritage recording and protection as part of its nature ction work. Some 50 sites identified, 22 legally protected	Since late 2000s geoconservation work considerably reduced due to inadequate institutional base and Governmental indifference. In recent years, the Agency of Environment, the Speleological Association, and the national television company involved in attempt to popularise geoheritage.
	First designated geosite, the Detunata basalt columns of Apuseni Mountains (1938). By late 1995, NPAN had >200 officially protected geosites. Board established National List of Geotopes - 26 sites of international interest. Hateg promoted as a European Geopark (2005).
nal Council for the Geoheritage of Serbia and enegro initiated geoheritage inventory project (1996) er for the Conservation of the Movable Geoheritage of a at the Natural History Museum (2007)	
1995 c.100 geosites of national importance (excepting norphosites) were registered - 25 protected as geological mplex natural monuments.	2004 > 700 geosites, 450 of national importance.

Albania's geoheritage

Albania's geology has been described in some detail (Meco and Aliaj 2000; Robertson and Shallo 2000) and is only briefly discussed here. Due to its geological position within the eastern Alpine Mediterranean Belt and a country of neo-tectonic movements, Albania's rocks provide striking evidence of crustal evolution, especially convergent plate collision with the ophiolite belt (part of the Earth's oceanic crust and the underlying upper mantle that has been uplifted and exposed above sea level and often emplaced onto continental crustal rocks). In the 'External Albanides' (see Serjani 1996 for a summary), a belt of deformed folded and thrust rocks, the exposed rocks are Mesozoic and Cainozoic neritic (formed in the relatively shallow water, less than 200m deep, on the continental shelf) and pelagic (formed in the deeper water, with a mean depth of 3,680m and maximum depth of 11,000m above the ocean floor) carbonates formed on shallow platforms and in oceanic basins. Albania's rich mineral resources, with workable reserves of chrome, copper, nickel-iron and coal, contribute significantly to its economy. Mineral exploration, exploitation and processing developed rapidly under its former Communist regime.

The first outline geological map of Albania, and the first such in the Balkans, was completed in 1922. Detailed geological mapping was undertaken between 1950 and 1990. Largely an agricultural economy for much of its history, it was in the post-war Communist period that Albania industrialised, with rapid economic growth and hitherto unprecedented progress in public education and health. Albania's industrialisation and consequent urbanisation meant that, for much of the 20th century, little consideration was given by the state to landscape and geoheritage protection.

Nature conservation legislation in Albania

Albania's first nature protection legislation was enacted, as in much of the Balkans, in the latter half of the 20th century; general Acts for the protection and conservation of forests, as utilitarian measures, focused on protecting hunted wild animals and waterfowl (especially rare species) and their habitats. By Special Laws, specific woodlands and hunting lands were established. Of Albania's 28,748km², almost 11,000km² (or 37% of its area) is forested; some 136 Natural Monuments were listed to protect these natural areas. The Council of Ministers proclaimed six National Forest Parks: Thethi (230km²); Lura (128km²); Dajti (330km², extended in 1994 to 731.2km²); Boz Oveci (138km²); Divjaka (125km²); and Llogara (101km²). In 1966, by Special Decision No 102, five new National Forest Parks were proclaimed: Tomorri Mountain (400km²) for its alpine habitats (400km²); Hotova (120km²) for its fir-tree forest; and Qafe Shtama (200km²), Valbona (800km²) and Zall Gjoçaj (14km²) for their biodiversity.

By Special Decision No 413 (1994), the Divjaka Forest Park and Laguna of Karavasta Lagoon were proclaimed a 'Natural Ecosystem'; the latter was protected from 1975 as a wetland under the auspices of the Ramsar Convention. Within the parks designated in the mid-20th century, significant geomorphological features were not discretely recognised; nor were the various outstanding rock formations and geosites that later constituted recognised geo-monuments. In the National Forest Parks (such as Valbona, Thethi, Lura, Dajti and Llogara), the recognised biodiversity and the geo-monuments are of national and international importance; some of the forest parks or preserves are now potential geoparks. Law No 8906 (2002) on 'The Protected Areas' defined their categories as: a strict natural reserve; a national park (incorporating the already designated National Forest Parks); a natural monument; a cave as a natural monument; a managed natural reservation; protected landscapes; a protected area of managed resource; and forests, waters and other natural resources within the protected areas. The criteria set by the International Union for the Conservation of Nature (IUCN) were the basis for the various area categorisations, their status and their level of protection. The Decision of Council of Ministers No 676 (2002), 'For Proclamation of Protected Areas of Albanian Natural Monuments', was the country's response to the 'Environment for Europe' strategy.

Geoconservation in Albania

Arising from the Decision of Council of Ministers No 676 (2002), amongst the 669 areas and sites protected as Natural Monuments are almost 300 areas of geoheritage interest, in the form of 100 karst caves; 58 cold-water springs, mineral-water and thermal springs; 28 wetlands – mainly karst and glacial lakes and marshes; and 195 geological sites of various origins. The list of National Monuments surprisingly omits numerous geosites of local, national and international importance that have been listed as Natural Monuments; nowhere does the list specifically include or define the terms 'geosite' and 'geopark'. Indeed, Albania has no specifically geoheritage or geoconservation Albanian Law or Decision.

However, some aspects of geosite protection are indirectly covered in laws pertaining to the rational exploration of mineral ores and raw materials, and later in the 'Mining Law of Albania'; the most important of these are the 1971 Decree No 4874 (1971) and Law No 4927 (1971), 'The Protection of Cultural Monuments and Rare Natural Wealth'. Within Albania, as elsewhere in south-east Europe, geosite selections from the national and international Balkan (South-East Europe) lists generated by ProGEO are based upon general criteria of regional significance; these encompass scientific, educational (didactic), aesthetic and geotouristic values. The subsequent Comparative Documentation derived from the selected geosites was presented within an agreed ProGEO WG-1 classification that had been discussed in Belogradchik (in 1998), Athens (in 2001) and Ankara (in 2002). It recognised the following interpreted categories: stratigraphical sites; petrological sites (igneous, metamorphic and sedimentary rocks, petrology, textures and structures); sites of mineralogical and economic geological significance; sites of structural (tectonic/structural) importance; complex geosites (geomorphologic, erosion, karst, glacial and neo-tectonic processes); continental or oceanic-scale geological features and the relationship of tectonic plates; sites historically important in the development of geology; and palaeoenvironmental sites. Geomorphosites of neo-tectonic origin, lakes and lagoons are presented in a separate group; some can be included in cross-border geoparks.

Albania's geosites were selected on the basis of a desktop study of numerous publications on Balkan geology and the eastern Mediterranean, plus presentations from various ProGEO meetings; the 'Balkan Framework List of Geological Heritage' (made up of various published contributions from Albania, Bosnia and Herzegovina, Bulgaria, Greece, Serbia and Montenegro, and Turkey) was, as a first attempt, a major guide. For each selected geosite, the draft documentation included its index and number in the First National Inventory; topographic coordinates; status, mainly based upon the Decision No 676; a short description (including its main values, defining especially its scientific, educational, exploration and geotourism features); and photo-illustrations.

Recognising the national geoconservation shortcoming, the Geological Survey of Albania and ProGEO-Albania presented to the Council of Ministers a 'Project-Proposition on Geosites and Geoparks in Albania'; a provisional list of geosites proposed for protection had already been prepared (Serjani and Cara 1996) and their basis partly outlined (Serjani 1996). From 1998 to 1999, both bodies undertook the first official geoconservation project; a geoheritage inventory was completed and the 1:50,000 *Map of Geological Sites of Albania* was published in 2001. The 'Regional, National and Local Lists of Geological Sites of Albania' study was completed in 2006 (Hallaci and Serjani 2007).

Albania's geoparks are territories of varying geographical extent with special flora and fauna (biodiversity), as well as special geological and geomorphological phenomena (geodiversity). The definition of Albania's National Parks was reviewed to include geo-monuments, so they could then (where appropriate) be proclaimed as National Geoparks. Some, following support by governmental bodies to complete their documentation, management and infrastructure, could eventually be designated and then accepted into the European Geoparks Network.

The present situation in Albania

By the early 21st century, Albania's geo-monuments, geosites and geomorphosites had been inventoried and digitised in GIS format, facilitating recognition of potential geoparks. Several Albanian state institutions have within their remit natural monuments and environmental protection; for example, the Committee of Environment Preservation and Protection, the Centre of Geographical Studies of the Academy of Sciences of Albania, and the Tourist Committee of Albania (Serjani *et al* 2005). What Albania perhaps shows best is the influence of a specific, if relatively short-lived (compared to its mixed Ottoman and Habsburg roots) political system. That centralised system has influenced the country's focus on landscapes as an economic resource, particularly for mineral extraction, which has in turn influenced the emerging geoheritage and geoconservation concepts.

TURKEY: A GEOCONSERVATION CASE STUDY

Geography and geology of Turkey

Turkey is the most easterly of south-east Europe's countries. Covering an area of 779,450km^2, it is bounded by the Mediterranean Sea to the south, the Black Sea to the north, and the Aegean Sea to the west. The Sea of Marmara, the Bosphorus and the Dardanelles (which jointly form the Turkish Straits) separate Europe and Asia and also form the boundary between Thrace and Anatolia. Turkey is bordered by eight countries: Armenia, the Azerbaijani enclave of Nakhchivan, Bulgaria, Georgia, Greece, Iran, Iraq and Syria. Turkey has long been a cultural and geographical crossroads between Europe and Asia.

Geologically, Turkey lies on the boundary of the Arabian portion of the African plate and the Asian branches of the Alpine geosyncline. Anatolia demonstrates evidence of extensive volcanic activity, particularly during the Neogene to Quaternary; it has the dormant strato-volcanoes (the typically conical form composed of layers of hardened lava and volcanic ash) of Mt Erciyes, Mt Karacadag, Mt Hasandag, Mt Nemrut and Mt Agro, calderas, volcano and tuff cones, and pyroclastic covers (Kazanci *et al* 2005). The Taurus Mountains are an important element of the Alpine-Himalayan orogenic belt (intensely folded and thrust sedimentary rocks intruded by volcanic material, created by the mountain-building process of colliding tectonic plates) and traverse the entire country from east to west; they show many typical alpine features, including thrusts, nappes and major faulting, together with simple and complex anticlines and synclines. Many of Anatolia's geological structures are in carbonaceous (coal-bearing) and evaporate (salt-bearing) rocks; hence, karst landforms and underground features such as caves, dolines (shallow, usually funnel-shaped, depressions in the ground caused by the dissolving away and subsequent collapse of underlying limestone) and poljes (large, flat-floored depressions underlain by lime-stone) are very common. Turkey as a whole has some 2,500 caves and more than 300 lakes.

Geological inquiry within modern Turkey dates from 1935, with the establishment of the Institute of Mineral Research and Exploration. The Institute was partly a consequence of the influx and availability of some talented western European geologists, some of whom were fleeing the period's nationalistic regimes. Working for the Institute as field geologists, and at the universities in Istanbul and Ankara, they supported the training and development of a generation of domestic geologists who also undertook training outside Turkey. The result was a major national geological mapping programme (at scales of 1:100,000 and 1:250,000) which, when completed in the 1950s, showed that Anatolia had unique structures, landforms, formations, stratigraphical sections and fossil faunas. Many of the programme's publications appeared under the auspices of the Geological Society of Turkey (but this ceased operations in 1985), together with numerous meetings, panels and workshops. Some publications were particularly significant in the development of modern geological thought; for example, the term 'mélange' (a mappable body of rock characterised by its lack of continuous bedding and including fragments of rock of all sizes) with its modern meaning was introduced into the literature by Bailey and McCallien (1950) as one of the results of these Turkish studies (Şengör 2003).

Geoconservation in Turkey: the background

Unlike wildlife or nature protection, geoheritage and geoconservation are relatively new concepts in Turkish geological literature, and for the public. Interest in geoconservation has developed gradually (Ketin 1970; Öngür 1976), and the first populist Turkish geo-journal, *Yeryuvarı ve İnsan*, was released by the Geological Society of Turkey to draw attention to 'natural monuments' (Arpat 1976; Arpat and Güner 1976). From the 1970s, due to a difficult domestic political situation, both cultural protection and geoconservation activities were curtailed until the 1990s. Presently, geoheritage and geoconservation terms are not specifically mentioned within Turkish legislation; the most common terms employed are 'natural matter', 'natural object', 'natural monument', 'natural objects to be protected', 'natural parks', and the two geological terms, 'fossil' and 'mineral'.

Turkey's rich and complex geology meant that the registration and protection of its numerous natural sites, even if just for the most scientifically significant aspects of its geoheritage, was really thought to be a way of preventing natural resource exploitation by various administrations with a possible negative impact on the country's economic development. However, in the mid-20th century its growing population and industrial expansion (with the need for construction materials) led to increasing losses of woodlands and farmland, together with their associated geosites. Consequently, in 1961 an Article was added to the country's constitution about the maintenance of forest boundaries. A fortunate, if unintended, by-product of this was the conservation of numerous geosites within forested areas. Many areas had also been designated as Natural or National Parks, so they already had protection. Further, geosites in or adjacent to registered archaeological sites were generally fortunate enough to be protected; for example, the Pamukkale Quaternary travertines near Hierapolis (Denizli), and the tuff cones and maar in Cappadocia (Nevşehir). A recent project has shown the importance of National Parks for wildlife and nature conservation and geoconservation in Turkey (Kazancı *et al* 2007; 2008; 2009).

Geoconservation in Turkey: its administration

The management of registered Turkish sites is the responsibility of the institution that accepted their selection and registration. Two national bodies were established in 1961 to authorise nature conservation site selection, registration and conservation: the Superior Council for the Conservation of Natural and Cultural Property (SCC) – with revised remits in 1983 and 2005 – and the Regional Conservation Council (RCC) as, respectively, advisory and technical committees of the Ministry of Culture. The Environmental Protection Agency for Special Areas was established in 1989 by Cabinet Decision, to examine and protect important areas like islands, lakes, peninsulas, coastal zones and deltas. From 2003, responsibility for natural heritage protection was entrusted to the Ministry of the Environment and Forestry; the Ministry of Culture was then amalgamated with the old Ministry of Tourism. The Environmental Protection Agency for Special Areas was, until 2011, under the control of the Ministry of the Environment and Forestry; however, it was renamed in 2011 as the General Directorate for the Conservation of Natural Properties and placed under the Ministry of the Environment and Urban Planning. The General Directorate of Culture and Museums has the most authority and even direct responsibility for conservation of natural resources. It replaced the General Directorate for the Conservation of Cultural and Natural Property in 2004. Also in that year, the General Directorate of Nature Conservation and National Parks was formed under the then Ministry of the Environment and Forestry (now the Ministry of Water and Forestry). It is the second most involved body for conservation, but until the latter half of the 2000s it mainly focused its activities on National Parks.

Some institutions that have no direct responsibility for nature conservation contribute to environmental protection. The General Directorate of State Hydraulic Works seeks and protects surface- and ground-waters, alongside water assurance, consumption and management roles. The General Directorate of Afforestation and Erosion Control establishes suitable plants on mountains and in rural areas, particularly in burned areas; indeed, fires are a common environmental hazard in the Mediterranean region that can trigger rapid erosion, a major threat to Turkey's geosites and geoheritage localities (Kazancı et al 2005). Other bodies without official or legal responsibility for geoconservation are important in researching geosites and geoheritage, and in their promotion to government ministries and the public. The General Directorate of Mineral Research and Exploration (MTA), the country's geological surveying organisation, produces maps, sections and geophysical information, and also owns Turkey's largest natural history museum, founded in 1935 but only opened to the public in 1968, in Ankara. The Turkish Petroleum Corporation works on sub-surface exploration, albeit focused on petroleum; staff on its fieldwork programmes have discovered and published on numerous geosites and geo-monuments. Some 96 universities, with their 26 geology departments, are an important contributing group.

There are at least 50 variously sized NGOs supporting efforts to conserve Turkey's natural environment, wildlife, specific habitats, landscapes, soil and water. The Geological Society of Turkey (1945–85) and the Chamber of Geological Engineers (founded in 1974) are specifically geological NGOs which, since 1946, have organised annual geological congresses that have promoted geoconservation. However, only the Turkish Association for the Protection of Geological Heritage (JEMIRKO), an off-shoot of ProGEO, is specifically engaged in geoheritage and geoconservation matters. It was founded in 2000 in Ankara, as an initially amateur group of a few Ankara University lecturers and their students who were interested in geoconservation. JEMIRKO organised a series of meetings and conferences between 2000 and 2005, including the ProGEO WG-1 Ankara meeting. With JEMIRKO's encouragement, some other university geology departments have formed student working groups on natural heritage, which organise activities for their local communities. The MTA has also established a project group for professional research on geoparks and natural heritage; Ankara University then undertook to offer a graduate programme on the subject. Meanwhile, the Kızılcahamam-Çamlıdere Geopark in Ankara, scheduled to open in July 2015, comprises 23 geosites, some of which are endangered (Kazancı 2012); it will be the first in Turkey to be realised as a joint project of the state bureaux and various NGOs.

The contributions of these various bodies are significant, as they bridge the gulf between Turkish society and state bureaucracy to enact nature conservation. Through them, domestic nature and landscape tourism, particularly geotourism, have developed. Privatisation and the hiring out of caves and other interesting sites seem to be an emerging government strategy to service geoconservation – although this approach has not been without the emerging issue of geo-exploitation (Hose 2007b). However, there is no national systematic management or maintenance of geosites; most do not have any management plan specifically focused on their geological and/or geomorphological interest. Nevertheless, the management of some of those dozen or so sites designated by the Ministry of Culture and Tourism is very good.

Geoconservation in Turkey: its protected sites

Presently, Turkey has 41 National Parks covering an area of 8,977km², 34 nature parks covering 7,905km², 31 nature conservation areas covering 4,658km² and 105 natural monuments covering 529km². All, except for some of the natural monuments, belong to the Turkish state. However, only two geosites, at Pamukkale (carbonate terraces) and Cappadocia (erosional earth features), are officially protected and included on the UNESCO World Heritage List. Of the registered nature monuments, most are old, individual trees and/or patches of forest with old trees. The sites registered by the Ministry of Water and Forestry seem to have been selected purely for their bioconservation interest. Some 176 of the 211 registered natural sites are also sites listed by the Turkish Association for the Protection of Geological Heritage (JEMIRKO).

Natural sites may be declared and registered as 'A Special Environmental Protection Area'. Since the main objective of these areas is to protect nature and the landscape – particularly from inappropriate housing development – each covers a large area, and although there are many individual geosites within them, they are not cited as the intrinsic values behind protection. Fourteen areas have been registered and given first-order importance by the Turkish Environment Protection Agency (now the General Directorate for the Conservation of Natural Properties). With a combined total area of some 12,111km², all are famous tourist sites such as Köyceğiz, Gökova, Datça and Belek.

Some impressive archaeological and nature sites have been designated as open-air museums by the Ministry of Culture and Tourism, changing them from ordinary sites into tourism destinations. The number of natural sites (1,166) on the Ministry of Culture and Tourism's list is relatively high, the majority being geomorphological or simple geographical sites – landforms, waterfalls, valleys, beaches, lakes and wetlands. Whilst they are natural formations, most are atypical geosites (see Wimbledon 1996; Barretino and Gallego 1999). JEMİRKO identified over 450 geosites (and eight potential geoparks) through its procedures; about a quarter (107) of these geosites are cited in the official list. Though unofficial, JEMIRKO's list is probably a more reliable geosites inventory of Turkey's officially registered 1,166 sites.

Unsurprisingly, JEMIRKO's approach is different to that of the official natural site registration procedure. All geologists, whether members of JEMIRKO or not, may contribute to its selection process. A detailed, descriptive form is used to propose a geosite that is then widely distributed to colleagues for comment. Proposals are evaluated by the relevant subject committees, each of which has five appropriately experienced geologists who, if necessary, visit the site. The results are discussed in JEMIRKO's general assembly before a final decision is made and formally announced. Proposers follow the categorisation of geosites in ProGEO's geosites classification scheme (see Theodosiou-Drandaki et al 2004). JEMIRKO's list differs from that of the Ministry of Culture and Tourism, because the proponents interpret geosite and geoconservation issues differently. Currently, only around 70 geosites on JEMIRKO's list have been conserved to international standards by the Turkish government.

Geoconservation in Turkey: statutory provision

Turkish conservation and protection legislation provides for the registration of three site types for cultural and natural sites. First-order sites are closed to any usage or activity. Second-order sites are closed to new construction, but previous usage may continue. Third-order sites are open for new activities, provided that the appropriate permissions are obtained from the various ministries. Apart from these orders, sites are put into different categories by the related ministries' authorised bodies. Four major Turkish laws authorise some institutes to protect the environment, lands and some sites.

Law No 2863 (1983) on the Conservation of Natural and Cultural Property, enacted by the Ministry of Culture, mainly targets archaeological sites and structures, together with historical and natural material owned by the Turkish state; consequently, it forbids their private use, or the trade/export of cultural and natural assets. The Law provides six categories of site protection: Archaeological (7,766 sites); Urban archaeological (35 sites); Urban (an unknown number of sites, due to multiple listings); Historical (142 sites); Natural (1,166 sites); and 'Superimposed', covering two or more categories (393 sites). Arising from Law No 2863, the Regulation on Identification and Registration of Immovable Cultural and Natural Property to be Protected (1987) requires that their registration procedure must begin with the identification of a known relevant site. A team of three or more experts then prepares a detailed site report for the Regional Conservation Council (RCC), adding an assessment of the likelihood of successful registration. When accepting their report, the RCC considers the suggestions and/or objections from different stakeholders and then publicises the registration. This decision is open to objections raised by the Superior Council for the Conservation of Natural and Cultural Property (SCC), but it rarely challenges them. There are 28 RCCs across the country and they work rather like a court.

The selection and registration of sites, as set out in Law No 2872 (1983 and revised 2006), the Law for the Environment, is roughly the same for natural and cultural sites. The complexity, and an evaluation, of this law have been examined by Yildiz (2010). The Ministry of the Urban and Environment identifies natural heritage sites, discusses them with other relevant governmental bodies and then registers the sites. The Law requires that all activities and investments in rural areas, which require the permission of either the Ministry of Water and Forestry or the Ministry of the Urban and Environment, have to be separated by some distance from such registered protected sites. The Ministry of Water and Forestry developed four nature conservation categories based on the Law: National Park; Nature Park; Site for Nature Conservation; and Nature Monument. The site registration procedure starts with a notification letter from a naturalist, a local authorised person and/or even a villager about a site's aesthetic appeal and significance. Experts from the relevant bureaux then make a site visit to identify the registration type and category. This can be a lengthy process, due to different technicalities and a shortage of experts in the relevant offices. The designation of a first-order area currently in private ownership, or a National Park decided by the Cabinet, is consequent upon the suggestion of the relevant ministries.

Law No 2873 (1983), the Law for National Parks, was enacted by the Ministry of Agriculture and Forestry to conserve culturally, scientifically and naturally significant parts of lands, forests and mountains. It specified that such areas can be designated as 'National Parks' due to the perceived endangerment of agriculture, nature and wildlife by rapid urbanisation and accelerating industrial activities. Law No 5403 (2005), the Law for (the Protection of) Soil and Land-use, was enacted by the Ministry of the Environment and Forestry; it requires that all medium- and large-sized projects, plus all developments in rural areas, have to provide evidence that they do not pose a threat to nature conservation. Other laws outline the responsibilities of institutions and prescribe limits to anthropogenic activities to ensure nature conservation; for example, Law No 6831, the Law for Forests (1956); Law No 3194, the Law for Construction (1985); Law No 3261, the Law for Coasts (1990); and Law No 5177, the Law for Mines and Mining (2004). They provide significant legal underpinning for the protection of the biotic and abiotic environments. However, other laws promote anthropogenic activities and land use. For example, Law No 2634, the Law for the Encouragement of Tourism (1982, revised 2003), enlarged the authority of the Ministry of Culture and Tourism to encompass all wildlife, eco-tourism and geo-tourism activities.

Geoconservation in Turkey: final thoughts

Whilst Turkey's nature conservation legislative and administrative systems seem robust yet bureaucratic, with three ministries (Culture and Tourism, Water and Forestry, and Urban and Environment) and many institutes involved, their weakness is that they (and the laws) can easily be changed and even abolished when a new national government takes office. For example, two previous ministries (for Culture and Tourism) were merged in 2002, and in 2011 the Ministry for Environment and Forestry was changed to the Ministry of Water and Forestry, and the Ministry of the Urban and Environment was formed.

Turkey has a rich and diverse natural heritage due to its complex geological evolution and geomorphological development. However, none of this richness has yet been compiled into a complete national geoheritage inventory. A national action plan for heritage protection was prepared in 1997 (Bademli 1997), but only a small part of it has so far been realised. This is perhaps because much of the funded scientific research has ignored the natural heritage character of geosites. Recently, some NGOs and amateur sports groups interested in hiking, cycling and hunting have begun working with local administrations on natural site management and promotion; leaflets and booklets have been produced, but these have yet to reach the professional interpretative standards seen elsewhere in southern Europe, especially in Spain and Italy.

The sheer abundance of Turkey's cultural and historical heritage, together with a heavy burden of official regulations (mostly serving archaeological sites), has seemingly had a negative impact on geoconservation. The government and its ministries are very active in conserving cultural sites, perhaps because of their potential tourism significance, but they are not similarly keen on caring for registered natural sites – excluding the National Parks and some special areas – that in themselves could impede tourism developments. However, in all fairness, this promotion of cultural heritage protection and the relative neglect of natural heritage protection is something of a Europe-wide issue. Unsurprisingly, not every site can be efficiently managed, because of the high number of sites and various additional problems (such as limited budgets, insufficient number of experts, and authorisation problems within offices). It might actually be the case that the abundance of legislation, ministries and NGOs, coupled with enforcing bodies with different remits, has created an unwieldy and ineffective nature conservation situation in general, let alone geoconservation, bureaucracy; perhaps one rather reminiscent of its Ottoman forebears.

CONCLUDING THOUGHTS ON THE REGION

The landscapes of this chapter's region range from vast river plains to high alpine mountains. Whilst countries such as Italy were pivotal early in geology's development, others such as Albania are new to this role. Perhaps one geopolitical element that binds much of the region is the past influence of two of Europe's lost (and these days lesser known) empires and the influence of their dominant religions – the Ottoman and Hapsburg empires. They have left a legacy of political and legislative approaches to landscape appreciation and understanding (whilst essentially under-pinned by geodiversity, their recognition and appreciation are influenced by geocultural elements – becoming the modern geoheritage), as well as in matters of land ownership and management. Whilst Turkey was at the centre of the Ottoman Empire, particularly noted for its bureaucracy, Albania was at its periphery. Whilst the former moved towards secular statehood, with religion accepted but separated from the state, the latter spent much of the second half of the 20th century as part of the strictly secular Communist bloc. Italy spent part of the 20th century under centralised Fascist regimes, following which it moved to almost federal government; a consequence has meant delays in developing national legislation as each semi-autonomous region maintains its measure of independence. The past empires have shaped modern cultures, influencing public perceptions of and concern for the use and protection of landscapes – from the strictly utilitarian to the recreational and aesthetic.

Hence, if the region's future generations are to inherit geosites and geomorphosites that are fully representative, then the development of a better appreciation of the many profound socio-political and natural agents that impact upon them is an obvious prerequisite. Threats to the region's geoheritage are mainly a consequence of industrial and urban development, coupled with (especially for geomorphosites) agricultural land-use changes and extractive industry activities. The former are particularly evident along the Mediterranean coast of Italy and Spain. The latter are particularly evident in forests and wetlands of Albania and Turkey. Across the region, the emphasis of protection and conservation measures is principally on the biotic, with some emphasis on sport hunting, rather than the abiotic natural heritage components. There is a huge difference in the volume of studies on biodiversity compared with geodiversity. Whilst biodiversity conservation already has international conventions and numerous outstanding well-funded projects, geoconservation is relatively young and poorly funded.

Thus, present efforts must be intensified in order to achieve better recognised and well-funded promotion and geoconservation of the region's rich geoheritage. These will require some standardisation of the geoheritage and geoconservation concepts and their terminology; an inventory and atlas of the region's geoheritage; the development and promulgation of effective geoheritage interpretive and promotional approaches; and finally the enactment of an adequate, and most importantly enforced, local to international geoheritage protection legislative framework. These various measures would benefit from the development of at least a draft South-East European Geoheritage Convention.

References

Alcala, L, Espilez, E, Mampel, L, Kirkland, J I, Ortiga, M, Rubio, D, Gonzalez, A, Ayala, D, Cobos, A, Royo-Torres, R, Gasco, F, and Pesquero, M D, 2012 A new Lower Cretaceous vertebrate bone-bed near Arino (Teruel, Aragon, Spain); found and managed in a joint collaboration between a mining company and a palaeontological park, *Geoheritage* 4 (4), 275–86

Andrea, M D, and Zarienga F, 2000 Italian national actions for nature preservation and geological sites, in *Geological Heritage: Its Conservation and Management* (eds D Barretino, W A P Wimbledon and E Gallego), Instituto Tecnológico Geominero de España, Madrid, 157–63

Arpat, E, 1976 İnsan ayağı izi fosilleri; yitirilen bir doğal anıt, *Yeryuvarı ve İnsan* 1 (2), 45–9

Arpat E, and Güner, Y, 1976 Göktaşı çukuru mu, çökme çukuru mu? *Yeryuvarı ve İnsan* 1, 12–13

Bademli, R R, 1997 *National Action Plan for Environment: Protection of Natural, Historical and Cultural Sites and Matters*, State Planning Organisation, Ankara (in Turkish)

Bailey, E B, and McCallien, W J, 1950 The Ankara mélange and Anatolian thrust, *Bulletin of Mineral Research and Exploration* 15, 12–22

Barca, S, and Di Gregorio, F, 1991 *Proposta metodologica per il rilevamento dei monumenti geologici e geomorfologici*, Bollettino dell'AIC, n. 83, Todi (in Italian)

Barettino, D, and Gallego, E (eds), 1999 *Towards the Balanced Management and Conservation of the Geological Heritage in the New Millennium*, ProGEO/Sociedad Geológica de España, Madrid

Barettino, D, Wimbledon, W A P, and Gallego, E (eds), 2000 *Geological Heritage: Its Conservation and Management*, ProGEO/Sociedad Geológica de España, Madrid

Becker, S O, Boeckh, K, Hainz, C, and Woessmann, L, 2011 The empire is dead, long live the empire! Long-run persistence of trust and corruption in the bureaucracy, *CEPRE Discussion Paper No 8288*, The Centre for Economic Policy Research, University of Warwick, Warwick

Burek, C V, 2012 The role of LGAPs (Local Geodiversity Action Plans) and Welsh RIGS as local drivers for geoconservation within geotourism in Wales, *Geoheritage* 4 (1–2), 45–63

Burek, C V, and Potter, J A, 2002 Minding the LGAPs – a different approach to the conservation of local geological sites in England? *Geoscientist* 12 (9), 16–17

Burek, C V, and Prosser, C D, 2008 *The History of Geoconservation, Special Publication No 300*, The Geological Society, London

Di Gregorio, F, and Ulzega, A, 2002 The state of the knowledge in the conservation and the valorisation of geomorphological sites in Sardinia, in *Proceedings of the Workshop 'Geomorphological Sites': Research, Assessment and Improvement* (eds P Coratza and M Marchetti), 19–22 June 2002, Modena, Italy

Doglioni, C, and Flores, G, 1997 *An Introduction to the Italian Geology*, Lamisco, Potenza

Ferrero, E, Giardino, M, Lozar, F, Giordano, E, Belluso, E, and Perotti, L, 2012 Geodiversity action plans for the enhancement of geoheritage in the Piemonte region (north-western Italy), *Annals of Geophysics* 55 (3), 487–95

Ghiraldi, L, Coratza, P, Biaggi, E de, Giardino, M, Marchetti, M, and Perotti, L, 2009 Development and usage of geosites: new results from research and conservation activities in the Piemonte region (Italy), *Geologia* 54 (2), 23–6

Giardino, M, and Mortara, G, 2001 Carta geomorfologica degli elementi di interesse paesaggistico del Parco Nazionale del Gran Paradiso, *Revue Valdotaine d'Histoire Naturelle* 53, 5–20

— 2004 *I geositi nel paesaggio della Provincia di Torino*, Pubblicazione del Servizio Difesa del Suolo della Provincia di Torino, Torino

Gisotti, G, and Burlando, M, 1998 The Italian job: earth heritage conservation in Italy, *Earth Heritage* 9, 9–11

Hallaci, H, and Serjani, A, 2007 Database of the geological sites of regional importance in Albania: selection and completion with database of geological sites of regional importance in our country – a project of geological survey of Albania, in *Geological Heritage in South-Eastern Europe: Book of Abstracts* from the 12th Regional Conference on Geoconservation and ProGEO Working Group 1 Annual Meeting, Ljubljana, Slovenia, 5–9 September 2007 (eds B Hlad and U Herlec), Agencija RS za okolje, Ljubljana, 19–20

Hose, T A, 2000 European geotourism – geological interpretation and geoconservation promotion for tourists, in *Geological Heritage: Its Conservation and Management* (eds D Barretino, W A P Wimbledon and E Gallego), Instituto Tecnológico Geominero de España, Madrid, 127–46

— 2005 Geo-tourism – appreciating the deep side of landscapes, in *Niche Tourism: Contemporary Issues, Trends and Cases* (ed M Novelli), Elsevier, Oxford, 27–37

— 2007 Geoconservation versus geo-exploitation and the emergence of modern geotourism, in *Abstracts Volume, Geo-Pomerania Szczecin 2007 Joint Meeting PTG-DGG: Geology Cross-Bordering the Western and Eastern European Platform, 24–26 September 2007* (eds H G Röhling, C T Breitkreuz, T Duda, W Stackebrandt, A Witkowski and O Uhlmann), University of Szczecin, Szczecin, 122

Hose, T A, and Vasiljevic, D A, 2012 Defining the nature and purpose of modern geotourism with particular reference to the United Kingdom and south-east Europe, *Geoheritage* 4 (1–2), 25–43

Inaner, H, Tokcaer, M, Kaya, T, Somuncu, M, Calapkulu, K, and Akkoc, S N, 2005 A potential geopark area: Kula (Katakekaumene) volcanic region in Western Turkey, in *Proceedings of the ProGEO WG-1 Sub-Regional Meeting and Field Trip* (ed Anon), ProGEO, Tirana

Kazancı, N, 2012 Geological background and three vulnerable geosites of the Kızılcahamam Çamlıdere Geopark Project in Ankara, Turkey, *Geoheritage* 4 (4), 249–61

Kazancı, N, Saroglu, F, Kirman, E, and Uysal, F, 2005 Basic threats on geosites and geoheritages in Turkey, in *Proceedings of the Second Conference on Geoheritage of Serbia* (ed Anon), Institute for Nature Conservation of Serbia, Belgrade, 149–54

Kazancı, N, Suludere, Y, Mülazımoğlu, N S, Tuzcu, S, Mengi, H, Hakyemez, Y, and Mercan, N, 2007 *Geosites of Soğuksu National Park and its Surroundings*, General Directorate of Nature Conservation and National Parks/Turkish Association for Protection of Geological Heritage (JEMIRKO), Ankara

Kazancı, N, Suludere, Y, Mülazımoglu, N S, Tuzcu, S, Mengi, H, and Hakyemez, Y, 2008 *Geosites of Hattuşaş ve Alacahöyük National Parks and their Surroundings*, General Directorate of Nature Conservation and National Parks/Turkish Association for Protection of Geological Heritage (JEMIRKO), Corum

— 2009 *Geosites of Yozgat Çamlığı National Park and its Surroundings*, General Directorate of Nature Conservation and National Parks/Turkish Association for Protection of Geological Heritage (JEMIRKO), Yozgat

Ketin, İ, 1970 Türkiye'de önemli jeolojik aflormanların korunması, *Türkiye Jeoloji Kurumu Bülteni* 13 (2), 90–3 (in Turkish)

Meco, S, and Aliaj, S, 2000 *Geology of Albania*, Gebruder Borntraeger, Berlin

O'Halloran, D, Green, C, Harley, M, Stanley, M, and Knill, J, 1994 *Geological and Landscape Conservation: Proceedings of the Malvern International Conference 1993*, The Geological Society, London

Öngür, T, 1976 Doğal anıtlarin korunmasında yasal dayanaklar, *Yeryuvarı ve İnsan* 1 (4), 17–23

Robertson, A, and Shallo, M, 2000 Mesozoic-Tertiary tectonic evolution of Albania in its regional eastern Mediterranean context, *Tectonophysics* 316, 197–254

Santos V F dos, Silva, C M da, and Rodrigues, L A, 2008 Dinosaur track sites from Portugal: scientific and cultural significance, *Oryctos* 8, 77–88

Saroglu, F, Kazanci, N, and Inaner, H, 2005 Last ten years of geoheritage in Turkey, in *Proceedings of the ProGEO WG-1 Sub-Regional Meeting and Field Trip* (ed Anon), ProGEO, Tirana

Şengör, A M C, 2003 The repeated rediscovery of mélanges and its implications for the possibility and the role of objective evidence in the scientific enterprise, *Geological Society of America, Special Paper 373*, 385–445

Serjani, A, 1996 Geological sites of the external zones of Albania, *Geologica Balcanica* 26 (2), 11–14

Serjani, A, and Cara, F, 1996 List of geological sites of Albania, *Geologica Balcanica* 26 (1), 57–60

Serjani, A, Neziraj, A, and Hallaci, H, 2005 Ten years of geological heritage in SE Europe, in *Proceedings of the ProGEO WG-1 Sub-Regional Meeting and Field Trip* (ed Anon), ProGEO, Tirana

Theodossiou-Drandaki, I, Nakov, R, Wimbledon, W A P, Serjani, A, Neziraj, A, Hallaci, H, Sijaric, G, Begovic, P, Todorov, T, Tchoumatchenco, P, Diakantoni, A, Fassoulas, C, Kazanci, N, Saroglu, F, Dogan, A, Dimitrijevic, M, Gavrilovic, D, Krstic, B, and Mijovic, D, 2004 IUGS Geosites project progress – a first attempt at a common framework list for south-eastern European countries, in *Natural and Cultural Landscapes – The Geological Foundation: Proceedings of a Conference 9–11 September 2002* (ed M Parkes), Royal Irish Academy, Dublin, 81–90

Vai, G B, and Caldwell, W G E (eds), 2006 *The Origins of Geology in Italy (Special Paper 411)*, Geological Society of America, Boulder

Vera, J A (ed), 2004 *Geología de España*, Sociedad Geológica de España, Madrid

Wimbledon, W A P, 1996 National site selection; a stop on the road to a European geosite list, *Geologica Balcanica* 26, 15–27

Wimbledon, W A P, and Smith-Meyer, S (eds), 2012 *Geoheritage in Europe and its Conservation*, ProGEO, Oslo

Yildiz, S, 2010 The model of Turkey in legal protection of cultural heritage, in *International Archives of the Photogrammetry, Remote Sensing and Spatial Information Sciences*, Vol XXXVIII, Part 5, Commission V Symposium, Newcastle-upon-Tyne, UK, 627–32

Zagorchev, I, 1996a Geological heritage of the Balkan Peninsula: geological setting (an overview), *Geologica Balcania* 26 (2), 3–10

— 1996b Information about the first sub-regional meeting: 'Conservation of the Geological Heritage in South-East Europe', *Geologica Balcania* 26 (1), 3–6

Zouros, N, 2006 The European Geopark Network: geological heritage protection and local development – a tool for geotourism development in Europe, in *4th European Geoparks Meeting – Proceedings Volume* (eds C Fassoulas, Z Skoula and D Pattakos), European Geoparks Network – Psiloritis National Park, Anogia, Crete, 15–24

Zunino, M, Cavagna, S, Clari, P, and Pavia, G, 2012 Two examples of the geoheritage potential of Piedmont (NW Italy): Verrua Savoia for the present, Valle Ceppi for the future, *Geoheritage* 4 (3), 165–75

FURTHER READING

Goodwood, J, 1998 *Lords of the Horizons: A History of the Ottoman Empire*, Henry Holt and Company, New York

Maran, A, 2008 Geoconservation in the Balkan region: practices and legal instruments, *Bulletin of the Natural History Museum in Belgrade* 1, 41–63

Sharples, C, 1995 Geoconservation in forest management: principles and procedures, *Tasforest* 7, 37–50

UNESCO, 2004 *Operational Guideline for National Geoparks Seeking UNESCO's Assistance*, Global UNESCO Network of Geoparks, Paris

Wimbledon, W A P, Ishchenko, A, Gerasimenko, N, Alexandrowicz, Z, Vinokourov, V, Liscak, P, Vozar, J, Vozarova, A, Bezak, V, Kohut, M, Polak, M, Mello, J, Potfai, M, Gross, P, Elecko, M, Nagy, A, Barath, I, Lapo, A, Vdovets, M, Clincharov, S, Marjanić, L, Mijović, D, Dimitrijević, M, Gavrilović, D, Theodossiou-Drandaki, I, Serjani, A, Todorov, T, Nakov, R, Zagorchev, I, Perez-Gonzales, A, Benvenuti, M, Constantini, E, D'Andrea, M, Gissoti, G, Guaddo, G, Marchetti, M, Massoli-Novelli, R, Panizza, M, Pavia, G, Poli, G, Zarlenga, F, Satkunas, J, Mikulenas, V, Suominen, V, Kananoja, T, Lehtinen, M, Gonggrijp, G, Look, E, Grube, A, Johansson, C, Karis, L, Parkes, M, Raudsep, R, Andersen, S, Cleal, C, and Bevins, R, 1998 A first attempt at a geosite framework in Europe – an IUGS initiative to support recognition of World Heritage and European geodiversity, in *Special Issue 'Geological Heritage of Europe'* (eds I Zagorchev and R Nakov), *Geologica Balcanica* 28 (3–4), 5–47

Zouros, N, 2004 The European Geoparks Network: geological heritage protection and local development, in *Episodes* 27 (3), 165–71

10

Geoheritage Case Study:
The Isle of Wight, England

MARTIN C MUNT

The Isle of Wight lies off the south coast of England and encompasses around 380km², approximating to a diamond shape, measuring some 37km east to west and 21km north to south. It is separated from the mainland county of Hampshire by the Solent. The highest point, St Boniface Down above the town of Ventnor, is 240m in height. Whereas 'the island', as it is known, has probably been an island, or perhaps a peninsula in the recent geological past, linked to sea-level changes during the Quaternary, the exact date of separation from the mainland is unknown. Palaeolithic people certainly lived there, and there is a rich archaeological heritage recording human occupation throughout much of prehistory (Basford 1980). Today it is home to a population of approximately 120,000 people, with the main town of Newport situated almost in the centre of the island. Famously a past home of Queen Victoria and Prince Albert (Osborne House was built between 1845 and 1851 as a summer retreat), it is perhaps better known today for annual rock concerts, yacht racing at Cowes, and as Europe's richest dinosaur locality.

AN OVERVIEW OF THE ISLAND'S GEOLOGY

The Isle of Wight has long been considered a simple layer cake of Cretaceous (144–65 million years ago) rocks, overlain by Palaeogene (65–23 million years ago) sediments, with a scatter of Pleistocene superficial deposits; these are interpreted as formed across a central east-west trending monocline and an anticline running out to sea to the south-west. Oil exploration in the early 2000s revolutionised understanding of the structural origin of the island. The Chalk backbone sits astride a great east-west trending fault, the legacy of Palaeozoic plate tectonics, known to have dominated the landscape of the dinosaurs and to have controlled sedimentation and development of the main structure. The island is further dissected into blocks by north-south trending faults. The island can be divided simply into three terrains: the central Chalk Downs, the northern Palaeogene and the Early Cretaceous area south of the Downs (see Figure 10.1).

In the south, the seaward face of the sequence between Bonchurch and Blackgang (Figure 10.2B) dips shallowly towards the English Channel; a series of terraces are formed by mass movement forming the so-called Undercliff. The Cretaceous Wealden rocks are the oldest and are world famous for dinosaurs. The overlying Lower and Upper Greensand and Gault Clay record the great Cretaceous marine transgression through to the Chalk. The eroded top of the Chalk is overlain by a sequence of Palaeogene sediments, including the London Clay, Bracklesham and Barton Groups and unique Solent Group, which record both local and regional sea-level change. Throughout much of this sedimentary pile, fossils are abundant and diverse. The Wealden rocks

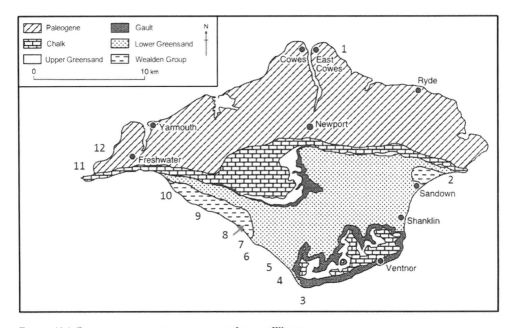

Figure 10.1 Outline geological map of the Isle of Wight
Numbered locations on map are: 1. Osborne; 2. Yaverland; 3. St Catherine's Point; 4. Blackgang
Chine; 5. Whale Chine; 6. Atherfield Point; 7. Shepherd's Chine; 8. Dinosaur Farm;
9. Brighstone; 10. Hanover Point; 11. Alum Bay; 12. Totland Bay [map adapted from Insole et al
1998]

yield dinosaur bones, teeth and footprints, other reptiles (such as turtles), mammals, fish, invertebrates and plants. The Marine Cretaceous rocks are a magnet for collectors of ammonites,
and also yield a huge diversity of invertebrates, plants and vertebrates. The Chalk is Britain's
main continuous sequence, but it is not fully accessible. The Palaeogene contains many unique
horizons only comparable with the Paris Basin, and contains rich diverse plant, invertebrate and
vertebrate assemblages – notably, globally important mammal faunas that record Palaeogene
climatic change. Superficial deposits, mainly considered to be Pleistocene, are enigmatic, but
principally relate to ancient and existing rivers and sea-levels.

The island's landscape and visitors

Perhaps it was merely the fact that, in the 1840s, Queen Victoria and her husband Prince Albert
built Osborne House near East Cowes that caused the Isle of Wight to become a popular resort
and holiday destination. In 1853 Alfred Lord Tennyson (1809–92) began his 40-year ownership of Farringford House at Freshwater, forming a circle of intellectuals that included his near
neighbour, the pioneering photographer Julia Margaret Cameron (1815–79). Cowes Week and
its yacht racing brought the rich and fashionable to the island, and others would follow. Brinton
(2006) suggests that the island's popularity was a result of its discovery by the gentry, coupled
with the natural beauty and charm of the coast – particularly the chines, the steep-sided valleys

Figure 10.2 Three iconic Isle of Wight geosites

A. Dinosaur Isle Museum – opened in 2001, it is Britain's first purpose-built dinosaur museum; it was funded by the Isle of Wight Council and the Millennium Commission.

B. Gore Cliff – standing above Blackgang Chine spectacularly exposes the Albian Upper Greensand (lighter shaded) overlain by the Cenomanian Grey Chalk (darker shaded) that represents a change in the area to deeper marine conditions some 100 million years ago.

C. Hanover Point – part of a fairly extensive trackway of dinosaur footprints (ascribed to Iguanodon) damaged by irresponsible collecting with a power saw, with one damaged footprint still *IN SITU* and another clearly removed in this view from July 1996 (scale = 30 cm).

D. Hanover Point – the so-called 'Pine Raft' of fossilised araucaroid gymnosperm logs (scale = 30 cm), derived from the local sandstone and considered to be some 125–130 million years old.

where rivers flow to the sea. Whatever the reasons, by the 1950s the island had become a mass tourism destination, with resorts such as Sandown and Shanklin attracting bucket-and-spade holidaymakers. Despite the rise of the affordable Mediterranean holiday, somehow the island has managed to retain its place as one of the UK's favourite family holiday destinations. Today, 'niche market destination' and 'off-season' are the buzzwords of tourism, and the island's marketing reflects these trends, with (for example) an annual walking festival. Maybe, despite the pull of the rich landscape and cultural heritage, it is the ferry trip across the Solent that creates the sense of being away – somewhere and somehow different – that gives the island its appeal to visitors.

The island is described as a microcosm of southern England; this can be attributed to one clear cause – geology. Just one hour's drive across the island can give one the thatched cottage feel of

Devon, downlands like Sussex, the wooded continuation of the New Forest, and a rugged coast-line where it is only by looking at the geology that the visitor can realise that this is not Cornwall. Two key tourist attractions, Blackgang Chine and Alum Bay, are built on more than geology, although the latter certainly sells itself on geology with souvenirs made from its coloured sands. Museums vie with each other as dinosaur attractions, only to be trumped in terms of visitor numbers by 'Jurassic' rides; however, the only Jurassic on the Isle of Wight are derived pebbles in Cretaceous rocks. Geology, its interaction with weather, waves and generations of people and plants, underpins the landscape that brings tourists to the island.

There could be little doubt as to the importance of geoheritage to the island when, in 2013, the 'Dinosaur Island' branding came to the fore, with the launch of *Walking with Dinosaurs: The 3D Movie*, which was premiered on the Isle of Wight. Eleven years earlier, Sandown had become the home of the UK's first purpose-built dinosaur museum, Dinosaur Isle Museum (Figure 10.2A), funded by the Millennium Commission; it has outlived many other rival attractions.

THE ISLAND'S LITERARY GEOHERITAGE

It is quite remarkable that, when the first official geological memoir on the island was published (Forbes 1856), there were already more than 130 publications relating to some aspect of its geology. When the only full memoir was published (Bristow *et al* 1889), there were more than 330 papers, books and monographs relating to the island; the authors listed in its bibliography read like a who's-who of 19th century geology. We can with certainty point to the year 1816 as a key year for the island's geological literature. The end of the Napoleonic Wars permitted a fresh start in travel, and in 1816 (a year after Waterloo), Thomas Webster's letters on the island's geology appeared in the first substantial tourist guide to the island, Sir Henry Englefield's *Picturesque Beauties*. However, it was hardly readily portable or affordable, because of its size and cost. The letters included superb aquatint images showing the coastline and strata. Webster (1773–1844) was the first geologist to recognise the presence of both fresh-water and marine in the island's Caenozoic strata. He was also appointed as the house-secretary and curator to the Geological Society of London in 1826; a skilled draughtsman, he edited and illustrated its *Transactions* for many years. Geology remained a feature of tourist guides to the island throughout the 19th century – Brannon (1855, one of numerous editions) and Barber (1834) being two of the best known. Munt (2006) continues the tradition by providing a geology chapter in a recent island tourist guide.

Popular guides to the island's geology form an important part of the published geoheritage. Foremost of these is Mantell (1854), describing also the geology of Purbeck, supported by one of the earliest geological maps of the island and plates of common fossils. By no means as grand as his *Fossils of the South Downs*, it remains a valuable source of information. With real local flavour, Norman's (1887) publication on the island is a treasure, with maps, sections and plates of fossils. Mark William Norman (1820–99) was a mason with a passion for fossils; he supplemented his income by selling fossils to tourists. The last such guide was that of Hughes (1922), again with plates of fossils. More recently, probably in a similar vein of tourist guide for the interested visitor, the Geologists' Association field guides (No 25) by Curry *et al* (1972, as a third edition, first published in 1958) and subsequently Insole *et al* (1998) have been the most notable additions to the literature. Today it is easy to forget how geology became accessible through geological maps; 2014 sees the latest version of the island's geological map by the British Geological Survey, which first appeared in print in 1856.

THE ISLAND'S GEO-HISTORICAL SIGNIFICANCE

It takes only the most superficial glance to realise why the Isle of Wight is of interest to geologists. With almost unbroken exposure around the coast and – for the most part – easy access, albeit frequently tide-dependent, the island has become a key training ground for geologists. But what made the island so significant? We have already recognised the role played by the monarchy in making the Isle of Wight fashionable during the 19th century. Famous geologists of the day, such as Sir Charles Lyell, were guests at Osborne House, and it is known that Prince Albert had an interest in geology. But to be important, the geology has to be special. The first memoir by Forbes (1856) focused on the Tertiary (that is, Palaeogene) deposits of the north coast – the most complete set of Palaeogene rocks in the United Kingdom. Unlike the Tertiary deposits of the Paris Basin (which are rich in gypsum – at the time, a strategically important resource), the island's Palaeogene proved to have very limited gypsum deposits of no strategic value. Nevertheless, the memoir sets a very high standard for the exploration of Britain's geology.

When fossil collectors, including Gideon Mantell, started to find the bones of large reptiles in Sussex, the geological structure of southern England was poorly understood, and it is well documented that the age of the bones was disputed. Similar bones were known to occur on the Isle of Wight, and were in fact so common that Wilkins (1859) was later to record that a 'wagon load of bones' was collected from the beach at Brook in 1854. Fitton (1847), in his work on the Lower Greensand, was to prove critical in verifying the antiquity of the bones; in this work, he demonstrated that the Wealden beds that contained the bones underlay and were therefore older than the Chalk, and hence of Cretaceous age. The Atherfield section, where Fitton worked and devised the sequence of Lower Cretaceous rocks, is regarded as a classic section in British geology; it records in just a few kilometres of exposure the onset of the Cretaceous marine transgression.

In 1841 Sir Richard Owen coined the word 'dinosaur'. This is the name he gave to unite the ancient reptiles whose bones came principally from the Cretaceous, Jurassic and (as later realised) Triassic strata. These were the very creatures to which Fitton's observations on the island's geology gave great antiquity. It was also the island that was to provide the key to uniting these creatures. Owen (1842) recognised that the sacrum of dinosaurs typically comprised five sacral vertebrae; he ascribed this character from a specimen collected from the island that is now housed in the Natural History Museum, London. The Isle of Wight can therefore rightly claim to be the birthplace of the antiquity and characterisation of the dinosaurs. Today it still holds the title of the richest dinosaur locality in Europe, and one of the top ten localities in the world.

Much of the work that followed the pioneering stage and the publication of the memoirs was very much focused on the detail. New species of dinosaur, such as *Polacanthus* and *Hypsilophodon*, were discovered by a local curate, the Reverend William Fox (1813–81) of Brighstone, and later footprints were recognised as well. In the 1920s the Geological Survey was back, with a review and update with the 'short account' by H J O White in 1921. Interest in the island as a key dinosaur location then started to fade, but never quite went away. Dinosaur discoveries had been made across the American West and in Africa, which had long since eclipsed the island's limited but productive Wealden rocks. Much of the Cretaceous had been effectively summarised and brought together in the stratigraphical memoirs by Jukes Brown and Hill (1900–04). For more extensive coverage of the island's role in the history of geology, the publications of Freeman (2004) and Rudwick (2005) are highly recommended. The history of dinosaur collecting, together with short biographies of principal characters, is provided by Martill and Naish (2001).

THE ISLAND'S RENAISSANCE

It would be easy to point to dinosaurs as the beginning of the renaissance of interest in the Isle of Wight. In the 1970s, a revolution in our understanding and interpretation of dinosaurs began. This was popularised in the UK by books such as Charig (1983). However, in the 1950s it was a new look at the ammonite fauna of the Lower Greensand by Raymond Casey (1961–78) that revitalised interest. Subsequently published, but unfinished, the Palaeontographical Society monograph on Lower Greensand ammonites represents the first modern systematic study of Isle of Wight fossils. But then there were the dinosaurs. One could even pick up their isolated bones on the beach; and then, a few new collectors started finding partial skeletons. Blows (1978) described some of the finds and introduced the personalities involved in dinosaur hunting on the island. New research on the Solent Group by Brian Daley (stratigraphy) and Alan Insole (mammals) provided a fresh understanding of the island's unique late Eocene to Oligocene strata. In terms of palaeontological interest, it became something of a three-horse race – ammonites, mammals and dinosaurs.

GEOHERITAGE AND THE DEVELOPMENT OF TOURISM ON THE ISLAND

It could be argued that the island is one big geoheritage site; however, in order to break it down into its component parts, it can be divided into museums and attractions, and principal field localities – that is, primary and secondary geosites (Hose and Vasiljević 2012). In 1810 the Isle of Wight Philosophical Society established the first geological collection on the island (Munt 2008); what remains of that collection now forms part of the Isle of Wight Council Geological Collection. Throughout a major part of that collection's history it was housed in the Museum of Isle of Wight Geology, located above the public library in Sandown. Opened in 1914, the collection grew, gradually incorporating collections from small local museums. Important acquisitions were made whilst J F Jackson (1894–1966) was employed as honorary curator in the 1930s and early 1940s. Significant growth restarted in the 1980s, when the museum became part of the Isle of Wight County Council Museum Service, with the employment of full-time staff. After 50 years of growth, exceeding its storage and display capacity, the museum (a free attraction) closed in 2001. It was reopened and rebranded as the Dinosaur Isle Museum, with an entry fee, located a short distance from the old site. Dinosaur Isle is now the ideal starting point, as indeed the old museum was, for exploring the island's geology. Expecting to greet its millionth visitor in 2014, Dinosaur Isle Museum is a modern building splendidly exhibiting the best of the local dinosaurs and fossils; its internationally important collections include type specimens of dinosaurs, mammals, reptiles and invertebrates. As an accredited museum, it is the only venue on the island where the secure future of the various geology collections is officially assured.

Dinosaur Farm near Brighstone was originally opened as an off-shoot of the old Museum of Isle of Wight Geology, but acquired a life of its own. The farm buildings are on land where, in the early 1990s, the remains of a still undescribed sauropod dinosaur were found and excavated. The venue was the focus of the BBC television event, *Live from Dinosaur Isle*, broadcast in 2001. The site has had a mixed history and still operates as a fossil-themed seasonal tourist attraction near to where the fossils are found. Numerous fossil shops have come and gone from the island, including Dinosaur Farm; other venues have included Blackgang Chine, and currently two shops in Shanklin and one at Isle of Wight Pearl, a tourist attraction not far from Dinosaur Farm.

The Godshill Natural History Centre has a small geology collection and also serves as a fossil shop.

For a short time in the 1990s and early 2000s, Ventnor was home to the Coastal Heritage Centre; operated by the Isle of Wight Council, it was an information centre focused on coastal management. Employing civil engineers, geomorphologists and interpretation staff, it provided a valuable service to developers, local people and schools. Funded primarily through European Union grants, the centre was closed in 2009 following funding cuts, and coastal engineering and management disbanded as a council operation.

Alum Bay is one of the island's premier tourist attractions. The cliffs are the site of the famous coloured sands, part of the London Clay, Bracklesham, Bournemouth and Barton Groups. For generations, tourists have filled jars of the sand as a memento of their stay on the island. The attraction has been built up around this activity. A chair-lift to the beach provides spectacular views of the cliffs of coloured sands and the Needles, sea stacks formed at the adjacent Chalk headland. Less fortunately, it is also a location for a 'Jurassic themed' ride, from which, on a clear day, one can see the Jurassic rocks of the Jurassic Coast World Heritage Site in the distance.

Blackgang Chine is a theme park made famous, to its many past visitors, by its huge fibreglass dinosaurs, installed by helicopter in the 1970s. The year 2014 saw fresh life for Blackgang, with the installation of new animatronic dinosaurs. It affords dramatic views along the south-west coast of the island, including the Atherfield section. 'Chine' is a local word (also used in east Dorset) for a ravine-like gully. The chines have formed as a result of rapid coastal retreat and the shortening of valley lengths of rivers that once fed a river system to the west, now lost below the sea. There are about ten such features on the island; from the geomorphological perspective, Shanklin Chine (also a tourist attraction) and Shepherd's Chine are of interest, but there are also abandoned chines to the north of Blackgang (Walpen Chine). Whale Chine has historically been the principal point of access to the Atherfield section. The scatter of chines along the coast provides access points to the coastal geology.

The 'Pine Raft' at Hanover Point (Figure 10.2D) was recognised in the 19th century as a log-jam dating from the time of the dinosaurs. It probably exposes the oldest surface rocks on the island, part of the Wealden Group. It is the focal point of most field geotourism on the island, with hundreds of schoolchildren and public field trips visiting each year. The site is managed by the National Trust, and is famous for its many dinosaur footprints and some gigantic footprint casts. Groups operated by Dinosaur Farm and fossil shops, and occasionally the Dinosaur Isle Museum, can be seen there daily at low tide during the tourist season. Hanover Point was previously marked by a conical concrete pillar, a range finder for the Victorian Needles gun battery, but that was lost to the sea during the winter storms of 2013 to 2014. The site has also had a controversial history, having been one of the first sites in the world to have dinosaur footprints destroyed by being cut out of the bedrock using power saws (Figure 10.2C). The National Trust has requested that visitors respect the importance of the footprint casts and do not remove them from the beach.

Other sites used for geotourism include Yaverland, St Catherine's Point and Atherfield. As with Hanover Point, school parties and public groups are given an introduction to geology and fossil collecting by staff from the attractions and the museum. The island's coastal (and a few inland) sites are designated as Sites of Special Scientific Interest (SSSIs), with 16 named sites for geological interest. Many sites are also included in the national Geological Conservation Review (GCR); these categories, along with Regionally Important Geological Sites (RIGS) and Local

Geological Sites (LGS), give advisory protection, at least in principle. The history of geoconservation on the island was reviewed by Munt (2008).

Coastal mass movement is as much a part of the geoheritage as the rocks themselves. The Ventnor area has been described as the largest urbanised landslip in Europe. In the 1990s, the Ventnor area and the Undercliff to the west were extensively mapped for geomorphic features, and hazard maps and predictions were developed. The area is now an open textbook for mass movement features and management techniques. Coastal defence works, frequently tied to landward mass movement, are features of the Ventnor area and other locations around the island. There is also a longer term legacy of coastal defence projects, one of the earliest of which (from the 1970s), at Totland Bay, has recently failed. The wet winter of 2013 to 2014 took a heavy toll on the area, as management of sites and roads stopped. There is rapid erosion of the south-west coast, as much as one metre per year, with major threats to road links. Non-invasive coastal management, where the intrinsic landscape and geological heritage interests outweigh transport links and land values, will prove to be a major challenge to geoheritage features in the near future.

GEOHERITAGE AND ITS VALUE TO THE ISLE OF WIGHT

So what is the value of the island's geoheritage? Current tourist numbers to the island vary between 2.3 and 2.4 million visitors annually, bringing approximately £336 million to the local economy in 2012 (source: Isle of Wight Tourism Research reports). Research carried out for Natural England (Webber *et al* 2006) found that geodiversity generated between £2.6 and £4.9 million of tourism industry revenue for the island in 2004/5, in a year when tourism generated £352 million in total. So, at best estimate, geoheritage on the island generates 2.8% of tourism income. Specifically, this supports the operation of one permanent museum (Dinosaur Isle), one seasonal exhibition (Dinosaur Farm), four shops, field trip operators and school agencies offering geological education.

CONCLUSION

The Isle of Wight, with one of the best geoheritage pedigrees in Britain, stands out as a truly special place, with fossils in leading museums, hundreds of publications (including highly sought-after historic books), readily accessible cliffs and fossils. Today that heritage pays its way, contributing to the local economy – so much so that Blackgang Chine (one of the island's leading tourist attractions) has made a new investment in animatronic dinosaurs.

The visitor and collector can enjoy the geology and fossils freely within the terms of protected SSSI status. The island as a whole does have a unique selling point, its dinosaurs, as one of the world's richest localities; the casual visitor can still find loose pieces of bone and other fossils on the beach. One clear issue is that the island has never seriously considered the importance of the land ownership rights, but this is no different to much of the UK; this may be why the UK has a thriving fossil-collecting community, which feeds clear benefit to British science and geoheritage.

REFERENCES

Barber, T, 1834 *Barber's Picturesque Illustrations of the Isle of Wight, Comprising Views of Every Object of Interest on the Island. Engraved from Original Drawings, Accompanied by Historical and Topographical Descriptions*, Bather, London

Basford, H V, 1980 *The Vectis Report: A Survey of Isle of Wight Archaeology*, Isle of Wight Archaeological Committee

Blows, W T, 1978 *Reptiles on the Rocks*, Isle of Wight County Council, Sandown

Brannon, G, 1855 *Picture of the Isle of Wight, or The Expeditious Traveller's Index to its Prominent Beauties & Objects of Interest. Compiled especially with Reference to those Numerous Visitors who can spare but two or three Days to make the Tour of the Island*, George Brannon, Wootton

Brinton, R, 2006 *Isle of Wight: The Complete Guide*, Dovecot Press, Wimborne

Bristow, H W, Reid, C, and Strahan, A, 1889 *The Geology of the Isle of Wight*, Memoir of the Geological Survey of England and Wales, HMSO, London

Casey, R, 1961–78 *A Monograph of the Ammonoidea of the Lower Greensand*, Palaeontographical Society Monograph, London

Charig, A, 1983 *A New Look at the Dinosaurs*, British Museum (Natural History), London

Curry, D, Daley, B, Edwards, N, Middlemiss, F A, Stinton, F C, and Wright, C W, 1972 *Geologists' Association Guide No 60: The Isle of Wight*, Geologists' Association, London

Daley, B, 1999 Palaeogene sections in the Isle of Wight: a revision of their description and significance in the light of research undertaken over recent decades, *Tertiary Research*, 19 (1–2), 1–69

Englefield, H C, 1816 *A Description of the Principal Picturesque Beauties, Antiquities and Geological Phaenomena, of the Isle of Wight. With Additional Observations on the Strata of the Island, and their Continuation in the Adjacent Parts of Dorsetshire, by Thomas Webster, Esq.*, Payne and Foss, London

Fitton, W H, 1847 A stratigraphical account of the section from Atherfield to Rockenend, on the south-west coast of the Isle of Wight, *Quarterly Journal of the Geological Society of London* 3, 289–327

Freeman, M J, 2004 *Victorians and the Prehistoric: Tracks to a Lost World*, Yale University Press, New Haven

Forbes, E, 1856 *On the Tertiary Fluvio-Marine Formation of the Isle of Wight*, Memoirs of the Geological Survey of Great Britain, Longman Green and Longman, London

Hose, T A, and Vasiljević, D A, 2012 Defining the nature and purpose of modern geotourism with particular reference to the United Kingdom and south-east Europe, *Geoheritage* 4 (1–2), 25–43

Hughes, J C, 1922 *The Geological Story of the Isle of Wight*, Stanford, London

Insole, A N, Daley, B, and Gale, A S, 1998 *Geologists' Association Guide No 60: The Isle of Wight*, Geologists' Association, London

Jukes Brown, A J, and Hill, W, 1900–04 *The Cretaceous Rocks of Britain*, Memoir of the Geological Survey of Great Britain (3 vols), Memoirs of the Geological Survey of the United Kingdom, London

Mantell, G A, 1854 *Geological Excursions Round the Isle of Wight, and Along the Adjacent Coast of Dorsetshire; Illustrative of the Most Interesting Geological Phenomena and Organic Remains*, Henry G Bohn, London

Martill, D M, and Naish, D, 2001 *Palaeontological Association Field Guides to Fossils, No 10: Dinosaurs of the Isle of Wight*, The Palaeontological Association, London

Munt, M C, 2006 The geology of the Isle of Wight, in *Isle of Wight: The Complete Guide* (ed R Brinton), Dovecot Press, Wimborne, 50–1

— 2008 A history of geological conservation on the Isle of Wight, in *The History of Geoconservation: Special Publication No 300* (eds C V Burek and C D Prosser), The Geological Society of London, London, 173–9

Norman, M W, 1887 *A Popular Guide to the Geology of the Isle of Wight*, Knight's Library, Ventnor

Owen, R, 1842 Report on British fossil reptiles, part ii, *Reports of the British Association for the Advancement of Science, Plymouth 1841* (11), 60–204

Rudwick, M A, 2005 *Bursting the Limits of Time: The Reconstruction of Geohistory in the Age of Revolution*, University of Chicago Press, Chicago

Webber, M, Christie, M, and Glasser, N, 2006 *The Social and Economic Value of the UK's Geodiversity: English Nature Research Report No 709*, English Nature, Peterborough

White, H J O, 1921 *A Short Account of the Geology of the Isle of Wight*, Memoirs of the Geological Survey of Great Britain, HMSO, London

Wilkins, E P, 1859 *A Concise Exposition of the Geology, Antiquities and Topography of the Isle of Wight*, T Kenfield, London

11

A Geoheritage Interpretation Case Study: The Antrim Coast of Northern Ireland

Kevin R Crawford

Understanding and conceptualising tourist experiences has been a major tourism research focus since the 1960s. Even after more than 50 years, the nature of contemporary tourist experiences continues to be an important research focus amongst tourism researchers (see Ryan 2002; Uriely 2005; Sharpley and Stone 2010). Today, the focus of research into tourists' expectations and experiences largely examines and explores the concept of postmodern, or post-mass, tourism (see Cohen 1995) and responses to transformations in the dynamic socio-cultural world of tourism. This has led to a focus on responsible tourism (Harrison and Husbands 1996) and ecotourism (United Nations World Tourism Organisation 1997), all forms of special interest tourism (Hall and Weiler 1992) into which geotourism could be subsumed as a form of niche tourism (Hose 2005a; Novelli 2005). One area where this has particular resonance has been in tourism to UNESCO World Heritage Sites (Shackley 1998). Studies into visitor provision, experiences and expectations at cultural heritage settings have been a key research focus in heritage tourism studies since the early 1990s (for example, Pocock 1992; Fladmark 1994; Light 1995). In many cases, research has focused on improving the understanding of tourists' experiences, motivations and expectations, and on understanding their opinions of the nature of the interpretation provision (for example, Poria 2010; Poria *et al* 2006a; 2006b; 2009; 2013).

Interpretation was defined (by the leading mid-20th century influential figure in interpretative practice in the USA National Parks Service) as: 'An educational activity which aims to reveal meaning and relationships through the use of original objects, by first-hand experience and by illustrative media, rather than simply to communicate factual information' (Tilden 1977, 8). Ballantyne (1998) summarises the overlap between environmental education and environmental interpretation, and their role in educating future citizens (as schoolchildren) about environmental issues; he diagrammatically (in his Figure 6.1) summarises the variables involved, and his account is useful today when many geoheritage interpretation programmes include references to current climate change concerns. Badman (1994) provides both a definition of environmental interpretation as 'the art of explaining the meaning and significance of sites visited by the public' (Badman 1994, 429) and an approach to its employment at geological sites (that is, geosites). However, Uzzell (1993) cautions that visitors to interpreted sites, especially those with visitor centres, might actually leave them with little new knowledge.

In addition to providing the first definition of interpretation, Tilden also focused on the methods of interpretation, including the development of key principles of interpretation. However, Tilden's principles, like much else in the field, have seldom been subjected to empirical investigation; indeed, he actually 'shied away from any kind of analytical or systematic approach to understanding visitors. Uncharacteristically, he did not call for a foundation of empirical facts or for research on visitors' (Machlis and Field 1992, 4). Similarly, interpretative practitioners have also been reluctant to empirically assess the efficacy of their work; this might be because, 'Surprisingly, many interpreters and interpretive planners cannot specify exactly what it is they are trying to do' (Field and Wagar 1992, 22). Consequently, assessing the efficacy of their provision is hindered by a lack of measurable objectives for their broad – albeit laudable – aims. With regard to specific geotourism interpretative provision, the first such attempts at evaluation were only undertaken in the UK in the early 1990s (Hose 1994a; 1994b), although such work had begun in the USA in the mid-1980s, with a report (Patterson and Bitgood 1986) on the innovative interpretive provision (now completely removed) at the Red Mountain Road Cut Natural Landmark in Birmingham, Alabama.

Tilden's seminal work, plus developments by others such as Uzzell (1989a; 1989b) and Uzzell and Ballantyne (1998) on heritage interpretation, and Ballantyne (1998), Ham (1992) and Machlis and Field (1992) on environmental interpretation, provide key theoretical contexts for interpretation. Therefore, for the purpose of this chapter, interpretation is defined as 'the provision of on-site information in the form of interpretive media that is designed to inform visitors of the site features, in addition to engaging visitor interest in the site'. Interpretation affects the satisfaction derived from a visit by adding value and meaning, thereby helping visitors appreciate the site (Gilmore *et al* 2007). Interpretation can also be used as a promotional tool to attract visitors to a site, especially if there is competition between sites and tourist attractions (Poria 2010).

Previous studies on interpretation at heritage settings commonly involved an evaluation of the content of the interpretation (for example, Hose 2005b), the display or the educational contribution (for example, Hose 1994a; 1994b), and the nature of the visitors (for example, Hose 1996; 1998), rather than on the tourists' expectations of interpretation. This has been particularly true in studies that have taken place in geological heritage (that is, geoheritage) settings. Hose (2006) has provided an historical and theoretical overview of UK geological interpretative provision from the 19th century to the present day. In relation to interpretation at geoheritage sites, most studies focus on raising awareness and understanding of geological information (for example, Burek and Davies 1994; Dias and Brilha 2004; Hose 1996; 1997; 1999; 2006; Pralong 2006), through reporting on the engagement of the general public with the geoheritage or on the development of interpretation, rather than on expectations and experiences of the on-site interpretation itself.

These studies are important, since from a geoheritage and geotourism perspective raising awareness is a key requirement of any visit to a geosite, given that public understanding of geology and geoheritage issues is generally poor and most people are unfamiliar with geological terminology (Dias and Brilha 2004). Hose (1994a; 1996; 1997; 1999; 2006) suggests that appropriate geological interpretive provision – that is, geointerpretation – is important in not only informing visitors about geosites but also ensuring that they will understand and recall the information relating to the geosite. Pralong (2006) explores this further by indicating that on-site interpretation at geosites needs to be more adapted to the visitors' expectations, and more striking and original ways of communication have to be developed in order to improve their overall experience of the site. In order to explore many of these points, a case study on tourists' expectations and experiences of a key geoheritage site in the UK – the Giant's Causeway and Causeway Coast World Heritage Site – is presented and discussed in this chapter.

THE GIANT'S CAUSEWAY AND CAUSEWAY COAST WORLD HERITAGE SITE

The Giant's Causeway and Causeway Coast World Heritage Site (popularly known as the Giant's Causeway) has an area of 7.1km² and occupies a thin, approximately 5-kilometre wide strip of coast, between Causeway Head and Benbane Head on the north coast of County Antrim in Northern Ireland (Figure 11.1). The main Giant's Causeway feature – the Grand Causeway (Figure 11.2A) – is owned and managed by the National Trust (a charitable NGO) which first acquired it in 1961. The name 'Giant's Causeway' is connected to a local Irish myth that it was built by a giant named Finn MacCool.

It is globally significant for its scientific geoheritage because of its Tertiary lava sequences and their associated fossil soils. It is also significant as a cultural geoheritage location, because it has been a notable tourist destination for domestic and European tourists since at least the 17th century, with numerous literary and artistic accounts of the area. For tourists, it is the spectacular basalt columns of the Giant's Causeway and its associated mythology that are the main attractions. International acknowledgment of the geological importance of the Giant's Causeway came in 1986 with its inscription on the UNESCO World Heritage List as a natural site under the World Heritage Convention, and it is the only such designated locality in Ireland. Two cultural

FIGURE 11.1 LOCATION MAP OF THE GIANT'S CAUSEWAY AND CAUSEWAY COAST WORLD HERITAGE SITE

FIGURE 11.2 GIANT'S CAUSEWAY – TWO KEY SITES AND A TABLE SHOWING INFORMATION IMPROVEMENTS

A. VIEW OF THE SEA-LEVEL PROMONTORY OF THE GIANT'S CAUSEWAY (GRAND CAUSEWAY) FROM THE UPPER CLIFF PATH VIEWPOINT (KNOWN AS AIRD SNOUT). CLIFF HEIGHT IS APPROXIMATELY 100M ABOVE SEA LEVEL. FROM THIS VIEWPOINT, IT IS POSSIBLE TO SEE THE EFFECTS OF MARINE EROSION ON THE COLUMNS – IN PARTICULAR, THROUGH THE CREATION OF THE ERODED PLATFORM OF THE GRAND CAUSEWAY.

B. ROTATING HEXAGONAL COLUMNAR INTERPRETATION PANEL (AS SEEN IN MARCH 2009), LOCATED NEXT TO THE GRAND CAUSEWAY (APPROXIMATELY 100CM IN LENGTH). NOTE THAT THE FORM OF THE PANEL MIMICS THE NATURAL POLYGONAL BASALT COLUMNS AND THAT EACH FACE DISPLAYS DIFFERENT INFORMATION ABOUT THE GIANT'S CAUSEWAY AND CAUSEWAY STONES.

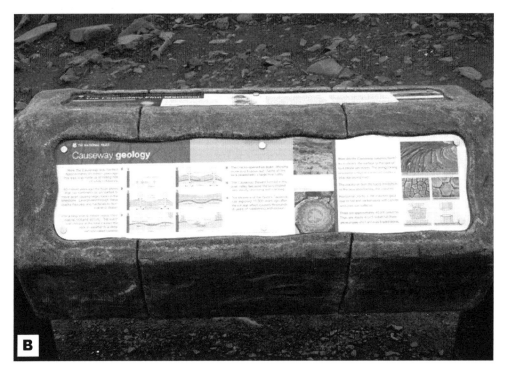

sites in the Republic of Ireland (Brú na Bóinne and Skellig Michael) are also designated as World Heritage Sites, and together with the Giant's Causeway are the only UNESCO World Heritage Sites on the island of Ireland.

The Giant's Causeway is actively promoted and marketed as a top tourist destination in Northern Ireland. As a result, it has become a 'must-see' attraction for anyone visiting the country, despite the scathing comment by the 18th century lexicographer Samuel Johnson: 'Worth seeing? Yes; but not worth going to see' (Boswell 1998, 1038). Consequently, it is a popular tourist attraction and was Northern Ireland's number one such attraction in 2014, with 788,000 people visiting the site (Department of Enterprise, Trade and Investment 2015). Such a high number of visitors brings with it the challenges of accommodating them (and their vehicles) and meeting their physical and interpretive needs, whilst also managing their impact on the site; these are examined in the site's original World Heritage Site management plan (Environment and Heritage Service, National Trust and Moyle District Council 2005), which specifically notes that: 'The most significant single issue facing the Site is the effective management of large numbers of visitors. The accommodation and management of visitors needs to be achieved in a manner that delivers a high quality visitor experience without compromising the conservation values of the Site or contravening any statutory designations that apply to the Site' (Environment and Heritage Service, National Trust and Moyle District Council 2005, 3); these site management issues are not discussed in any detail here.

The large number of visitors each year, and its popularity as an attraction, therefore present the Giant's Causeway's managers with significant challenges in terms of visitor management – including, for example, visitor movements, overcrowding, zoning and accessing interpretation (see Shackley 1999). Satisfactorily managing the visitor experience (as highlighted by Cochrane and Tapper 2006), for which dedicated interpretative and visitor management provision is essential (see Bromley 1994; Goulding 1999), is also a challenge. However, until July 2012, when a new visitor centre opened, the visitor experience had been based around temporary on-site visitor facilities after a fire all but destroyed the original visitor centre in 2000. Visitor facilities in this interim period included a visitor information centre, a shop selling literature on the site, and a series of on-site interpretative panels. It was acknowledged in the *Giant's Causeway and Causeway Coast World Heritage Site Management Plan* (2005–11) and by Gilmore *et al* (2007) that the temporary facilities were inadequate for a World Heritage Site – a situation that could have potentially threatened its designation.

THE 2007 GIANT'S CAUSEWAY VISITOR SURVEY

The aim of this survey was to provide some evidence of visitors' opinions about the on-site interpretation, and what they believed could be added to improve their understanding and experience of the site. Tourists' usage of the interpretation was assessed in October 2007 in a small-scale face-to-face questionnaire visitor survey. Visitors were approached and invited to participate in the survey. It was undertaken outside the peak summer tourist season and did not aim to be representative of the full profile of visitors to the Giant's Causeway, but to provide an insight into visitor expectations and experiences. Tourists' understanding of the site's diverse geoheritage and geoconservation value was especially examined. Their opinions of the on-site interpretation were also sought, to gauge whether it aids visitor understanding and enhances their experience of the site.

A total of 150 visitors were interviewed, of whom 91 were male and 59 female. All interviewees were over 18 years old, with almost three-quarters (72%) aged 18 to 35 years old. In addition to age, other socio-demographic details were collected. In terms of educational qualifications, almost all had secondary or tertiary level educational qualifications. Of these, just under two-thirds (c. 59%) were university degree level educated. It is interesting to note here that the educational profile of the visitors surveyed closely resembles the common adult geotourist characteristic identified by Hose (1996; 1998; 1999) in that, at specifically geological attractions, they are above the national average for educational attainment. Almost three-quarters (71%) of the interviewees were in either full- or part-time employment, and just over a fifth (22%) belonged to an environmental or other similar organisation, such as the National Trust. The great majority, at just over four-fifths (82.7%), were visitors from either the United Kingdom or the Republic of Ireland.

Visitors' general opinions on the Giant's Causeway on-site interpretation

The survey was designed to gauge visitor opinions on the quality of the on-site interpretation (provided by interpretative/informational panels at selected locations around the site, plus some free leaflets providing general site information) and whether it met their expectations; it was not designed to comment on the specific content of the interpretation or on the management of the site. Firstly, the survey asked whether the interviewees had noticed any interpretation panels (an example can be seen in Figure 11.2B) during their visit to the site; whether they had gained any new information from the interpretation; and what information they had gained from the interpretation. There was an almost equal split between interviewees noticing and not noticing the interpretation panels. There was a clear relationship between interviewees' ages and the likelihood of noticing the panels; 18 to 25 year olds were less likely to notice the panels, whereas interviewees aged 56 years and over were more likely to have seen them.

With regard to gaining any new information from the interpretation, almost three-fifths (58.7%) of interviewees stated that they did not feel they had learnt anything new, and of these half were aged 18 to 35 years old. Again, there was a clear age relationship, with most 18 to 25 year olds stating that they had not learnt anything new, whereas interviewees aged 56 and over mostly stated that they had learnt something new from the information provided on the site. The interviewees indicated that the main information gained related to the geology/rock features and the formation of the Giant's Causeway (38%). Survey responses support the findings of Light (1995), who indicated that there is a tendency for older visitors to make greater use of interpretation to improve their understanding, and for those aged below 30 years old to engage less with interpretation.

Secondly, interviewees were asked to comment on the quality of the interpretation, and whether it met their needs and expectations. Presentation was a key consideration in the survey. Only two-fifths (38%) of interviewees agreed that the interpretation was well presented. Educational level had a strong relationship with how well the interviewees felt the material was presented; lower levels of agreement were associated with interviewees who had tertiary level educational qualifications (degree level and above). In terms of meeting needs and expectations, interviewees were asked to consider whether the interpretation could be better presented and what they would like to see included that would improve their experience of the site. Just over a third of the interviewees agreed that the interpretation could be better presented. The remaining respondents disagreed or did not know if it could be improved. Further analysis determined that age and education were significant with regard to agreement. Interviewees aged 36 to 45 years old, and those with tertiary level qualifications, had higher levels of agreement. When asked if there was anything they had wanted to know that was not on the information boards, around a quarter (26%) said 'yes' and three-quarters (74%) 'no' or 'don't know'. Those most in agreement were interviewees with higher degree qualifications. Light (1995) had similar findings and stated that there is a tendency for better educated visitors to seek out information during their visit. Particular aspects highlighted by the interviewees included more information on the geology and formation of the site (including an explanation of red fossil soil horizons between the lava layers), the mythology/legend of the site, and more general information on the surrounding wildlife and coastal features.

A key focus of the survey was to determine whether the visitors' informational needs were being met by the on-site interpretation. In particular, it was important to know whether they had received sufficient information on how the Giant's Causeway was formed, and whether this information was explained in a way they could understand. Such interpretation of geological features is termed geointerpretation, defined as: 'The art or science of determining and then communicating the meaning or significance of a geological or geomorphological phenomenon, event, or location' (Hose 2012, 17). On the presentation of the geological information relating to the formation of the Giant's Causeway, almost three-fifths (58.7%) per cent of interviewees agreed that it had been explained in a way they could understand, but just over two-fifths (41.3%) felt that it had not. Responses by those not in agreement included: 'it wasn't clear enough', 'it was too technical' and 'needs more general information'. This is an interesting finding, since Hose (1997) indicates that visitors to geosites access geology interpretive panels least when they are in competition with other subject matter. In addition, Hose (2012, 17) highlights that: 'Commonly, it is the human interest of the science of a geosite or geomorphosite, rather than the pure science itself, that most engages people's attention. Hence, geohistory can underpin efficacious geointerpretation.' The survey findings also reiterate the observation by Poria (2010) and Poria *et al* (2006a) that the nature of the interpretation plays a key role in the tourist experience of a site. This may also help to explain the significance of belonging to an environmental or similar organisation, with regard to interviewees feeling that they had not received enough information. The previous experiences of interviewees at other heritage sites, such as those owned by an organisation such as the National Trust, may lead to higher levels of expectation.

With regard to the level of information received from the interpretation material, just over a third agreed that they had received sufficient information, but the remaining two-thirds felt that they had not. Most dissatisfied comments related to a general lack of availability of information, both on-site and in the free leaflets, and the poor provision of detailed information was also a common issue cited by interviewees. The availability of detailed information on the geological features of the site as a whole, and on the geological significance of the site as a UNESCO World Heritage Site, was seen as something that needed to be remedied.

The last part of the survey aimed to determine from the interviewees what they would like to see on the site in order to improve their learning experience. In addition to verbal responses, they were asked to choose up to three options from a list of examples of ways that could improve their experience. The top four choices were: the provision of rangers/people to explain the features on the site; provision of guided walks; more information boards; and the provision of an audio-guide about the site. The percentage responses of the interviewees (Table 11.1) would seem to match the findings of Hose (1997). That is, that guided walks and accompanied field excursions are very popular amongst geotourists – especially casual ones – as ways of enhancing their experiences of geosites.

Information Source	%*
Rangers/Person to explain	23.3
Guided walks	15.3
Audio guide	14.0
More information boards	14.0
Free detailed leaflet	8.7
Detailed booklet to buy	8.7
Detailed map	6.0
Detailed leaflet to buy	2.0
Other	8.0

* % calculated on multiple answers cited by interviewees (n=150)

Visitors' opinions on the outdoor on-site panels

The results of the survey revealed a relatively high level of dissatisfaction amongst interviewees with regard to the on-site interpretation/information provided on the outdoor on-site panels. Interviewees expressed difficulties in finding information appropriate to their needs; simply noticing the presence of interpretation panels at key locations on the site was identified as an issue. With regard to the quality of the interpretation, the survey highlighted that interviewees had key concerns about the presentation of the information, and about it being at the right level to give them sufficient understanding of the site's features. These are significant concerns, which need to be addressed in order to improve the visitor experience. One solution to the problem of finding information might be for site managers to consider the location of the interpretation panels. Key considerations should relate to the most appropriate locations to maximise their impact, and to ensure that they can be seen or found when there are large numbers of visitors. Provision of information relating to the location of the panels, such as on a site map or on a leaflet, might improve visitor engagement with the site features and could contribute to improving the visitor experience of the site.

The quality of the interpretation panels, in terms of meeting the needs and expectations of the visitors, was highlighted through the survey. The format (a rotating hexagonal columnar drum, at the time of the survey – see Figure 11.2B), size, content and presentation were all commented on by the interviewees. The interpretation panels provided for their basic needs, by giving general information to aid a broad understanding of the site features and how the Giant's Causeway was formed, but interviewees indicated that there was room for improvement. It appeared that there was a clear need for the information to be beneficial for all levels of understanding and interest, through the provision of information for different age groups or knowledge levels (for example, leaflets for children or for those with a specialist geological interest). This is in agreement with the findings of Tilden (1977) and Pralong (2006), who stated that different target groups, such as adults and children, should be considered as specific markets for interpretation materials. Hose (1997) adds a note of caution with regard to this, since there is a key misconception that providing for schools and families meets the needs of all visitors. If such targeted materials were in place at the Giant's Causeway, then this may help to address some of the age and education level differences found in the survey responses. In addition, the lack of information in languages other than English was highlighted by international visitors. The interviewees were generally disappointed by the presentation of the on-site interpretation panels; many had higher expectations of the quality of the information. More discerning visitors expected the use of modern technology to add to their experience. Examples stated included the provision of audio guides, downloadable podcasts (mp3 downloads) or SMS (text message) information.

In order for the previous comments to be addressed, thereby improving the visitor understanding and experience of the site, an overhaul of the site interpretation is clearly required. The interviewee responses indicate that the interpretation should contain more specific information on the Giant's Causeway itself, and on the significance of the site as a World Heritage Site. This would seem to be a vital development, since Pralong (2006) highlights that the demand for explanatory commentaries is important for natural sites and landscapes with earth science features of interest.

The interpretation should be designed not only to meet the needs and expectations of the visitors, but also to accommodate the nature of the site. The suggestion by Hose (2005a; 2006) of adopting an appropriate site-specific interpretative strategy should help to meet geotourist requirements. Hose (2005b) states that: 'Such a strategy depends upon identifying and promoting its physical basis, knowing and understanding its users and developing effective interpretation materials' (2005b, 28). In undertaking this approach, both the tourism and (geo)conservation requirements of the site could be accommodated.

In terms of developing interpretive plans and signage for geotourism attractions, examples of good practice can be found in Hughes and Ballantyne (2010). If the interpretation contains an appropriate level of geological information, then the visitors should better understand the nature of the site and thus improve their experience. They should also be better able to understand and recall the information relating to the geosite, which Hose (1996; 1997; 2006) states is a key issue for visitors to geosites.

Improving the Giant's Causeway visitor experience

The survey highlighted that a key requirement would be to provide more human interaction. Provision of on-site rangers or guides was considered by interviewees as a good way of improving their experience. Having rangers available to answer questions, give guidance on what to see and visit on the site, or to give guided walks would greatly improve visitor understanding and experience of the site. They are also much appreciated by geotourists (Hose 1997) and improve the learning experience (Hose 2006). The ranger services could be linked to other forms of interpretation, to provide an integrated approach to enhancing the visitor experience.

It was clear from the survey that a significant improvement to the visitor experience would be the provision of more information, particularly within a visitor centre context or at a 'gateway' to the site. This has previously been highlighted as an issue by Smith (2005), Wilson and Boyle (2006) and Gilmore et al (2007). This information would need to accommodate all levels of understanding and interest, as well as being available in multiple languages, so visitors can fully enjoy and understand the Giant's Causeway. This is especially important given its status as a World Heritage Site. The lack of cohesion in the visitor facilities and the possible lack of cooperation between operators and stakeholders (National Trust and Moyle District Council) may have affected the quality of the visitor experience at the time of the survey. The temporary visitor facilities were disjointed and, on arrival, visitors did not really know where to go, what to do, or what to find on the site. This lack of coordination may also have affected the quality of the on-site interpretation and its effectiveness in meeting many of the visitor management and experience objectives, detailed in the 2005 to 2011 Site Management Plan.

With the development of the new visitor centre (opened in July 2012), sufficient information should now be available for visitors to fully engage with, and understand, the site features. With the visitor centre designed to give a world-class visitor experience (state-of-the-art facilities and interpretation), the visitor concerns and issues with the interpretation raised during the survey should now be addressed. This should also have an impact on visitor management, since the site should be better controlled and coordinated. Again this is pertinent to the site, given its designation as a natural World Heritage Site, whereby the conservation and management of the geoheritage features are so internationally important.

Conclusions

The visitor survey revealed that the on-site interpretation at the Giant's Causeway (up to July 2012) did not fully meet the needs and expectations of its visitors. Clear improvements needed to be made to both the interpretation and infrastructure, in order to improve the visitor experience of the Giant's Causeway and Causeway Coast World Heritage Site. Based on the survey findings, it is clear that, when designing on-site interpretation or visitor facilities, it would be beneficial to determine visitors' opinions on what they would like to see included, so that they are provided with the best on-site experience to meet their expectations. Healy and McDonagh (2009) see it as essential that visitor infrastructural planning should take on board visitor views prior to any development, to ensure visitor satisfaction with the development. In addition, there is a need to have the budget for post-installation improvements, following visitor evaluations. Finally, the survey findings and consequent suggested improvements have clear implications for the broader provision of interpretative materials at similar coastal geoheritage sites across the UK and Europe.

References

Badman, T, 1994 Interpreting earth science sites for the public, in *Geological and Landscape Conservation* (eds D O'Halloran, C Green, M Harley, M Stanley and S Knill), Geological Society, London, 429–32

Ballantyne, R, 1998, Interpreting 'visions': addressing environmental education goals through interpretation, in *Contemporary Issues in Heritage and Environmental Interpretation* (eds D Uzzell and R Ballantyne), The Stationery Office, London, 77–97

Boswell, J, 1998 *Life of Johnson* (Oxford World's Classics), Oxford University Press, Oxford

Bromley, P, 1994 *Countryside Interpretation: A Handbook for Managers*, E & F N Spon, London

Burek, C V, and Davies, H, 1994 Communication of earth science to the public – how successful has it been? in *Geological and Landscape Conservation* (eds D O'Halloran, C Green, M Harley, M Stanley and S Knill), Geological Society, London, 483–6

Cochrane, J, and Tapper, R, 2006 Tourism's contribution to World Heritage management, in *Managing World Heritage Sites* (eds A Leask and F Fyall), Butterworth-Heinemann, Oxford, 97–109

Cohen, E, 1995 Contemporary tourism – trends and challenges: sustainable authenticity or contrived post-modernity, in *Change in Tourism: People, Places, Processes* (eds R Butler and D Pearce), Routledge, London, 12–29

Department of Enterprise, Trade and Investment, 2015 *Northern Ireland Visitor Attraction Survey 2014*, Department of Enterprise, Trade and Investment, Northern Ireland, Belfast

Dias, G, and Brilha, J, 2004 Raising public awareness of geological heritage: a set of initiatives, in *Natural and Cultural Landscapes – The Geological Foundation* (ed M A Parkes), Proceedings of the Conference, 9–11 September 2002, Dublin Castle, Ireland, Royal Irish Academy, Dublin, 235–8

Environment and Heritage Service, National Trust and Moyle District Council, 2005 *Giant's Causeway and Causeway Coast World Heritage Site Management Plan (February 2005)*, Environment and Heritage Service, Belfast

Field, D R, and Wagar, J A, 1992 Visitor groups and interpretation in parks and other outdoor leisure settings (1973), in *On Interpretation: Sociology for Interpreters of Natural and Cultural History (Revised Edition)* (eds G E Machlis and D R Field), Oregon State University Press, Corvallis, 11–23

Fladmark, J M (ed), 1994 *Cultural Tourism*, Donhead, London

Gilmore, A, Carson, D, and Ascenção, M, 2007 Sustainable tourism marketing at a World Heritage Site, *Journal of Strategic Marketing* 15, 253–64

Goulding, C, 1999 Interpretation and presentation, in *Heritage Visitor Attractions: An Operations Management Perspective* (eds A Leask and I Yeoman), Continuum, London, 54–68

Hall, C M, and Weiler, B, 1992 What's special about special interest tourism? in *Special Interest Tourism* (eds B Weiler and C M Hall), Belhaven, London, 1–14

Ham, S H, 1992 *Environmental Interpretation: A Practical Guide for People with Big Ideas and Small Budgets*, North American Press, Golden

Harrison, L C, and Husbands, W (eds), 1996 *Practicing Responsible Tourism: International Case Studies in Tourism Policy, Planning, and Development*, Wiley, New York

Healy, N, and McDonagh, J, 2009 Commodification and conflict: what can the Irish approach to protected area management tell us? *Society and Natural Resources* 22, 381–91

Hose, T A, 1994a Telling the story of stone – assessing the client base, in *Geological and Landscape Conservation* (eds D O'Halloran, C Green, M Harley, M Stanley and S Knill), Geological Society, London, 451–7

— 1994b *Interpreting Geology at Hunstanton Cliffs SSSI Norfolk: A Summative Evaluation (1994)*, The Buckinghamshire College, High Wycombe

— 1996 Geotourism, or can tourists become casual rock hounds? in *Geology on Your Doorstep: The Role of Urban Geology in Earth Heritage Conservation* (eds M R Bennett, P Doyle, J G Larwood and C D Prosser), The Geological Society, London, 207–28

— 1997 Geotourism – selling the earth to Europe, in *Engineering Geology and the Environment* (eds P G Marinos, G C Koukis, G C Tsiamaos and G C Stournass), A A Balkema, Rotterdam, 2955–60

— 1998 Selling coastal geology to visitors, in *Coastal Defence and Earth Science Conservation* (ed J Hopke), The Geological Society, London, 178–95

— 1999 How was it for you? Matching geologic site media to audiences, in *Proceedings of the First UK RIGS Conference* (ed O Oliver), Worcester University College, Worcester, 117–44

— 2005a Geotourism – appreciating the deep time of landscapes, in *Niche Tourism* (ed M Novelli), Elsevier Butterworth-Heinemann, Oxford, 27–37

— 2005b Writ in stone: a critique of geoconservation panels and publications in Wales and the Welsh Borders, in *Stone in Wales* (ed M R Coulson), Cadw, Cardiff, 54–60

— 2006 Geotourism and interpretation, in *Geotourism* (eds R Dowling and D Newsome), Elsevier Butterworth-Heinemann, Oxford, 221–41

— 2012 3G's for modern geotourism, *Geoheritage* 4, 7–24

Hughes, K, and Ballantyne, R, 2010 Interpretation rocks! Designing signs for geotourism sites, in *Geotourism: The Tourism of Geology and Landscape* (eds D Newsome and R K Dowling), Goodfellow, Oxford, 184–99

Light, D, 1995 Visitors' use of interpretive media at heritage sites, *Leisure Studies* 14, 132–49

Machlis, G E, and Field, D R, 1992 *On Interpretation: Sociology for Interpreters of Natural and Cultural History (Revised Edition)*, Oregon State University Press, Corvallis

Novelli, M (ed), 2005 *Niche Tourism*, Elsevier Butterworth-Heinemann, Oxford

Patterson, D, and Bitgood, S, 1986 *The Red Mountain Road Cut: An Evaluation of Visitor Behaviour – Technical Report No 86–75*, Center for Social Design, Jacksonville

Pocock, D, 1992 Catherine Cookson country: tourist expectation and experience, *Geography* 77, 236–43

Poria, Y, 2010 The story behind the picture: preferences for the visual display at heritage sites, in *Culture, Heritage and Representation: Perspectives on Visuality and the Past* (eds E Waterton and S Watson), Ashgate, Aldershot, 217–28

Poria, Y, Biran, A, and Reichel, A, 2009 Visitors' preferences for interpretation at heritage sites, *Journal of Travel Research* 48, 92–105

Poria, Y, Reichel, A, and Biran, A, 2006a Heritage management: motivations and expectations, *Annals of Tourism Research* 33, 162–78

— 2006b Heritage site perceptions and motivations to visit, *Journal of Travel Research* 44, 318–26

Poria, Y, Reichel, A, and Cohen, R, 2013 Tourist perceptions of World Heritage Site and its designation, *Tourism Management* 35, 272–4

Pralong, J-P, 2006 Geotourism: a new form of tourism utilising natural landscapes and based on imagination and emotion, *Tourism Review* 61, 20–5

Ryan, C, 2002 *The Tourist Experience* (2 edn), Continuum, London

Shackley, M, 1998 *Visitor Management: Case Studies from World Heritage Sites*, Butterworth Heinemann, Oxford

— 1999 Visitor management, in *Heritage Visitor Attractions: An Operations Management Perspective* (eds A Leask and I Yeoman), Continuum, London, 69–82

Sharpley, R, and Stone, P, 2010 *Tourist Experience: Contemporary Perspectives*, Routledge, London

Smith, B J, 2005 Management challenges at a complex geosite: the Giant's Causeway World Heritage Site, Northern Ireland, *Géomorphologie: Relief, Processes, Environment* 3, 219–26

Tilden, F, 1977 *Interpreting our Heritage* (3 edn), University of North Carolina Press, Chapel Hill

United Nations World Tourism Organisation, 1997 What tourism managers need to know, *A Practical Guide for the Development and Application of Indicators of Sustainable Tourism*, United Nations World Tourism Organisation, Madrid

Uriely, N, 2005 The tourist experience: conceptual developments, *Annals of Tourism Research* 32, 199–216

Uzzell, D L (ed), 1989a *Heritage Interpretation Volume 1: The Natural and Built Environment*, Belhaven Press, London

— (ed), 1989b *Heritage Interpretation Volume 2: The Visitor Experience*, Belhaven Press, London

— (ed), 1993 Contrasting psychological perspectives on exhibition evaluation, in *Museum Visitor Studies in the 90s* (eds S Bicknell and G Farmelo), Science Museum, London

Uzzell, D, and Ballantyne, R (eds), 1998 *Contemporary Issues in Heritage and Environmental Interpretation*, The Stationery Office, London

Wilson, L-A, and Boyle, E, 2006 Interorganisational collaboration at UK World Heritage sites, *Leadership & Organization Development Journal* 27, 501–23

12

A Geoheritage Case Study: GeoMôn in Wales

John Conway and Margaret Wood

Anglesey (Môn in the Welsh language) is the largest of the Welsh islands, covering some 720km² and with approximately 200km of coastline. A measure of its rich geoheritage is that its rocks span four geological eras and 12 geological systems – 1,800 million years of Earth history have fashioned more than 100 rock types. The island was also formerly the world's richest source of copper; an extensive – and somewhat derelict – mining and industrial archaeology heritage is variously regarded by its residents and visitors alike as either a blight on the landscape, a hope for future mining activity, or a tourism opportunity. The coastline provides easy access to the majority of the island's geology, with rocky shores exposing older rocks and wide sandy beaches and muddy lagoons providing modern sedimentary environments. Anglesey became a geopark in 2009, calling itself the 'GeoMôn – Anglesey Geopark', to include the Welsh name for the island in accordance with both national and European Geopark Network (EGN) policies on bilingualism.

Anglesey has long been the focus of geological interest; it is speculated – although not documented – that Charles Darwin, prior to embarking on *HMS Beagle*, visited the island on his June 1842 tour of North Wales with Adam Sedgwick, since he had relatives living on the island and there is a day's gap in his diary. His tutor at Cambridge, John Henslow, had previously published the first map and account of the geology of Anglesey in 1822. Mapped by Ramsay in the 1840s, and most famously in tremendous detail by Edward Greenly in 1919, Anglesey is possibly one of the most geologically mapped areas in Britain; research has continued up to the present day. GeoMôn staff have accompanied and assisted various groups carrying out research and mapping; most recently, this was with the Universities of Leicester and Tokyo. Given Anglesey's interesting and well-exposed geology, coupled with its popularity as a holiday destination, there are surprisingly few modern field guides covering the island's geology (Bates and Davies 1981; Treagus 2008; Conway 2010a); however, the publication by the Geologists' Association (Bates and Davies 1981) has gone through at least four reprints, indicating that the demand for a technical publication is remarkably strong.

History of geoconservation

Geoconservation in Anglesey began in organised form when the Gwynedd and Môn Regionally Important Geological Sites (G&M RIGS) group was established in 1994 under the leadership of Dr Margaret Wood, area geologist for the Countryside Council for Wales (CCW), and Dr Stewart Campbell, Head of Earth Sciences at CCW. This group was one of the founding members of the Association of Welsh RIGS Groups (AWRG), which met annually and sponsored research projects as well as the identification and recording of regionally important geological sites under the RIGS banner. Wales was the first to set up a national RIGS movement and to hold national forum meetings on an annual basis, starting in Brecon in 1994; G&M RIGS, together with North East Wales RIGS (NEWRIGS) group, led the way for many years in Wales. The history of RIGS in Wales has been published by Burek (2008).

The Anglesey Geodiversity Partnership was set up in 2005 with the twin aims of delivering the Anglesey Local Geodiversity Action Plan (LGAP) and achieving geopark status. Its core members were the G&M RIGS group, CCW and Isle of Anglesey County Council. Letters of support were received from key politicians, including the Welsh Assembly Government First Minister, Rhodri Morgan; the Minister for Environment, Planning and Countryside, Carwyn Jones; the Welsh Assembly Member for Anglesey, Ieuan Wyn Jones; and the Westminster MP Albert Owen. Other organisations also indicated their support, including the British Geological Survey; the British Nuclear Group (which then ran Wylfa Power station and its visitor centre); and the University of Wales, Bangor – all detailed in the unpublished (Wood 2007) 'Application Dossier for Nomination to the European Geopark Network'.

The LGAP, in common with most such plans, set out to review all the information available about Anglesey's geodiversity and then to perform a full audit of all potential sites. The role of LGAPs in Wales is explained by Burek (2012). The Anglesey LGAP was supported between 2004 and 2007 by a detailed project funded by the Aggregates Levy Sustainability Fund (ALSF) to develop a methodology for site recording and evaluation, leading to the mapping of over 200 Gwynedd and Môn sites (Wood *et al* 2007a; 2007b); this acted as a 'pilot' project and led to further grant applications to extend the survey to NEWRIGS in 2007 and subsequently to the whole of Wales. Designated RIGS sites were reported to the Island Council and feature in their planning processes. This was a first step to raising awareness amongst the local community at all levels, involving local and national politicians.

Applying for geopark status involved documenting the geology and the landscape, as well as all features of 'geoheritage' – a term that can include any use of the geodiversity. In Anglesey's case this included the soil diversity, building on an earlier RIGS project to provide educational soil science for school fieldwork (Conway 2006). The resulting book covered North Wales, and provided the impetus to develop a soil trail around the Anglesey coastline. Geopark status is about more than just having outstanding geology; the geodiversity partnership had to show the development of 'green' tourism and sustainable community development advantageously using the geoheritage.

ANGLESEY'S GEOHERITAGE

Anglesey has an amazing diversity for such a small area of well exposed and easily accessible geological formations (Figure 12.1), and is perfect for a geopark and a centre for geological exploration, education and enjoyment. A large area of the island consists of exposed Precambrian rocks – or, at least, rocks that have been considered to be of Precambrian age, although accurate dating was impossible until very recently. This group, known as the Mona Complex (Mona is the ancient name for Anglesey), comprises green and blue schists, gneisses, pillow lavas and associated fragmentary rocks, the Coedana granite and gneisses, as well as unspecified meta-sediments and coarse quartzites; it probably underlies most, if not all, of the island. Boundaries with adjacent rocks are in most places obscured, and field relationships have to be inferred, permitting a

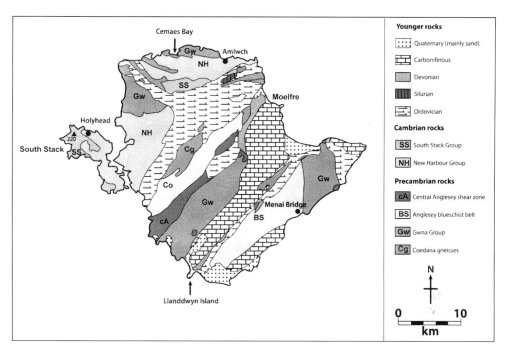

FIGURE 12.1 MAP SHOWING THE GEOLOGY OF ANGLESEY

range of interpretations which only adds to the interest. This is the largest area of Precambrian rocks in England and Wales.

Greenly (1919) published a two-volume description of the rock types, obvious structures and visible field relations, but his interpretation of age and sequence has been queried many times since. With the theory of plate tectonics applied to the island in recent years by geologists from the Universities of Leicester and Tokyo, Greenly's original succession in the Mona Complex has been found to be more accurate than that of the researchers in the 20th century. New dates of many rocks have validated many of Greenly's ideas, proving that detailed field research is still the best way to discover the geology of the island.

The occurrence of fossils in the limestone blocks of the Gwna Mélange (explained below) came to light when Margaret Wood describêd stromatolites (algal mats) from the north coast (Wood and Nicholls 1973); these have since been dated to around 860mya and are the oldest fossils in England or Wales.

The most recent work by a team of Japanese scientists has involved multidisciplinary studies. Japan lies on a subduction zone, and this team has spent many years studying such features in their own country. Some valuable new data and interesting interpretations are emerging. Perhaps a key conclusion is that, for all its tectonic dismemberment, Anglesey is in fact a classic example of a destructive plate margin sequence with marine sediments, blueschist (metamorphosed sediments typical of deep burial) mélange and an ophiolite (ocean crust) sequence. The Japanese team's work on ocean plate stratigraphy is best observed on Llanddwyn Island, where the green pillow lavas, red cherts and grey limestone provide spectacular exposures.

Anglesey's geology described

Precambrian

The Precambrian Gwna Group consists mainly of greenschists, but it features an extraordinary and spectacular mélange. Greenly coined the term 'mélange' to describe a massively fragmented rock, a product of a major undersea debris-slide of catastrophic proportions, accreted onto the underside of the overlying continental plate at a destructive plate margin; within the green silty matrix there are blocks, ranging from pebble-size up to rafts over a kilometre across, including stromatolitic limestone, basaltic pillow-lavas, bedded cherts, red mudstones and white quartzite. Wood (2004) studied the Precambrian of the Rhoscolyn Anticline. She has recently explored the origin and development of the term 'mélange' and explained the various outcrops of mélange within the Precambrian rocks of Anglesey (Wood 2012). The island is now placed at number seven in the top 100 significant exposures of the world by the *Japanese Journal of Geography*.

The Coedana Granite is a muscovite-granite, intruded into high-grade acid and basic gneisses. Mainly found in the interior of the island, there are some exposures on the coast south of Rhosneigr. Both appear to be related to ancient island-arc magmatism. This particular rock has been the subject of intense speculation, with interpretations ranging from an intrusion to the most highly metamorphosed member of a sequence of gneisses.

Gabbros and serpentinites (basic to ultrabasic intrusive rocks) are highly altered in places (with conversion of olivine to serpentine). They occur only in the New Harbour group of rocks and have not been dated. Locally, they have been quarried for ornamental serpentine and, at one locality, for chromite.

Blueschists form a linear belt in south-east Anglesey; these rocks include metabasic rocks containing a metamorphic mineral assemblage that shows that they have been subjected to unusually high pressures, such as those found deep within a subduction zone. They were originally the pillow lavas, extruded like toothpaste from a tube onto the deep ocean floor at a constructive plate margin, and when pushed sideways by continuing lava eruptions they make up the entire ocean crust.

Cambrian

The Cambrian period is represented by the South Stack and New Harbour groups of rocks (see Figure 12.2A). The South Stack group consists of alternating coarse and fine psammitic (sandy) layers – clastic sedimentary rocks, interpreted as proximal turbidites (marine sediments close to the shore) deposited by turbulent underwater currents laden with debris in a marine sedimentary basin. The New Harbour group consists of clay-rich beds alternating with thin fine sandy layers, interpreted as distal turbidites (marine sediments away from the shore and so finer grained); the sediment is chiefly the reworked product of island-arc volcanoes. The group also includes some bedded cherts, tuffs and basalts, metamorphosed into mafic greenschists. These rocks usually show spectacular, intense folding. They were formed at a destructive plate margin. Once thought to be some of the oldest rocks on the island, they have proved to be of Cambrian age and were emplaced below the older Gwna group by thrust faults.

Ordovician

Ordovician rocks are represented by coarse conglomerates, sandstones and shales; some of the latter have been metamorphosed to slates and they have been quarried along the coast near

Figure 12.2 Two iconic GeoMôn geosites

A. South Stack and its lighthouse; highly deformed Cambrian (South Stack Group) turbidites form this small island lying off Anglesey.

B. Porth Wen brickworks; this now derelict industrial site formerly worked decomposed Ordovician shales as a clay source and ground Precambrian quartzites as a silica source.

to Amlwch. Silurian rocks are very scarce; just a few graptolite-rich shales are found in the core of the syncline on Parys Mountain. They are best observed on the northern coast from Llanbadrig to Porth Wen, where they lie unconformably on Precambrian and Cambrian rocks (see Figure 12.2B).

Devonian

Devonian rocks are represented by a complex folded and faulted sequence of reddish mudstones, sandstones and limestones. These are generally interpreted as floodplain and temporary lake deposits of an intra-desert river valley (Allen 1965); their floodplain nature is attested by palaeosols (ancient fossil soils), possibly some of the oldest, since they are almost contemporary with the main emergence of plants onto land, a precursor for soil development.

Carboniferous

Carboniferous rocks are found on the eastern side of the island, having a major impact on the landscape with extensive, near-horizontal bedded, limestones forming quarried and natural cliffs and a distinctive offshore island. These rocks are perfect for demonstrating structural features such as bedding, dip and strike, and small-scale faulting, as well as various carbonate lithologies, a wide range of biological features and fossils. Perhaps the most intriguing features are the karst surfaces produced by weathering during the Carboniferous period, representing unstable land-sea levels, with the rocks being exposed above sea-level and subject to solution weathering. Coarse sandstones and conglomerates are present at several horizons, whilst exposed along one shoreline are beds exhibiting various lithologies (fine cross-bedded sandstone to coarse conglomerate) and plant fossils – they are assumed to be Millstone Grit. Elsewhere, Coal Measures rocks – once quarried for coal – underlie a low-lying area, which partially explains the lack of interest in this potential but largely unutilised resource; remnants of 19th century mining can be found, along with sparse spoil heaps demonstrating the rock types, but the twin problems of drainage and poor quality meant that the coal was never seriously exploited.

Dykes and sills

Palaeozoic dykes and sills are very common and can be seen in most bays along the coast; they were intruded when southern England collided with Scotland during the Caledonian orogeny. Much younger, Tertiary (olivine) dolerite dykes are less common and extensive, but interestingly they lie parallel to the older dykes; they are testimony to the plate movements leading to the opening of the Atlantic Ocean. These easily accessible intrusions are excellent for demonstrating igneous processes and rock types, as well as for extrapolating to modern plate tectonic movements and global geology.

Mesozoic rocks

Mesozoic rocks are largely absent on Anglesey, although underwater surveys show that extensive Permo-Triassic rocks floor the surrounding seas and explain the dominant red colouration of the Quaternary glacial deposits on the eastern side of the island. Small exposures reminiscent of Triassic mudstone are known from the Menai Strait shoreline, but recently Permo-Triassic red sandstone and conglomerate have been rediscovered in a disused quarry adjacent to the magnificent and aptly named Plas Coch ('the red mansion').

Quaternary and Holocene

Quaternary glaciation has played a major role in the evolution of Anglesey's present-day land-scape. Many rock surfaces show the characteristic smoothed and rounded landscape of glaci-ated areas, although one may have to overlook the roughness of subsequent weathering and erosion. Glacial sediments drape the eastern and southern coastlines in particular. The Irish Sea Till (formerly known by the more descriptive term 'red northern drift') contains erratics from southern Scotland and northern England; these include the famous Ailsa Craig microgranite (still used at source in Scotland to manufacture curling stones) and Shap Granite, together with flint and basalt from Northern Ireland. Outwash gravels and glacio-fluvial sands are also present.

Soils

Soil formation is varied across such a complex and varied solid geology and sediment cover. First mapped by Evan Roberts (1958) and David Ball (1963), then reinterpreted by Rudeforth *et al* (1984), Anglesey contains examples of most soil types found in the UK. Conway (1984a; 1984b) drew attention to the importance of soils and their relationship to landscape evolution. A RIGS project funded by CCW, designed to provide sites for school parties to study soils in the field with minimum effort, was reported in 1995 to the RIGS forum and published online in 1999; it was later translated into Welsh and eventually published in hard-copy (Conway 2006). This unique project attracted considerable attention, including presentations at the UNESCO Global Geopark Network conference and a number of subsequent conferences, the latest for the British Soil Science Society (Conway 2012a; 2012b). Most recently (Conway 2014), GeoMôn has produced a soil trail leaflet directing people to a number of sites around the coastline where soil types can easily be seen on the constantly eroding or refreshed cliffs.

Economic geology: Parys Mountain

Parys Mountain was historically a major source of copper, and in the late 18th century it was Europe's biggest producer of copper. Its owner, Thomas Williams, claimed to control the world price through its association with Cornish mines (Rowlands 1981). In the early 1960s, the first of several major exploratory drilling programmes commenced at Parys Mountain; currently, Anglesey Mining Plc is investigating the potential for re-opening the mine. As exploration has proceeded, a much better geological picture of the stratigraphic sequence and the nature of the mineralisation has emerged; stratiform lenses of massive sulphide mineralisation, containing percentage levels of copper, lead and zinc, with noteworthy concentrations of silver and gold, occur both to the west and north of the old mine. Several million tonnes of ore have now been proved by drilling and exploration in these areas. The ore deposit is thought to have been formed on the late Ordovician or early Silurian sea-bed from heated solutions percolating up through the rocks and exiting into the water – similar to the 'black-smokers' seen on present-day mid-ocean ridges. The mineralisation is accompanied by intense silicification and pyritisation of the surrounding sedimentary and igneous rocks. Numerous uncommon compounds of anti-mony, arsenic, bismuth, gold, lead and silver are also present. Amongst these is the lead sulphate mineral anglesite, first recognised in 1783 as a mineral species by the English botanist William Withering (1741–99), who discovered it at Parys Mountain; its name (honouring the island) was conferred on it by the French mineralogist François Sulpice Beudant (1787–1850) in 1832. The archaeological, environmental and social impact of mining at Parys Mountain has recently been noted (Conway and Wood 2012a); likewise, the relevance of Parys Mountain to the work of the geopark has been noted (Conway and Wood 2012b).

PRESENTING THE GEOHERITAGE: GEOTOURISM

Developing from the Gwynedd and Môn RIGS group, which had already completed a survey of geoconservation sites, GeoMôn focused on geotourism, producing a variety of leaflets and geotrails and establishing a geopark visitor centre, originally in the Pritchard Jones Institute before settling in the Watch House in the historic harbour at Porth Amlwch in 2010. The harbour had developed from a small fishing centre to become one of the major 18th century industrial ports in Wales, servicing the Parys Mountain copper mine. GeoMôn's major activity has been to develop a network of geosites and geotrails. These were largely based on the recently designated Coastal Footpath, a local authority project to link together the various sections of public footpath (with a guaranteed legal right of way) with permissive paths, to provide a complete island coastal footpath, some 125 miles (200km) long; it is now a component of the all-Wales coastal footpath. The Coastal Footpath project publishes a bi-annual magazine, *Swn-y-Mor*, in collaboration with the Anglesey Area of Outstanding Natural Beauty (AONB), which carries a geotrail in every issue. This magazine, freely available in all tourist outlets and available to download from the local authority website, brings geology to the attention of all on the island, resident or visitor alike; 14 such trails have now been published in *Swn-y-Mor* (see Annex 1).

These trails are short walks to specific areas of geological interest. There are also information boards in some 'honey-pots' such as Cemaes Bay, South Stack and Malltraeth; the latter encompasses a picnic spot. The development of the whole coastal path spurred GeoMôn to provide a guide to the entire coastal geology (Conway 2010a; 2010b). This publication pre-dated any official guidebook by four years, but is not a guidebook as such; rather it is a narrative, to accompany the walk around the entire coastline, describing the rocks and structures, the landscape evolution and geoheritage features. In the last three years GeoMôn, financed by the Isle of Anglesey County Council, has produced and installed 13 new information boards around the coast. Geosites include the classic geological localities of Llanddwyn Island (Precambrian pillow lavas and mélange), Marquess of Anglesey's Column (Precambrian glaucophane blueschists), Cemaes Bay (Precambrian mélange and fossils, Palaeozoic dykes) and Llanbadrig (Precambrian mélange and fossils). Where appropriate, the geological interpretation is complemented by historical and folklore details. For example, Llanddwyn is the 'home' of St Dwywen, the Welsh patron saint of lovers; Llanbadrig is the church founded by St Patrick after he was wrecked and washed ashore here. More information boards have been erected for geological SSSI and RIGS at Rhoscolyn, Rhosneigr, Parys Mountain, Traeth Lligwy, Red Wharf Bay, Lleiniog and Gallows Point. These boards showcase partnership geoconservation work between GeoMôn, Natural Resources Wales (NRW) and Isle of Anglesey County Council (IoACC) across the island.

There is always a debate on whether installing information boards is a benefit or an intrusion into the landscape. GeoMôn's boards are carefully produced and normally installed on a rock plinth in an unobtrusive location. Leaflets (especially paperless versions, freely available from websites) do not intrude on the landscape, but are far less available to casual visitors than to residents. Anglesey suffers from poor mobile phone coverage in many parts of the coastline – so, whilst online resources or smart-phone apps might be more useful in some places, they are not a priority for GeoMôn. Nevertheless, audio-trails have been prepared as part of an Anglesey County Council initiative for South Stack, Parys Mountain, Porth Amlwch and Cemaes. Both Cemaes and Llanbadrig Community Councils have long been enthusiastic partners and, as well as requesting geotrails for their community newsletter, they were keen to have a permanent feature. GeoMôn obliged by installing a 'rock walk' on the headland to the west of Cemaes Bay, displaying many of the island's rock types on plinths. To the east, geology and culture have been combined in a trail to St Patrick's cave and well on the cliffs below Llanbadrig church, where he is reputed to have stayed when wrecked on this coastline around 440AD.

Walking is one form of exploring the countryside, but cycling is also very popular. GeoMôn collaborated with the local authority in the development of the Copper Trail (national cycle route N566), which provides a 34-mile (55km) circular tour of northern Anglesey; GeoMôn provided information for boards at various stages of the trip, explaining the local landscape and geology visible along that section. More recently, a guide to accompany a network of cycle routes on Holy Island (Ynys Gybi) has been developed by Sustrans (a UK national cycling charity), and GeoMôn has written trail leaflets for the four circular routes. GeoMôn also has a partnership with B-Active Rhoscolyn, a sea-kayaking company who lead 'rock hopping geology by kayak' trips for the geopark (http://b-active-rhoscolyn.co.uk/course/rock-hopping-geology-by-kayak).

Both G&M RIGS and NEWRIGS have produced geology trails for their towns (Chester, Conway, Caernarvon, Bangor, Llandudno, etc), so it was logical for the GeoMôn development team to continue the trend with a trail for its most visited town, Beaumaris, which was launched during the Anglesey Walking Festival in 2008. The trail leaflet is available free of charge in the town, and features as a guided walk most years in the Anglesey Walking Festival. Work is currently in hand to produce a town trail for Holyhead, including local and exotic building stones, the New Harbour rock exposures and the enormous harbour wall (the latter is already part of the Breakwater Quarry trail). GeoMôn's latest publication, *Footsteps through Time* (Campbell, Wood and Windley 2014), is a lavishly illustrated, limited edition, comprehensive account of the geology of Anglesey.

The geopark has trained 12 geo-guides (using sponsorship from Plas Goch, a large hotel and holiday complex), who lead guided walks following some of the leaflet trails; the personal touch of a guide can explain so much more than a printed leaflet and can take advantage of the current state of nature or the landscape. Four more guides have been trained specifically for Newborough Forest and Llanddwyn Island, where GeoMôn is creating a super new trail at the request of the NRW warden and the Bro Aberffraw community (covering all of south-west Anglesey). Guides also take out parties by request, and act as local experts for visiting university groups and other interested parties. Individual guides have their own specialism, as well as geology, and can therefore enthuse groups on a range of topics from natural history through archaeology to shipwrecks. Specialist trips are also provided by kayak or rib-boat.

Attendance of geopark representatives at the European Geopark Network (EGN) and the UNESCO Global Geopark Network (GGN) conferences is a requirement for any geopark applying to be accepted into, and to remain within, the networks. Accordingly, various GeoMôn staff have attended many EGN and GGN conferences, and their lead spokespersons have presented conference papers at these and at related conferences (see Annex 2), building on the earlier geoconservation and geoheritage work carried out under the RIGS banner.

Developing the geopark requires the partnership and support of the local community, usually through local government. In the UK there are several levels of local government, starting with parish or community councils, which have little power but do reflect the interests and wishes of the local people. Both Cemaes and Llanbadrig Councils have supported the geopark from the start, recognising that their economy is built, at least in part, on tourism. In 2011 they requested that three trails, detailing the geology and landscape of the coastline to either side of the town, be written for their community newsletter.

Real local power is vested in the IoACC, which manages most aspects of business development, tourism and environment; it also administers EU funding through the Rural Development Plan. It has supported GeoMôn in various ways, through funding geopark signage at major transport nodes on the island, such as Anglesey airport and Holyhead rail and ferry terminal, and also funding more than 20 information boards for carefully selected geosites, with technical content and images provided by GeoMôn and NRW staff. IoACC staff have managed practical aspects, such as obtaining planning and landowner permissions.

The geopark has a network of local businesses supporting and promoting its activities; these include food producers, accommodation providers, activity holiday providers and local craftspeople. The aim is to use the geoheritage to stimulate the local economy, partially by extending the tourism offerings (eg guided walking holidays, kayaking trips etc), and partly by extending the tourism season beyond the summer school holiday period, by providing guided walks and activities in spring and autumn. GeoMôn contributes to various school and children-focused clubs. In the last two years, it has formed a partnership with B-Active, a kayaking venture in Rhoscolyn, and provides geo-kayaking sessions for schools.

Geopark status was originally awarded in 2009, initially by the EGN, followed by ratification by the GGN in 2010. As is customary, GeoMôn had to submit to the revalidation process after four years and was successfully revalidated in 2013. Further developments are planned: GeoMôn is a partner in various EGN projects promoting Europe-wide networks on fossils, on mining heritage, on coastal landscapes and, most excitingly, on food.

GeoMôn is one of two geoparks in Wales (alongside Fforest Fawr in the Brecon Beacons) and seven throughout the UK, all of which are members of the UK Geopark Forum (in association with the UK UNESCO team), which now encourages aspiring geoparks and vets proposals for new geoparks in the UK. Geoconservation has moved on from the early days of statutory bodies designating protected sites, through the grass roots RIGS voluntary designation scheme, to actively promoting the use and understanding of geoheritage through the international network of Global Geoparks, which is on the cusp of establishing full recognition within UNESCO. Anglesey is justifiably proud to be part of this international community.

REFERENCES

Allen, J R L, 1965 The sedimentation and palaeogeography of the Old Red Sandstone of Anglesey, *Proceedings of the Yorkshire Geological Society* 35 (2), 139–85

Ball, D F, 1963 *Soils of Bangor and Beaumaris: Memoirs of the Soil Survey of England and Wales*, Lawes Trust, Harpenden

Bates, D E B, and Davies, J R, 1981 *Geologists' Association Guide No 40: Anglesey*, Geologists' Association, London

Burek, C, 2008 History of RIGS in Wales, in *The History of Geoconservation, Special Publication No 300* (eds C V Burek and C D Prosser), The Geological Society, London, 147–71

— 2012 The role of LGAPs and Welsh RIGS as local drivers for geoconservation within geotourism in Wales, *Geoheritage* 4 (1), 45–63

Campbell, S, Wood, M, and Windley, B, 2014 *Footsteps through Time: The Rocks and Landscape of Anglesey Explained*, Isle of Anglesey County Council

Conway, J S, 1984a Soil – the most valuable geological resource, *OU Geological Society Journal* 5, 4–6

— 1984b Soil change in the later prehistoric period – progressive or quantum pedogenesis, in *Welsh Soil Discussion Group Report 26* (ed S Limbrey)

— 2006 *Soils in the Welsh Landscape (A Field-Based Approach to the Study of Soil in the Landscapes of North Wales) / Priddoedd yn Nhirwedd Cymru (dull ar raddfa maes o astudio pridd yn nhirweddau Gogledd Cymru)*, Seabury Salmon and Associates, Ludlow

— 2010a *Rocks and Landscapes of the Anglesey Coastal Footpath / Creigiau a thirweddau Llwybr Arfordirol Ynys Môn*, GeoMôn Geopark, in association with Seabury Salmon and Associates, Ludlow

— 2010b A soil trail? – a case study from Anglesey, Wales, UK, *Geoheritage* 2 (1), 15–24

— 2012a North Wales soil trails – soils RIGS in action. Invited talk at Soils, Geodiversity and Soils Educational Site Networks: SW England Soils Discussion Group, 20 March, Rothamsted Research, North Wyke

— 2012b Soils in public view … what do we want them to see? Invited talk at Welsh Soils Discussion Group, IBERS, 20 March 2012, Aberystwyth

— 2014 *The Anglesey Soil Trail*, GeoMôn Geopark, Anglesey

Conway, J S, and Wood, M, 2012a The legacy of mining – archaeological, environmental and social impact – at Parys Mountain, Anglesey, in *Proceedings of the 10th European Geoparks Conference: Sustainability through Knowledge Communicating Geoparks* (ed K Rangnes), Langesund, 86–92

— 2012b World's largest copper mine supports new geopark project, *Network: The European Geopark Magazine* 9, 24

Greenly, E, 1919 *Geology of Anglesey: Memoirs of the Geological Survey of Great Britain* (2 vols), Geological Survey of Great Britain, HMSO, London

Henslow, J, 1822 *Geological Description of Anglesea*, J Smith, Cambridge

Roberts, E, 1958 *The County of Anglesey, Soils and Agriculture*, Agricultural Research Council, HMSO, London

Rowlands, J, 1981 *Copper Mountain*, Anglesey Antiquarian Society, Anglesey

Rudeforth, C, Hartnup, R, Lea, J W, Thompson, T R E, and Wright, P S, 1984 *Soils and their Use in Wales, Bulletin 11*, Soil Survey of England and Wales, Harpenden

Treagus, J, 2008 *Anglesey Geology – A Field Guide*, Seabury Salmon and Associates, Ludlow

Wood, M, 2004 *Precambrian Rocks of the Rhoscolyn Anticline*, Seabury Salmon and Associates, Ludlow

— 2007 Application dossier for nomination to the European Geopark Network, unpublished, GeoMôn Geopark, Anglesey

— 2012 The historical development of the term 'mélange' and its relevance to the Precambrian geology of Anglesey and the Lleyn Peninsula in Wales, UK. The 100s: significant exposures of the world, *Japanese Journal of Geography* 7, 168–80

Wood, M, Campbell, S, Roberts, R, Brenchley, P, Conway, J S, Crossley, R, Davies, J, Fitches, B, Mason, J, Matthews, B, Treagus, J, and Williams, T, 2007a *Developing a Methodology for Selecting Regionally Important Geodiversity Sites (RIGS) in Wales and a RIGS Survey of Anglesey and Gwynedd, Volume 1: Methodology*, report to Welsh Assembly Government, Cardiff

— 2007b *Developing a Methodology for Selecting Regionally Important Geodiversity Sites (RIGS) in Wales and a RIGS Survey of Anglesey and Gwynedd, Volume 2: Site Survey*, report to Welsh Assembly Government, Cardiff

Wood, M, and Nicholls, G D, 1973 Precambrian stromatolites from Northern Anglesey, *Nature* 241, 65

Annex 1: Published Anglesey geotrails

Conway, J S, 2008 Geotrail along the South Stack Coastline, *Swn y Mor* 4, 12–13

— 2008 Holyhead Breakwater Country Park Geotrail, *Swn y Mor* 5, 12–13

— 2009 *Beaumaris Town Geology Trail*, G&M RIGS

— 2009 Cemaes – Llanbadrig Geotrail, *Swn y Mor* 7, 12–13

— 2010 Geotrail from Wylfa visitor centre to Cemaes Bay, *Swn y Mor* 8, 12–13

— 2010 Geotrail from Bodafon to Moelfre, *Swn y Mor* 9, 12–13

— 2011 Geotrail from Llanbadrig to Porth Wen, *Cemaes Voice* 8, 12–13

— 2011 Geotrail from Lleiniog to Beaumaris, *Swn y Mor* 10, 12–13

— 2011 Geotrail from Cemaes Bay to Llanbadrig Church, *Cemaes Voice* 6, 10–11

— 2011 Geotrail from Wylfa Visitor Centre to Cemaes Bay, *Cemaes Voice* 7, 12–13

— 2011 Red Wharf Bay Geotrail, *Swn y Mor* 11, 10–11

— 2012 Porth Amlwch Geotrail, *Swn y Mor* 12, 10–11

— 2013 Geotrail around Parys Mountain, *Swn y Mor* 13, 12–13

— 2013 Geotrail: Aberffraw to Porth Cwyfan, *Swn y Mor* 14, 16–17

— 2014 Geotrail: Penmon, *Swn y Mor* 15, 16–17

Conway, J S, and Wood, M, 2009 Geotrail across Newborough Forest to Llanddwyn Island, *Swn y Mor* 6, 12–13

Annex 2: Further papers presented by GeoMôn spokespersons at EGN/GGN conferences and at related conferences

Conway, J S, 2007 Raising the profile of soil with geoconservation: a case study from the Anglesey Geopark, paper presented at *European Geopark Network Conference*, 13–17 September 2007, Ullapool

— 2008 Raising the profile of soil – a case study from the Anglesey Geopark, paper presented at *3rd International UNESCO Conference on Geoparks*, June 2008, Osnabruck

— 2009 Raising the profile of soil within geo-conservation – a case study from the Anglesey Geopark, paper presented at *ProGEO – WG3: Geodiversity, Geoheritage and Nature and Landscape Management*, 19–23 April 2009, Drenthe, The Netherlands

— 2009 Developing a virtual geopark – how Google software can be used to provide a geopark experience for disabled people with mobility problems, in *New Challenges with Geotourism: Proceedings of the 8th European Geoparks Conference* (eds C de Carvalho and J Rodrigues), Portugal, 55

— 2010 Using long distance footpaths to promote geoheritage, paper presented at *GEOTRENDS International Conference on Geoheritage and Geotourism Research*, June 2010, Novi Sad, Serbia

Conway, J S, and Mabvuto-Ngwira, P, 2012 Geotourism opportunities for sustainable development and poverty alleviation in developing countries: the potential of Victoria Falls as a geopark, in *Proceedings of the 10th European Geoparks Conference: Sustainability Through Knowledge Communicating Geoparks* (ed K Rangnes), Langesund, Norway, 123–30 (refereed article)

Wood, M, and Conway J S, 2012 Cultural and archaeological heritage of the Parys copper mine, Anglesey, in *Proceedings of the 11th European Geopark Conference: Smart Inclusive Sustainable Growth*, 19–21 September 2012 (ed A Sa, D Rocha, A and V Correia), Arouca, Portugal, 303–4

— 2012 Developing an integrated geo-cultural trail, in *Proceedings of the 11th European Geopark Conference: Smart Inclusive Sustainable Growth*, 19–21 September 2012 (ed A Sa, D Rocha, A and V Correia), Arouca, Portugal, 301–2

13

A Geoheritage Case Study:
The Ruhrgebiet National GeoPark, Germany

VOLKER WREDE

THE RUHRGEBIET

The Ruhrgebiet ('Ruhr Area') is the most densely populated industrial heartland of Germany. Stretching eastwards from the River Rhine for some 100km along the valleys of its tributaries Ruhr, Emscher and Lippe, it is a continuous urban agglomeration covering an area of some 4,400km² and housing some five million people. Historically, this townscape is very young; 200 years ago, neither the term 'Ruhrgebiet' nor the structure existed. A few old, but small, towns were situated in the Ruhr Valley, using hydropower for mills and small-scale industries. Bochum, Essen and Dortmund grew up along an old trade route, called the Hellweg. This led from the Rhine, eastwards to central and eastern Germany. Other than these urban developments, most of the area was remote – a rural, sparsely inhabited, agricultural landscape.

The Ruhrgebiet is not easily defined as a distinct geographical unit. The southern part, including the valley of the Ruhr, belongs to the Rhenish Mountains, while the northern and western sections are part of the German Lowlands – the Münsterland Basin and the Northern Rhine Lowlands. Likewise, the area never formed a political unit and was always split into different territories; even today it is divided by administrative boundaries into a 'Westphalian' part in the east and a 'Rhenish' part in the west. Its 12 independent larger cities and four districts belong to three administrative regions within the State of North Rhine-Westphalia. However, a Ruhr Regional Council ('Regionalverband Ruhr') was originally founded in the 1920s to coordinate action on some of the common issues of the communities. Today its focus is, for example, on economic development, environmental protection, tourism promotion and public relations. The award of the title 'European Capital of Culture – Ruhr 2010' helped to form a stronger common community bond for the development of an integrative Ruhr metropolis (Figure 13.1A).

The fast-paced development of the Ruhrgebiet started with the industrial revolution at the beginning of the 19th century. Within a few decades, the rural area had developed into one of the largest urban agglomerations in Europe. In 1852 the Ruhr area was inhabited by some 375,000 people; in 1925 the number of inhabitants exceeded 3.8 million (Farrenkopf 2009), and today the Ruhrgebiet is home to 5.4 million inhabitants. This development was initiated by the increasing use of coal, which outcropped at the surface in the Ruhr Valley. The discovery of iron and other metal ores then stimulated steel mills and metal works. Limestone and dolomite became important resources for metallurgical processes in the iron and construction industries. Clays and mudstones, used for the production of bricks, along with sandstones and other building materials found in the region, enabled the erection of housing areas and infrastructure for hundreds of thousands of people within a very short time. The use of salt springs and salt mining is another example of mineral production in the area, with a long history dating back to prehistoric times.

Therefore, the development and identity of the Ruhrgebiet are actually based only on the occurrence and use of its natural resources. The term 'Ruhrgebiet' came into use only in the 1930s. Before then, the area was described as the 'Rhenish-Westphalian Industry District' or the 'Ruhr Coal Mining District'. Few other landscapes in Europe demonstrate so well the connection between their natural resources and the economic, socio-cultural development of humankind as clearly as does Germany's Ruhrgebiet area. Today, after the decline of the mining industry, which started in the late 1950s and is scheduled to close down finally in 2018, the area is confronted with considerable changes in its economic and social structures (Wrede 2006). Therefore, it is necessary to preserve community points of orientation and identity. With natural resources being the crucial point of identity and economic development, the foundation of a geopark in this urban area was a logical consequence of this development (Wrede 2003; Wrede and Mügge 2006).

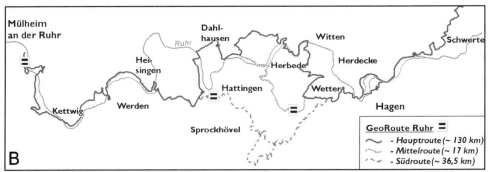

FIGURE 13.1 LOCATION OF THE
RUHR GEOPARK AREA
A. LOCATION OF THE RUHRGEBIET
NATIONAL GEOPARK IN GERMANY.
B. SKETCH MAP OF THE 'GEOROUTE RUHR'
HERITAGE TRAIL.

THE NATIONAL GEOPARK RUHRGEBIET

Aiming to preserve the awareness of its people for the roots of the area's development, the Ruhr Area GeoPark was founded in 2004 by the Ruhr Regional Council and the State Geological Survey of North Rhine-Westphalia. It is based upon the regulations and criteria of the German National GeoPark Scheme (Mattig *et al* 2003), which was stimulated by the Global and European Geopark Networks but is independent from these. The general aims of German National GeoParks are:

- Protection and conversation of the geoheritage of the area;
- Informing the public about geoscientific topics;
- Stimulation of local economic development, mainly by promoting geotourism activities.

The 'GeoPark Ruhrgebiet e.V.' was set up as an incorporated non-profit entity, with the mission to organise and run a geopark in the Ruhrgebiet.

Up to now, the working programme of the GeoPark Ruhrgebiet has concentrated on geotope conservation, carried out by volunteers and in close cooperation with the local environmental protection authorities; the outcomes have been the publication of guidebooks, information leaflets and maps, on-site information for a network of geotrails, and the organisation of field trips and congresses (Kirnbauer *et al* 2008; Mügge-Bartolović *et al* 2010; Wrede 2006; Wrede and Schmiedel 2009).

Considering the approaching closure of the last coal mines in the area, protection and conversation of the geoheritage – both on-site and in museum collections, as primary and secondary geosites respectively (Hose 2007) – has become particularly important for the area. In the past, mining geology in the Ruhr area contributed much to the scientific knowledge and understanding of the Carboniferous stratigraphy of central Europe (eg the term 'Westphalian' for a part of the Upper Carboniferous strata is derived from the area) and of coal geology in general. After 2018, underground exposures will no longer be accessible, and demonstration of the Carboniferous strata, including the coal seams, will be restricted to surface outcrops alone. Therefore, they have become more valuable than ever for university education and scientific research, and also for the public. The outcrops, where geological features can be studied *in situ* (at primary geosites), complement the rich collections of fossils and other samples preserved in the collections of numerous museums (secondary geosites). The GeoPark intends in future to intensify its collaboration with museums and collections, both public and private, to form the 'GeoArchives of the Ruhrgebiet', preserving as much information as possible on the coal-bearing strata which laid the cornerstone of the development of the region. This will become an outstanding example of geoconservation working towards enhancing the geoheritage of secondary geosites. Similar arguments have been put forward (Hose 2011; 2012; Hose and Vasiljević 2012) for the initial development of geotourism in the UK and Europe; these concepts by Hose (1995) were also incorporated into the initial UNESCO (2000) geopark scoping document.

Mass tourism does not play a primary role in the Ruhrgebiet. Few people come to the area for holidays, but there is a great demand for local recreation facilities. The large number of people residing within the GeoPark, many of them still with family roots in the mining industry, forms a large target group for geo-educational programs. As a public opinion poll in 2013 proved, there is a great public interest in issues like mining history, fossils, showcaves and mine visits (Elsner and Lawitzke 2013). They can easily be combined with outdoor activities, including hiking,

cycling, or canoeing on the River Ruhr. Significantly, the area is linked to other national recreation routes such as the Ruhr Valley Cycle Route; this 230km cycle trail follows the River Ruhr
from its source in Sauerland to its confluence with the River Rhine at Duisburg. The opportunity
to combine outdoor (geo-)activities with all the facilities offered by a vibrant metropolitan area
establishes a considerable potential for the further development of geotourism. These include
leisure boat trips or heritage railway lines leading directly to outstanding geosites, visits to thea-
tres, concerts and museums, or to historic sites in the area.

The GeoPark Ruhrgebiet is organised as a network of different members; they include local
authorities (town and district administrations), universities and museums, representatives of the
extraction and mining industry, environmental and conservation organisations, mining heritage
groups, scientific societies and touristic enterprises, as well as personal members. The members
generally act autonomously, but the GeoPark coordinates their activities and creates synergy.
Working groups on the conservation of geosites, on educational programmes and on museum
cooperation have been established for their respective purposes. 'GeoPark Ruhrgebiet' is the
common label – a marketing tool for the public presentation of the programmes and activities.

GEOLOGY OF THE GEOPARK

The three different geographic units, of which the GeoPark is part, reflect the geology of the
area. The southern part is dominated by Palaeozoic rocks of Devonian and Carboniferous age.
Surface geology in the north-eastern part is determined by Cretaceous and Pleistocene sediments.
The western part mainly has sediments of Tertiary, Pleistocene and Holocene age. Sediments
of Permian, Triassic and older Mesozoic age, covered by younger strata, are well known from
boreholes and mine shafts. Therefore, the geology of the Ruhr area represents, more or less
continuously, some 350 million years of Earth history (Wrede and Brix 2009) (see Table 13.1).

The oldest known strata in the GeoPark date from the transition between the Lower and the
Middle Devonian; they are found in the centre of the Remscheid Anticline, a large fold structure
in the southernmost part of the region. The clayish and sandy layers were deposited on the shelf
adjoining the coast of the 'Old Red Continent' in the north. The Brandenberg-beds of upper
Eifelian age represent a coastal situation; besides marine fauna (eg bivalves and fishes), early land
plants have been preserved and some sedimentary features indicate a terrestrial-fluviatile (river)
environment. During the late Middle Devonian (Givetian) and Upper Devonian times, the
sea-level rose and intensive growth of limestone reefs became dominant; the reef-limestones, in
parts more than 1,000m thick, are of high economic value. They underpin the dramatic karstic
landscapes in the areas south of the Ruhr Valley and beyond.

The 'Culm Facies' (named after a locality in Devon, England) sediments of the late Devonian
and Lower Carboniferous (around 330–380mya) indicate deposition in a relatively deep marine
basin with very low rates of sediment input. Since the Namurian (313–326mya), and initiated
during the Variscan Orogeny, or mountain-building period, which occurred 380 to 280 million
years ago far in the south, rivers transported more and more sediments into this basin. The
vast deltas created by these rivers gradually grew until their surface reached sea-level. On top of
these delta platforms, the first coal-forming swamps developed. In the Upper Namurian (around
313–318mya), marine and terrestrial sediments still alternated frequently, as can be seen by
marine fossils (eg brachiopods) and fossil land plants and insects. The Vorhalle Quarry in Hagen
provides the oldest known winged insects in the world, numerous in species and individuals,

TABLE 13.1 THE STRATIGRAPHY OF THE RUHRGEBIET NATIONAL GEOPARK.

Quaternary	Holocene	Fluvial deposits: sand, gravel locally bogs
	Pleistocene	glacial deposits, moraines, erratic blocks, loess
Tertiary		sands, clay (marine)
Cretaceous	Upper	limestones, marls, sands (marine)
	Lower	
Jurassic		(mining and borehole-exposures only)
Triassic		
Permian	Upper (Zechstein)	
	Lower (Rotliegend)	conglomerates, arid conditions
Carboniferous	Pennsylvanian	transition from marine to paralic / terrestrial facies: sandstones, clay- and siltstones, coal seams; plants, early insects and reptiles; Variscan Orogeny
	Mississippian	carbonate platform in the west, deep water sediments (Kulm facies) in the east
Devonian	Upper	limestones, sand- and siltstones marine fossils
	Middle	reef limestones
	Lower	coastal silt- and sandstones, marine fauna, early plants

and most of them excellently preserved; they date from Namurian B times (Hendricks 2005) around 315mya (Figure 13.2E).

During the Westphalian (304–313mya), the number of marine ingressions decreased and a huge basin developed; sedimentation of the coal-bearing strata in this basin was controlled by three factors: the subsidence of the basin (which averaged 0.4m/1000y), sedimentary input, and changes of the sea-level. The combination of these factors led to cyclic sedimentation with regular coarsening-up or fining-up sequences. The total thickness of the coal-bearing strata in the Ruhr basin from Westphalian B to Westphalian D is some 4,000m of clays, siltstones and sandstones. Embedded in the sediments are some 250 coal seams, ranging in thickness from a few centimetres to more than two or three metres. However, the net coal content of the stratigraphic column is only between 2% and 4%. During the late Westphalian and Stephanian, the front of the Variscan Orogen reached the Ruhr basin, but stopped within it. Consequently, in the southern part the strata are intensely folded and thrusted, causing an orogenic shortening of the Earth's continental crust of around 50%. In the northern part of the area, in contrast, orogenic shortening is less than 5% (Drozdzewski and Wrede 1994).

FIGURE 13.2 SOME KEY LOCATIONS IN THE RUHR GEOPARK

A. ZOLLVEREIN COLLIERY, ESSEN, UNESCO WORLD HERITAGE SITE.

B. FORMER ENGINE-HOUSE OF NIGHTINGALE MINE AND DÜNKELBERG BRICK-KILN, WITTEN, NOW AN INDUSTRIAL MUSEUM AND INFORMATION CENTRE FOR THE GEOPARK.

C. BOCHUM: GEOLOGICAL GARDEN – VARISCAN UNCONFORMITY (MARKED WITH WHITE LINE AND NAMED ON IMAGE) OF UPPER CRETACEOUS SANDSTONE OVERLAYING FOLDED CARBONIFEROUS SANDSTONES AND SHALES.

D. MÜLHEIM SOIL SCIENCE TRAIL – SOIL EXPOSURE OF A CAMBISOL WITH AGRICULTURAL USE AND A RIVER-TERRACE GRAVEL AT BASE.

E. VORHALLE QUARRY IN HAGEN – *LITHOMANTIS VARIUS*; NOTE THE PRESERVED COLOUR PATTERN ON THE WINGS (WING-SPAN OF THE INSECT IS C.12CM).

In the Lower Permian (around 260–270mya), evidence of the Variscan Orogen as seen in surface outcrops was completely eroded. For most of the GeoPark, a huge gap in the geological record exists for the time interval between the Carboniferous, Permian and the Upper Cretaceous (of around 190 million years). However, Upper Permian strata (including economically important rock-salt deposits) and Triassic and Jurassic (145–200mya) sediments have been revealed by boreholes in the basement of the Northern Rhine Lowlands, overlying the folded Carboniferous basement rocks and overlain by Cretaceous or Tertiary strata. These exposures provide evidence for strong fault-tectonic activities in these time intervals. In the Münsterland part of the GeoPark, marine sediments of the Upper Cretaceous directly rest on the folded strata of the Carboniferous – a time interval of nearly 200 million years.

The Cretaceous rocks (around 66–145 million years old) found in surface exposures typically are marls, limestones and less consolidated sands; all were deposited in a shallow marine environment and they are partly rich in fossils. During the Tertiary Period, marine sands and clays were deposited in the newly forming Lower Rhine Embayment, being part of the Central Graben system of the North Sea.

The Pleistocene Saale Glacier covered most of the GeoPark area. Its terminal moraine is found on the western bank of the River Rhine, forming a chain of hills within the flat lowlands. In the other parts, the basal till and aeolian (loess and cover sands) Pleistocene sediments lie atop the older strata. The youngest sediments in the GeoPark are the Pleistocene and Holocene alluvial deposits of the Rhine and its tributaries.

Geoheritage and geoconservation

As an expression of the versatile geology, and being favoured by topography and historic mining operations, a large number of both natural and artificial geosites are to be found in the GeoPark. Some 400 primary geosites are listed in the geosites catalogue of the State Geological Survey. Three of them have been awarded the status of 'National Geosites': the already mentioned Vorhalle Quarry in Hagen; the bizarre karst area of Felsenmeer near the town of Hemer; and the historic mining district of Muttental near Witten. This latter location is considered to be (one of) the cradle(s) of coal mining history in the area. Mining is first documented from the 14th century, but probably started much earlier.

Today many relics of the mining activities and primary geosites displaying the coal-bearing strata in the Muttental area have been restored and made accessible by heritage trails. The former Nightingale Mine, which had closed down at the end of the 19th century, serves as an industrial museum. At the mine there is not only an authentic underground mine, accessible to visitors, but also a historic brickworks built later on the premises; brick production was based on the use of Carboniferous claystones quarried nearby. One of the historic buildings now houses the central information centre of the GeoPark (Figure 13.2B).

Cliffs and flowstone caves in Devonian limestones, several sites exposing the Variscan unconformity, and impressive outcrops of post-Variscan rocks and structures are only some examples of the natural inventory of the GeoPark. More than 20 museums display regional geology and mining history. Numerous examples of mining and industrial heritage sites from all ages have been developed as tourist attractions or are re-used, for example, as event locations; they are linked by the 'Industriekultur' heritage route. After being closed down in 1986, the former Zollverein Mine in Essen, founded in 1834 and completely rebuilt in the 1920s and 1930s in the Bauhaus style (Figure 13.2A), was listed as a UNESCO World Heritage Site. Today it houses, amongst other attractions, the Ruhr Museum – the largest collection dealing with the social and natural history of the region. With the intention to extend the World Heritage status of the Zollverein Mine, and with the support of the State Government of North Rhine-Westphalia, an application to include the industrial landscape of the Ruhrgebiet in the World Heritage list was suggested (Mügge-Bartolović et al 2012; Stiftung Industriedenkmalpflege und Geschichtskultur 2012). The GeoPark Ruhrgebiet is an integral part of this proposal, which has, however, been postponed for the present.

A major input into the GeoPark is from the activities of mining heritage groups which were, perhaps unsurprisingly, mostly founded by former miners focused on the preservation of their local mining history. Today, membership of these groups is recruited from people with a range of professional and artisan backgrounds. With the support of local authorities and private sponsors, their enthusiasm and their knowledge about the history and details of sites have enabled them to create heritage trails and local museums, and even to restore an authentic 18th to 19th century mine near Dortmund – now operated as a visitor attraction.

The 'Muttental Mining Heritage Trail', established in 1972 by the German Mining Museum in Bochum, the town administrations of Witten and Herbede and private initiatives, was the first of more than 20 geology and mining heritage trails which already existed when the GeoPark was founded in 2004. Most of them are located in or near the Ruhr Valley, where mining ceased in the 1970s. A high concentration of mining relics from several centuries is still to be found here. Meanwhile, parts of the region have been de-industrialised and completely re-landscaped, becoming attractive for tourism – much as can be seen across other similar areas in Europe, in which industrial heritage preservation has become incorporated within cultural tourism provision (see Alfrey and Putnam 1992, Chapter 1). This is very similar to what has happened in northern England (see Hose and Vasiljević 2012), especially in County Durham (Durham County Council 1994) and South Wales (Wanhill 2000) in a similar timeframe. Besides mining history, the heritage trails cover a wide range of topics, similar to those that have been developed at other classic German geosites (see Look, Quade and Muller 2007). Classic geotrails can be found in Essen and Hagen, initiated by members of the Geological Department of Essen University and by the Department for Environment of the Hagen town administration respectively. A soil science trail was installed in Mülheim (Figure 13.2D). Near Essen-Kupferdreh, the 'Deilbachtal Heritage Trail' presents reminders of early copper and iron production, while the 'Energy Trail' in Herdecke points to the changes in energy technology over time.

Because of their decentralised origin, most of the trails had their own individual design and different types of interpretation, and they pursued different marketing strategies. The impact of most of these trails was limited to local residents. The GeoPark Ruhrgebiet has developed a proposal to integrate these individual trails into an all-embracing network. A new long-distance track, 'GeoRoute Ruhr', leading from Mülheim to Schwerte, was laid out to connect most of the individual trails and, supported by uniform trail-marking, to enable supra-regional marketing (Wrede and Mügge-Bartolović 2012). All institutions involved in the management of the pre-existing trails, regional authorities and the officially appointed hikers association were persuaded to support the proposal. Wikinger Reisen GmbH, one of Germany's leading holiday-operators, based in the Ruhrgebiet, generously sponsored the project.

Due to the distribution of the existing trails, it was not possible to link them all by one single route. In the central part of the area, where trails had been established both north and south of the Ruhr Valley, the main track was split into up to three parallel routes. All in all, the length of GeoRoute Ruhr reached 180km, all following public footpaths. A guidebook was published, with detailed maps of the route and descriptions of some 150 sites along the trail (Mügge-Bartolović 2010). Taking into account that the interest of most visitors is not focused on geology alone, the sites described in the guidebook include not only geological/geomorphological and mining heritage objects, but also cultural sites such as historic churches, castles and manor houses. For the future, packages of hiking trips on the GeoRoute Ruhr, combined with trips on the historic Ruhr Valley Steam Railway or the pleasure boats run on some sections of the river and its lakes, are under consideration. Some of the sites have been provided with new information panels; however, their number is limited due to the problem of sustainable maintenance. The urban setting of the GeoPark creates not only a potentially large number of visitors to its installations, but unfortunately also an enlarged risk of vandalism. Therefore, investments in on-site installations require people or organisations to adopt these installations for maintenance. Members of the mining heritage associations or local citizens' groups undertake much of this care. Together with the pre-existing heritage trails, the GeoRoute Ruhr forms a complex network of trails, some 300km in length and including more than 500 sites (Figure 13.1B). The description of the complete network can be found on the internet (www.geopark-ruhrgebiet.de).

The GeoRoute Ruhr was officially opened in 2010. Within three years, it was broadly accepted by the public. One of the findings of the public opinion poll in 2013 (previously mentioned) was that the GeoRoute Ruhr is well known to around a fifth of the inhabitants of the Ruhrgebiet. The increasing interest in the primary geosites along the trail leads to surprising discoveries. For example, in 2012 a Dortmund family inspected a small sandstone quarry, abandoned decades ago; they recognised the footprints of a tetrapod animal on one of the bedding planes, which had been overlooked by generations of visitors. This animal track proved to be the oldest record of a tetrapod animal in central Europe, and it has been salvaged in a joint effort between several institutions, coordinated by the GeoPark (Wrede 2014). Another long-distance heritage trail – the 'GeoRoute Lippe' – is planned in the northern part of the GeoPark (Abels *et al* 2012). Distances between the individual geosites are larger here than on the GeoRoute Ruhr, and the landscape is less hilly and therefore more appropriate for cycling. The GeoRoute Lippe is thus designed to be developed as a combined bicycle and hiking track.

Some more heritage trails, which are not linked to these networks, can be found in the eastern and westernmost parts of the GeoPark; for example, the 'Weg der Geotope' ('Geosites Trail') near Unna and Fröndenberg, or the 'Sonsbeck Geotrail', focusing on the glacial landforms of the Saale epoch. Other locations, which are not linked to any of the heritage trails, have been furbished and made accessible for the public. One such example is the 'Bochum Geological Garden', one of the most instructive exposures of the Variscan unconformity in Europe. Originally established as a quarry for the extraction of Carboniferous clay for the production of bricks, it was in operation until the late 1950s. It was saved from being infilled by the initiatives of the staff of the former Bochum School of Mines (now a University of Applied Sciences, called the Technische Hochschule Georg Agricola), who used it in their geology tuition. Meanwhile, legally protected as a Nature Monument, the 'Geological Garden' (Figure 13.2C) is embedded in a public park and maintained by the Parks Authority of the City of Bochum (Ganzelewski *et al* 2008).

Looking to the future

As already mentioned, structural changes are a determinant in the history of the Ruhrgebiet. The important role that the GeoPark will play as the 'GeoArchives' for science and education has already been noted. Today, the development of the Ruhrgebiet away from heavy industries to a more diversified economic structure enables a gradual development of tourism, which did not exist at all few years ago. The contrast between industrial history and modern metropolitan structures on one side and green, natural landscapes full of sights on the other side attracts visitors, not only from inside the area but also in increasing numbers from outside. The further development of geotourism will play a part in this development; the extended geotourism infrastructure will contribute to the touristic attractiveness of the region, and the increasing number of visitors to the Ruhrgebiet will enlarge the number of potential participants in GeoPark activities.

References

Abels, A, Mügge-Bartolović, V, and Wrede, V, 2012 Georoute Lippe – Erlebnisradroute durch den Norden des Nationalen GeoPark Ruhrgebiet, *SDGG* 79, 14–15

Alfrey, J, and Putnam, T, 1992 *The Industrial Heritage: Managing Resources and Uses*, Routledge, London

Drozdzewski, G, and Wrede, V, 1994 Faltung und Bruchtektonik – Analyse der Tektonik im Subvariscikum, *Fortschritte in der Geologie von Rheinland und Westfalen*, 38, 7–187

Durham County Council, 1994 *Durham Geological Conservation Strategy*, Durham County Council, Durham

Elsner, U, and Lawitzke, P, 2013 Marktanalyse GeoPark Ruhrgebiet, *GeoPark News*, 2/2013, 4–7

Farrenkopf, M, 2009 Zur Geschichte des Ruhrbergbaus, *EDGG* 238, 30–41

Ganzelewski, M, Kirnbauer, T, Müller, S, and Slotta, R, 2008 Karbon-Kreide-Diskordanz im Geologischen Garten Bochum und Deutsches Bergbau-Museum, *Jahresberichte und Mitteilungen des Oberrheinischen Geologischen Vereins Band* 90, 93–136

Hendricks, A (ed), 2005 *Als Hagen am Äquator Lag: Die Fossilien der Ziegeleigrube Hagen-Vorhalle*, Westfälisches Museum für Naturkunde, Münster

Hose, T A, 1995 Selling the story of Britain's stone, *Environmental Interpretation* 10 (2), 16–17

— 2007 Geotourism in Almeria Province, South-east Spain, *Tourism: An International Interdisciplinary Journal* 55 (3), 259–76

— 2011 The English origins of geotourism (as a vehicle for geoconservation) and their relevance to current studies, *Acta geographica Slovenica* 51 (2), 343–60

— 2012 3G's for modern geotourism, *Geoheritage* 4 (1–2), 7–24

Hose, T A, and Vasiljević, D A, 2012 Defining the nature and purpose of modern geotourism with particular reference to the United Kingdom and south-east Europe. *Geoheritage* 4 (1–2), 25–43

Kirnbauer, T, Rosendahl, W, and Wrede, V (eds), 2008 *Geologische Exkursionen in den Nationalen GeoPark Ruhrgebiet*, GeoPark Ruhrgebiet, Essen

Look, E-R, Quade, H, and Muller, R (eds), 2007 *Faszination Geologie: Die bedeutendsten Geotope Deutschlands*, E Schweizerbartsche Verlagsbuchhandlung, Stuttgart

Mattig, U, Look, E-R, and Röhling, H G (eds), 2003 Richtlinien Nationale GeoParks in Deutschland, *SDGG* 30, 1–34

Mügge-Bartolović, V, 2010 *GeoRoute Ruhr: Durch das Tal des Schwarzen Goldes*, GeoPark Ruhrgebiet, Essen

Mügge-Bartolović, V, Linke, J, Mehrfeld, U, Pfeiffer, M, and Wrede, V, 2012 Nationaler GeoPark Ruhrgebiet – Vom Montanindustriellen Herzen Europas zum Welterbe 'Industrielle Kulturlandschaft Ruhrgebiet', *SDGG* 80, 126

Mügge-Bartolović, V, Röhling, H G, and Wrede, V (eds), 2010 Geotop 2010 – geosites for the public, *SDGG* 66, 1–244

Regionalverband Ruhr, 2013 GeoPark Ruhrgebiet – Regionalumfrage 2013, Essen (unpublished)

Stiftung Industriedenkmalpflege und Geschichtskultur, 2012 *Weltweit einzigartig. Zollverein und die industrielle Kulturlandschaft Ruhrgebiet*, Dortmund

UNESCO, 2000 *UNESCO Geoparks Programme Feasibility Study*, UNESCO, Paris

Wanhill, S, 2000 Mines – a tourist attraction: coal mining in industrial South Wales, *Journal of Travel Research* 39 (1), 60–9

Wrede, V, 2003 Probleme des Geotopschutzes im urbanen Ballungsraum Ruhrgebiet,. *SDGG* 27, 32–43

— 2006 The Ruhr Area National GeoPark – a geopark project in an urban area, *Regionalwiss. Forsch.* 31, 71–4

— 2014 Deutschlands älteste Wirbeltierfährte wurde im Nationalen GeoPark Ruhrgebiet entdeckt, *SDGG* 84, 100–2

Wrede, V, and Brix, M R, 2009 Geologie des Ruhrgebiets, *EDGG* 238, 30–41

Wrede, V, and Mügge, V, 2006 Die Einrichtung des GeoParks vor dem Hintergrund des Strukturwandels im Ruhrgebiet, *SDGG* 42, 100–5

Wrede, V, and Mügge-Bartolović, V, 2012 GeoRoute Ruhr – a network of geotrails in the Ruhr Area National GeoPark, Germany, *Geoheritage* 4 (1–2), 109–14

Wrede, V, and Schmiedel, S (eds), 2009 Nationaler GeoPark Ruhrgebiet – eine Bergbauregion im Wandel, *EDGG* 238, 1–100

14

A Geoheritage Case Study: Andalucía, Spain

Thomas A Hose

An introduction to Spain's geoheritage and its protection

Spain has some of the best exposed geology in Europe, due to its mountainous nature, extensive coastline and somewhat arid climate. Gibbons and Moreno (2002, 1) note that: 'The geology of Spain is remarkably diverse. It includes one of the most complete Palaeozoic sedimentary successions in Europe.' In addition, Spain's Mesozoic and Cainozoic rock record is almost as complete; the case study area is noteworthy for its Neogene (2.5–23mya) and almost globally unique lamproitic volcanism characterised by rocks, derived from material from the mantle and brought to the Earth's surface at colliding plate margins, and unusually potassium rich. The physical core of the Iberian Peninsula is the Iberian Massif, a block of rocks representing the eroded roots of a mountain chain that stood in the area some 300–360mya. Its north-eastern margin is bounded by the Pyrenean fold belt and its south-eastern margin by the Betic Cordillera fold belt; these are composed of highly folded sedimentary rocks intruded by volcanic rocks. The block's west edge is delimited by the continental boundary formed by the opening of the Atlantic Ocean. In the east it is mostly buried by Mesozoic and Cainozoic sediments. This solid geology has been beautifully mapped out since the 1970s (in various editions and scales) by the Instituto Geológico y Minero de España, the national geological survey; for example, on the 1:2,000,000 *Mapa Geológico de España* (2004a) and the accompanying 1:2,000,000 *Mapa Tectónico de España* (2004b). Concomitantly. the survey and the Sociedad Geológica de España co-published *Geología de España* (Vera 2004), which is supplied with both maps as well as a CD of these and the book's superb diagrams and illustrations – something that should be emulated by publishers of other national geological surveys! Unlike the Geological Society's English text (Gibbons and Moreno 2002), which takes a stratigraphical approach, the Spanish text adopts a regional approach, which is more useful for geotourism purposes. The major Spanish text on the country's landforms is *Geomorfología de España* (Elorza 1994), a well illustrated regional account. The information in this was much updated in the English text, *Landscapes and Landforms of Spain* (Gutiérrez and Gutiérrez 2014), which was also illustrated in colour.

The country was slow to offer any protection to its rich biological and geological heritages. This was perhaps partly due to the significance of mineral exploitation to the national and regional economies, where geoconservation might be considered an impediment. Indeed, in the closing decades of the 20th century, geoconservation matters were generally addressed as somewhat minor elements within general nature conservation and landscape protection measures. It was noted that: 'Analysis of the country's protected natural areas reveals that a quarter of the total area has been selected for Earth heritage features, but paradoxically, almost all are geomorphological in character, i.e. selection has been based more on landscape than rocks' (Alcala 1999, 14). However,

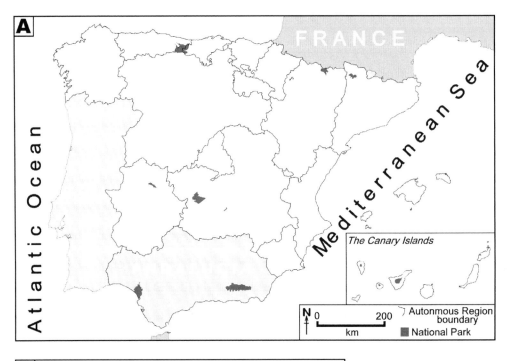

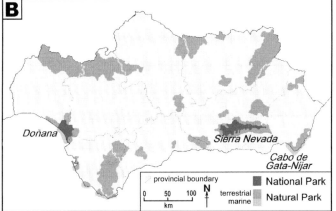

FIGURE 14.1 SPAIN AND
ANDALUCÍA MAPS
A. MAP OF SPAIN AND ITS
NATIONAL PARKS.
B. LOCATION MAP OF
ANDALUCÍA'S NATIONAL AND
NATURAL PARKS.

in the 21st century's opening decade, Law No 42 (2007) on Natural Heritage and Biodiversity mentioned the conservation of geodiversity and geological heritage as one of its sources of inspiration; it replaced Law No 4 (1989) on the Conservation of Wild Flora and Fauna, which had been the main legal framework for nature conservation in Spain for almost 30 years, and under which many of the areas protected for nature conservation purposes were initially recognised as Natural Monuments, Nature Reserves, Natural Areas, Natural Parks and National Parks.

In the Spanish context, Natural Monuments are natural spaces or specific elements that merit special protection because of their rarity or beauty. Nature Reserves are primarily

TABLE 14.1 TABLE OF ALMERÍA PROVINCE'S PROTECTED NATURE SITES.

Designation	Name
Natural Park	Cabo de Gata-Níjar, Sierra María-Los Vélez
Natural Area	Desierto de Tabernas, Sierra de Alamilla, Karst en Yesos de Sorbas, Punta Entinas-Sabinar, Alborán
Natural Reserve	Albufera de Adra
Natural Monument	Arrecife Barrera de Posidonia, Isla de Terreros e Isla Negra, Isla de San Andrés, Piedra Lobera, Sabina Albar

biodiversity-focused areas with ecosystems, communities or biological elements whose rarity, fragility, importance or uniqueness merits special protection. National Parks are larger than Nature Reserves and Natural Parks; they can include areas transformed by human exploitation, and usually focus on particular landscapes. Natural Parks are natural areas, little transformed by human exploitation and occupation; the beauty of their landscapes and the representativeness of their ecosystems, flora, fauna or geomorphological formations possess ecological, aesthetic, educational and scientific value or regional significance that merit protection. National Parks are designated on similar grounds to the regional Natural Parks, but specifically because their areas are particularly representative of the country's natural heritage, which includes some of its principal natural systems.

Natural Areas are technically registered in Andalucía as *Parajes Naturales*, and are protected because of their unique wildlife and/or landscape. Natural Reserves are small enclaves conserving local fragile ecosystems, commonly within wetlands; access to them is usually restricted by special permits, obtainable from the Junta de Andalucía Medio Ambiente (environment) offices in each provincial capital. Natural Monuments are typically exceptional natural features, such as a centuries-old tree or a distinctive rock formation. Many habitat types registered as Natural Reserves and Natural Monuments do have an underlying active geomorphological process, rarely acknowledged in the designation and completely overlooked by most visitors. Most notified geosites (some 25) are registered as Natural Monuments. An excellent range of publications, mainly in Spanish, has been prepared by Andalucía's government, describing its natural heritage by site type and specific location; for example, Natural Parks (Consejería de Medio Ambiente de la Junta de Andalucía 2003a), Natural Monuments (Consejería de Medio Ambiente de la Junta de Andalucía 2003b), and the Gypsum Karst of Sorbas (Calaforra 2003).

Spain now has 15 National Parks (Figure 14.1A), latterly defined under Law No 41 (1997), which established a new approach to their shared management between the national and regional governments. At the regional and sub-regional level, they are reinforced by legislation designating other protected areas, such as Natural Reserves and Natural Monuments, as the regions of Andalucía (Figure 14.1B) and Almería exemplify (Table 14.1); their management is entrusted to the environmental authorities of the autonomous regions.

DESIGNATING SPAIN'S GEOHERITAGE

Spain's 1975 Natural Areas Act required the competent authorities to develop inventories of protected nature conservation areas. These were developed under the auspices of the national Nature Conservation Institute (ICONA), which created the 1976 to 1977 Inventory of Outstanding Landscapes and subsequently the 1977 to 1980 Inventory of Natural Open Spaces Warranting Special Protection (see Cortes, Barettino and Gallego 2000). Whilst the latter included some geological elements, its chief focus was on areas of biological significance. From 1978, the Geology and Mining Technology Institute of Spain (ITGE) undertook a 'National Inventory of Geological Interest' (Elizaga 1988), but progress was very slow and in 1987 the work was included within the remit of the 1:50,000 Geological Cartography project. Concomitant with the national inventory, some regional ones were also prepared; for example, the *Geological Heritage of the Autonomous Community of Madrid* (1973), the *Sites of Geological Interest in the Province of Valencia* (1983), and a catalogue and inventory for Andalucía's Málaga Province (see Cortes, Barettino and Gallego 2000). A major overview of Spanish geosites was published in 2009 (García-Cortés *et al* 2009), the Spanish culmination of the Global Geosites project promoted by the European Association for the Conservation of Geological Heritage (ProGEO), within the International Union of Geological Sciences (IUGS) and sponsored by UNESCO. The project was intended to develop subsequent geoconservation activities and also to support other programs such as UNESCO's Global Geopark Network.

Geoparks are something of a natural successor and development to Spain's protected natural areas, especially those designated under the Law of the Historic Heritage (1985) and the Law of the Conservation of the Wild Flora and Fauna (1989). Section II, Article 15, of the Law of the Historic Heritage notes that a 'historical site is [a] place or natural area linked to events or memories of the past, to popular traditions, cultural creations or creations of Nature and the works of mankind that possess an historical, ethnological, palaeontological or anthropological value'. Its Section V, Articles 41 and 42, refer to 'geological and palaeontological elements relating to the history of mankind and to his origins and background', and include archaeological excavations 'carried out with a view to describing and researching all kinds of historical and palaeontological remains and geological components relating thereto'. Significant fossil sites can be protected by the Law of the Historic Heritage as a cultural resource, with their protection entrusted to the cultural agencies of the autonomous regions in which they are located.

The Law of the Conservation of the Wild Flora and Fauna (1989) recognised two geoheritage elements. Its Article 13 states: 'Parks are natural areas that because of their beautiful landscapes and geomorphological formations have aesthetic, educational or scientific values meriting conservation'. Article 16 states: 'Natural Monuments are those geological formations, palaeontological deposits and other elements of the geosphere that are of special interest due to their uniqueness or the significance of their landscape and scientific worth.' It also recognised National Parks, which – although they are designated by the Spanish parliament, latterly by Law No 41 (1997) – are jointly managed by the state administration and the autonomous communities within which they are situated.

SPANISH GEOPARKS AND GEOTOURISM

Spain has embraced geoparks more than any other country in Europe. Its geoheritage promotion concentrates on areas of palaeontological (especially dinosaur – Fuertes-Gutiérrez *et al* 2015), karstic and volcanic interests (Calaforra and Fernández-Cortés 2006). Its geoparks meet the original UNESCO requirement of 'a territory with well-defined limits that has a large enough surface area for it to serve local economic development. The Geopark comprises a number of geological heritage sites of special scientific importance, rarity or beauty' (Anon 2000, 43). Its geoparks have been at the forefront of innovative interpretative provision. They fully engage with the European Geopark Network's (EGN) programme, promoting scientific endeavour and territorially-focused geoconservation to meet broader societal needs, on the presumption that geotourism could provide sustainable economic development (Zouros 2006). They are required to participate actively in the economic development of their area, by working with locally based enterprises to develop new products and services; these range from interpretative and educational provision (such as visitor centres and trails) to souvenir manufacture (such as pottery and replica fossils – the latter since the sale of original geological material is not permitted within geoparks), and to leisure and amenity services. Services can include adventure activities such as mountain-biking, climbing, kayaking and rafting. Geoparks are required, as part of their designation, to provide educational facilities.

Prior to the establishment of a geopark, a sustainable geotourism audit is needed; in Andalucía, and specifically Almería Province, this was undertaken by the Spanish Geosites Project. This had been initiated by the International Union of Geological Sciences Geosites Working Group, with UNESCO's support (Wimbledon 1996a; 1996b; 1998). It recognised important geosites, especially as a tourism resource, in relation to their physical accessibility and proximity to urban areas. It was supported by the European Union's LEADER-IIC programme. The recognition and enforcement of measures to protect wildlife and geological areas vary in the consistency of their application across Spain, and a limited number of evaluations have been published (see, for example, Vallejo and Duran 1999). It is therefore useful to examine one autonomous region and one of its provinces to explore their efficacy.

ANDALUCÍA'S PROTECTED LANDSCAPES

Geologically, Andalucía can be split into three regions – namely, the Iberian Massif, the Cordillara Betica and the Neogene Depressions. Whilst the first two outcrop across the rest of Spain, the latter (which has unique volcanism and readily identifiable volcanic cones and columnar lava flows) is generally restricted to eastern Andalucía. Almost a fifth of Andalucía (Figure 14.1B) is statutorily protected – the largest proportion of any autonomous region, reflecting the unspoilt nature of its countryside. The environment department of the Andalucían government, the Consejería de Medio Ambiente, is responsible for overseeing the protected areas and has offices in each provincial capital. National Parks and Natural Parks also have a local headquarters, the Oficina del Parque, located inside the protected area. Examination of Andalucía's National Parks and designated protected sites (see Figure 14.1B) in its south-eastern province of Almería (see Table 14.1) shows some of the complexity of geoheritage protection and management in a major Spanish autonomous region, and one especially popular with domestic and foreign tourists.

Andalucía's two National Parks, the Doñana and Sierra Nevada, are remarkably different in their characteristics; the former is coastal and the latter mountainous. The region's 22 Natural Parks make up the bulk of its protected areas. They show a great range of geology, climate and habitat type; these include coastal sand dunes, beaches, semi-desert steppe, mountain forests, Mediterranean woodland, salt marshes and marine zones. Natural Park legislation is designed to protect cultural and architectural traditions as well as the natural environment. Almost all of the Natural Parks offer unrestricted access, excepting a few areas where a special permit to visit is required, due to the risk of forest fires or disturbing nesting birds.

The Doñana National Park

The Doñana region, in the provinces of Huelva and Seville, shows the complexity of Spanish protected areas in action. The region includes both the Doñana National Park, established in 1969, and the Doñana Natural Park, created in 1989. The National Park is a designated UNESCO biosphere reserve, a Ramsar Wetland Site, and a UNESCO World Heritage Site. It was one of the World Wildlife Fund's first projects in 1969, when (in partnership with the Spanish government) it purchased a large tract of the Guadalquivir River delta wetlands to help create the park. The Natural Park was expanded in 1997 to create a protection buffer zone managed by the regional government. The two parks are now classified as a single natural landscape. In June 2006, responsibility for maintaining them was transferred by Royal Decree to the government of Andalucía; the functions and services of the Nature Conservation administration thus transferred to the Andalucían state were widened, and Doñana National Park and the Natural Park became the Natural Area of Doñana, a single territory divided into areas with different levels of environmental protection.

The Doñana National Park covers 543km², of which 135km² are in a protected area. Its coastline is dominated by 38km of pristine beaches on one of Spain's few remaining large stretches of undeveloped coastline. The surrounding area of the park's single natural landscape has several major geomorphosites. These lie within what are termed the pre-parks, with their associated habitats. These vary from sand-dune belts and marshes to extensive pine and eucalyptus forests. Doñana reflects the geomorphological development over several hundred thousand years of a deep underground aquifer, and surface features such as dunes and salt marsh of a major river delta system. After the end of the last glacial period, it was covered by freshwater and brackish marshes, ponds and sand dunes. However, there were some high energy marine incursion events (tsunamis and large storms). At the beginning of the Flandrian interglacial (around 12,000ya), which marked the beginning of the present-day climate regime, the melting of the old glaciers led to a period of comparatively rapidly rising global sea-levels that reached their maximum about 6,500 to 7,000ya. At that time, the Doñana National Park and the surrounding areas were flooded and a lagoon formed, later named Lacus Ligustinus by the Romans; it gradually began to infill with sediment washed in by the Guadalquivir River. The pace of infilling of the lagoon somewhat increased from then on to the present; this has been associated with the consequent accelerated growth of sand-spits and the creation of new inland marshes and wetlands.

Today, the Park's seaward extension is characterised by the flat topography of salt marshes, whilst on its landward side there are some depressions filled by temporary and/or permanent wetlands, locally called *lucios*. The whole complex is protected by the Doñana spit, a wide sandy littoral barrier with mobile dune systems growing toward the south-east. The vegetation of the mobile dunes, rarely found in Spain outside this region, is formed by the prevailing south-westerly winds and is particularly noteworthy; the gradual burial of vegetation, especially of trees that develop and are then killed by the slow encroachment of sand, is an aspect of the Park's dynamic geomorphology readily appreciated by visitors. However, very little if any of the geomorphological information is interpreted and presented to Doñana's visitors. Instead, the main focus is on presenting the park's wildlife and habitats.

Several centres with interpretative provision attempt to meet visitors' needs. The two main visitor centres are at El Acebuche and La Rocina. The former, in the incorporated Matalascañas Dune Park and Marine World Museum, is an old farmhouse; from there, visitors can access its bird hides and trails. The Bajo de Guía visitor centre, located in Sanlúcar de Barrameda, was purposefully developed for visitors to the original Natural Park. The Acebrón Palace, built in the 1960s as a private residence and hunting lodge, now serves as a park visitor centre. The Ice House visitor centre is a former early 20th century structure, sympathetically converted to its new use; located in Sanlúcar de Barrameda, it lies alongside the dock from which a boat takes visitors up the Guadalquivir River to the town of La Plancha, from which they can access the Llanos de Velázquez and Llanos de la Plancha, where there are nature observatories. There are numerous park publications and some interpretative and informational panels across its area.

Access beyond the beaches, trails, bird hides and lagoons near the visitor centres to almost all of Doñana is strictly by guided tour. Trips, usually in boats or four-wheel drive vehicles, are run from the main visitor centre in El Acebuche and the tourist office in Sanlúcar de Barrameda. The northern part of the park can be entered from the José Antonio Valverde visitor centre, about 30km south of the town of Villamanrique de la Condesa; this is down the Raya Real drovers' track between the visitor centre and El Rocío, but access is only for non-motorised traffic (horses and bicycles) and pedestrians.

The Sierra Nevada National Park

The Parque Nacional Sierra Nevada (the Sierra Nevada National Park), founded in 1999, lies within Granada and Almería. Spain's largest National Park, it covers some 858.8km^2 and has 20 peaks rising above 3,000m – the highest being Mulhacén (3,479m), Veleta (3,396m) and Alcazaba (3,371m). The rivers rising on its northern face feed into the Guadalquivir River basin and hence, ultimately, into the Doñana National Park and the Atlantic Ocean. The rivers rising on its western and southern faces (all tributaries of the Guadalfeo River) flow into the Mediterranean. There are almost 50 high-level mountain lakes. Much of the mountainous land-scape, particularly above 2,400m (the perpetual snow line, prior to and during the last Ice Age), was shaped by the action of glaciers, resulting in characteristic U-shaped valleys.

The entire Park is open to the public, but many routes are restricted to pedestrian and bicycle traffic. The area's popularity with skiers, hikers, mountain bikers and paragliders has created a market for suitable publications. An excellent full-colour guide to the Park's natural and cultural heritage was published by the government of Andalucía (Consejería de Medio Ambiente de la Junta de Andalucía 1996) and recently updated by the park authority (Parques Nacionales 2012); the latter is available in both hard-copy and online pdf formats. Both texts are in Spanish.

ALMERÍA PROVINCE AND ITS MAJOR GEOHERITAGE LOCALITIES

Almería Province provides most of Andalucía's readily accessible geoheritage attractions (Figure 14.2), with excellent geotourism provision (Hose 2007). As the prologue to a major educa-tional field-guide to the region makes clear: 'The arid landscapes of Almería are well known

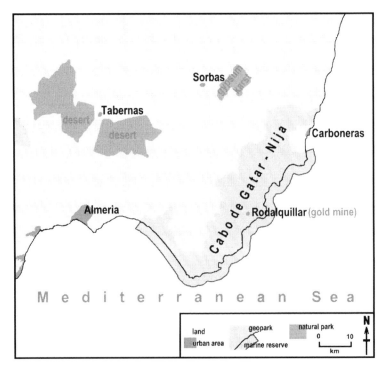

FIGURE 14.2 MAP OF SOME OF ALMERÍA'S MAJOR GEOHERITAGE SITES

amongst professors of geology ... of an astonishingly high number of European universities, who appreciate it and use it as a huge natural laboratory to carry out practical field investigation' (Villalobos 2003, 10); to the region's renowned landscapes can be added the nature and history of mineral exploitation, ranging from iron to gold and from lignite coal to uranium. Apart from the mines and quarries, there is also the infrastructure of mineral processing and transport. A most spectacular example of the latter is the cast-iron arched El Cable Inglés in the coastal resort of Almería, the regional capital. Originally constructed between 1902 and 1904 by the Alquife Mines and Railway Company Limited (a company based in Glasgow – which the Spanish seemingly thought was in England), it operated until 1970. It was used to load ships with iron ore brought in by the Linares to Almería railway from the Alquife mines. In 1998 it was protected as an Asset of Cultural Interest. In late 2010, a €2.8million project began to repair and conserve the structure. The four main geoheritage interest areas are all within an hour's drive of Almería.

Acosta and Cabo de Gata coastlands

This coastal strip, with several small villages (including Cabo de Gata, Almadraba de Monteleva and La Fabriquila), is generally known as Cabo de Gata. There are salt works at Almadraba de Monteleva, with the lagoons of Las Salinas de Cabo de Gata (Figure 14.3B) behind. Carboneras, near the easternmost border of the park, is dominated by a large cement factory working the local Miocene limestone. Its sandy Cabo de Gata beach is a major tourism asset; the adjacent Playa de los Muertos is a light pebbly beach.

Cabo de Gata-Níjar Natural Park and Geopark

The name Cabo de Gata is partly derived from the silica mineral agate that was formerly mined in the area. The Cabo de Gata-Níjar Natural Park, covering some 456.63km^2, is Andalucía's largest protected coastal area. Noteworthy for its wild, rugged and somewhat isolated landscape, it has Europe's only warm desert climate. The mountainous Natural Park is characterised by mainland Spain's most extensive (Neogene) volcanic rock formations; these include volcanic domes (Figure 14.3E), volcanic calderas and lava flows. The mountain range of the Sierra del Cabo de Gata has El Fraile (493m) as its highest peak. The range's sharp red and ochre-coloured peaks and crags abruptly fall into the Mediterranean; its 100m high cliffs are cleft by gullies that provide difficult access to some white sandy beaches. Between Cabo de Gata headland and the village of San Miguel lie salt flats, the Las Salinas de Cabo de Gata, a Ramsar site. The flats are separated from the sea by a 400m sand bar. There are numerous tiny rock islands and offshore underwater coldwater coral reefs. Almost at the park's western extremity is the Amoladeras Wash, where Quaternary fossil beaches can be seen (Figure 14.3A). The area was designated in 1987 as a UNESCO Biosphere Reserve of some 380km^2 of land and 120km^2 of Marine Reserve. In 2001 it was included among the Specially Protected Areas of Mediterranean Importance. In 2006 it joined UNESCO's Global Geoparks Network; it is also a member of the EGN. Most surprisingly, in 2010 it was proposed as a site of a dump for nuclear waste; less surprisingly, the proposal was rejected.

The Park is popular with walkers and mountain bikers. Its visitor centre at Las Amoladeras houses displays on the area's volcanic geology and natural history. There are also information points at Las Sirenas on the southernmost tip of Cabo de Gata, in La Isleta and at the Punta de los Muertos. Additionally, the main stops and viewpoints around the park have colourful information panels.

Cabo de Gata formerly had Spain's only gold mines at Rodalquilar (Figure 14.3F) (López-García et al 2011). Although gold was discovered at the end of the 19th century, its large-scale underground exploitation did not start until the mid-20th century, and the mine closed in 1996. After that, a town of 1,400 inhabitants virtually disappeared, leaving fewer than 100 residents, mainly engaged in farming and tourism. The dangerous state of the underground mine workings and the derelict surface treatment works, coupled with the isolated location and poor roads, suggest that a show mine visitor attraction is unlikely in the near future. However, it has some on-site information panels.

Tabernas Desert Natural Park

With its wadi-like river valleys and rounded eroded hills, the landscape of Tabernas is more like that of North Africa than Europe (Figure 14.3C). Its geoheritage interest is due to events that occurred in the Middle Miocene (around 8mya), when the area was flooded by a warm shallow sea that deposited mainly marls and sandstones. At the edges of this sea, roughly the present-day hills, coral reefs formed and these can be seen today. Its landscape is shaped by limited rainfall that is usually torrential and causes flooding and landslips, resulting in badlands topography.

Sorbas Gypsum Karst Natural Area and cave tourism

Around one fifth of Spain's land area is karstic, and Almería Province is especially noted for its unique gypsum karst (Figure 14.3D). Both old gypsum mines and natural limestone caves were opened up as tourist attractions from the mid-19th century. The whole area provides 'a pioneering example of how geo-resources can become an asset within the socio-economic development of rural areas, which benefit from the possibilities offered by environmental legislation' (Romero 2003, 7). A study of the area's unexploited gypsum karst caves (Calaforra and Fernández-Cortés 2002) noted the threats posed by changes in the local water regime, caused by visitors and by the services (such as lighting, in terms of both light and the heat generation) provided for them, resulting in the ingress and spread of micro-flora and fauna, which alters the appearance of the caves and damages their unique ecosystem. They noted that geotourism, even when managed as part of an ecotourism venture, requires 'monitoring and assessing in order to facilitate more comprehensive decision-making ... before opening the cave to tourists. This also means that the

visitor regime can be adapted if a change is observed relative to the conditions existing before tourism' (Calaforra and Fernández-Cortés 2002, 119).

Such a monitoring programme (Calaforra *et al* 2004) was installed in the Sorbas gypsum caves, and its data was broadcast in real time to the University of Almería, before being streamed onto the internet as an overview of microclimatic response to visitors. Visits to the Sorbas Gypsum Karst Natural Area were surveyed from an environmental economics perspective (Contreras and Calaforra 2002), and a value of €279,350 per year for tourist use of the area was identified; however, with measures put in place to restrict access, this would decrease to €41,340 per year. Three-fifths of the survey's respondents indicated that they would accept reduced gypsum exploitation in favour of its conservation. This suggests that the area's long term economic future, especially with its overall focus on mass tourism, is best assured by sustainable (geo)tourist resource development. It already has a small, privately managed, field study centre at Sorbas.

Promoting geotourism in Andalucía and Almería

In Andalucía, considerable support is evident for wildlife and geological protection and promotion measures. A range of technical and populist publications focused on wildlife, landscape and geology has been prepared under the auspices of the Tecnología de la Naturaleza (TECNA) and the Consejería de Medio Ambiente of the government of Andalucía. The latter announced in 2001 the Declaration on the Conservation of the Geodiversity of Andalucía (Anon 2001), and in 2003 the Andalucían Strategy for the Preservation of Geodiversity (Villalobos 2007, 20–22); the latter led in 2004 to the Inventory of Andalucían Geo-resources, innovatively published as a DVD. Almería Province's geotourism potential was particularly explored and promoted by a major regional conference (Primeras Jornadas Técnicas sobre Conservación y Uso Sostenible de la Geodiversidad de Andalucía), held in Almería in May 2004 (Benetez 2004). Geotourism promotion in the Province has benefited from publications developed under the auspices of TECNA.

FIGURE 14.3 SOME OF ALMERÍA'S MAJOR GEOHERITAGE SITES

A. ALMERÍA BAY – A FOSSIL BEACH (OF CEMENTED PEBBLES WITH SOME LARGE SHELLFISH FOSSILS) IN THE AMOLADERAS WASH, FORMED DURING THE QUATERNARY WHEN WARM INTERGLACIAL PERIODS MELTED THE ICE, RAISING SEA LEVELS AND FLOODING THE LAND.

B. THE SALT WORKS AT ALMADRABA DE MONTELEVA – THE LAGOONS OF LAS SALINAS DE CABO DE GATA CAN BE SEEN ON THE LANDWARD SIDE.

C. TABERNAS DESERT NATURAL PARK – THE HILLS, COMPOSED CHIEFLY OF MARLS AND SANDSTONES, COVERED WITH SPARSE VEGETATION, AND THE SEASONALLY DRY RIVER BED, WITH TREES SUCH AS THE DATE PALM, ARE TYPICAL OF THIS SEMI-DESERT OR 'BADLANDS' AREA.

D. SORBAS GYPSUM KARST NATURAL AREA – THE SPARSE SCRUB VEGETATION IS INDICATIVE OF THE LIMITED SURFACE WATER. OLD GYPSUM MINE WORKINGS AND THEIR SPOIL HEAPS CAN BE SEEN TO THE LEFT OF THE MODERN ROAD BRIDGE.

E. CABO DE GATA-NÍJAR NATURAL PARK – THE EXTINCT NEOGENE VOLCANIC CONE OF THE VELA BLANCA DOME FORMS THE HEADLAND; THE PANEL ON THE LEFT SHOWS VISITOR RESTRICTIONS WITHIN THE PARK, AND THAT ON THE RIGHT EXPLAINS (IN SPANISH) THE VOLCANIC ROCKS OF THE HEADLAND.

F. CABO DE GATA-NÍJAR NATURAL PARK – THE RODALQUILAR GOLD MINE COMPLEX IS AN EXTENSIVE DERELICT GOLD MINE, AND ITS TREATMENT PLANT (INCLUDING ONE OF THE CIRCULAR SETTLING TANKS ON THE BOTTOM RIGHT) REQUIRES CONSIDERABLE RESTORATION TO MAKE IT SAFE FOR CASUAL GEOTOURISTS.

Two major Spanish texts outline Andalucía's geodiversity and geoconservation strategy (Anon 2001) and provide an inventory of its Natural Monuments (Anon 2003a). The first text is supported by a technical DVD, *Inventario, diagnóstico y valoración de la geodiversidad en Andalucía*, and a populist DVD, *Patrimonio Geológico de Andalucía*, released by TECNA in 2004 and 2003 respectively. The second text is supported by a CD-ROM by TECNA, released in 2003. There is a tourist guide to its Natural Parks (Anon 2003b). Two videos, *Desiertos de Almería: Paisajes Geológicos Excepcionales* and *Karst en Yesos de Sorbas: Un mundo subterráneo entre el desierto y el humedal*, were produced in 2004 by TECNA, complementing commercial offerings from some of the geo-attractions, such as the DVD *La Cueva del Gato; Las Cuevas de Sorbas*, released in 2004.

A bilingual guide (Calaforra 2003) to the Sorbas Gypsum Karst was published by the Consejería de Medio Ambiente in 2003 and meets the needs of dedicated geotourists. Another major publication (Villalobos 2003) is a full-colour illustrated, if somewhat technical, guide to the Sorbas area, meeting the needs of dedicated geotourists. Villalobos (2007) is a fairly comprehensive educational full-colour geology field-guide for the whole of Andalucía, with Chapters 7 and 8 focusing on the Cabo de Gata and the Sorbas Gypsum Karst respectively. A range of populist booklets and leaflets (in Spanish) was published in the early 2000s by Consejería de Medio Ambiente, meeting the needs of casual geotourists. Almería has also attracted for academic geotourists a few research papers (see, for example, Martin *et al* 2004) and field-guides (Mather *et al* 2001).

CONCLUSION: THE GEOHERITAGE OF ANDALUCÍA AND ALMERÍA

Superb and readily accessible rock exposures, landforms, show caves and mines make Andalucía, and especially Almería Province, ideal for both informal and informed geoheritage appreciation. Their setting in dramatic and aesthetically appealing landscapes, coupled with a generally mild climate, provides ideal geotourism development opportunities. These are promoted to dedicated and casual geotourists; the former already visit the area in increasing numbers from universities in the UK and Germany, and the numbers of the latter are rising due to the inclusion of geotourism activities within general tourism offerings from the major coastal resorts, such as Almería. The economic impact of this provision has yet to be fully assessed, although it has been considered for the Sorbas Gypsum Karst Natural Area (Contreras and Calaforra 2002); further studies of that area, together with the Cabo de Gata-Níjar Geopark, could provide one of Europe's best geoheritage economic impact case studies.

The fragile nature of many geosites, coupled with their susceptibility to irreparable damage by hammering and over-collecting, requires that active protection and geoconservation measures are implemented. The subterranean geomorphosites are susceptible to unintentional damage from the presence of visitors and the facilities provided for them; restricting access through permits and shorter opening hours can help mitigate these issues. Exhortations alone, no matter how carefully presented in interpretive media (whether in publications, on-site panels or displays), are unlikely to change visitors' behaviour enough to prevent on-site damage. The widespread availability of fairly priced good-quality texts and multi-media materials in Spanish and English promotes the region's geoheritage and recognition of its uniqueness, and might well result in greater public support for its protection by enhanced nature protection legislation and planning procedures. Geoheritage protection and promotion in Andalucía, along with geotourism's development in Almería Province, has developed rapidly with some measure of success, in terms of geosite/geomorphosite designations, protection and interpretation.

Together the sites provide an excellent case study of a varied geoheritage that has only comparatively recently been studied and protected, compared with countries in northern and western Europe. Unlike countries such as the UK and Italy, its legacy geoheritage of past geological study and collections is very limited; most of its secondary geosites are newly created visitor centres found in the National Parks and Natural Areas. Consequently, its geotourism emphasises primary geosites, supported by a range of recent materials published over the past 25 or so years, explored in the field. The combination of superbly exposed and interpreted geosites and geomorphosites with a generally amenable climate makes it a prime future geotourism destination.

Acknowledgments

Thanks to Miguel Villalobos for much help in the field, and for freely supplying numerous field-guides and other publications on Andalucía's geology.

References

Alcala, L, 1999 Spanish steps to geoconservation, *Earth Heritage* 11, 14–15

Anon, 2000 *UNESCO Geoparks Programme Feasibility Study*, UNESCO, Paris

Anon, 2001 *Propuesta de Estrategia Andaluza de Conservación de la Geodiversidad*, Consejería de Medio Ambiente, Junta de Andalucía, Granada

Anon, 2003a *Monumentos Naturales de Andalucía*, Consejería de Medio Ambiente, Junta de Andalucía, Granada

Anon, 2003b *Parques Naturales de Andalucía: Guía del Viajero*, Consejería de Medio Ambiente, Junta de Andalucía, Granada

Benetez, A, 2004 La Junta apuesta por la conservación de recursos y por un turismo geológico, *La Voz de Almería*, 21 May, 15

Calaforra, J M, 2003 *El Karst en Yeso de Sorbas* [*The Gypsum Karst of Sorbas*], Consejería de Medio Ambiente, Junta de Andalucía, Calle Mayor

Calaforra, J M, and Fernández-Cortés, A, 2002 Cave management: what to do before making suitable a tourist cave, paper presented at *The 4th Samcheok International Cave Symposium: The Sustainable Management of Cave: Academic and Policy Implications*, 10 July 2002, Gangwon, Korea, and published in their proceedings, 118–27

— 2006 Geotourism in Spain: resources and environmental management, in *Geotourism, Sustainability, Impacts and Opportunities* (eds R Dowling and D Newsome), Elsevier, Oxford, 199–220

Calaforra, J M, Fernández-Cortés, A, and Gázquez-Parra, J, 2004 Low-cost telemetry monitoring of the cave environment: Sorbas Gypsum Karst, Spain, *Transactions of the British Cave Research Association* 31 (1), 37–41

Consejería de Medio Ambiente de la Junta de Andalucía (ed), 1996 *Guía del Parque Natural Sierra Nevada*, Consejería de Medio Ambiente de la Junta de Andalucía, Granada

— 2003a *Parques Naturales de Andalucía: Guía del Viajero*, Consejería de Medio Ambiente de la Junta de Andalucía, Granada

— 2003b *Monumentos Naturales de Andalucía*, Consejería de Medio Ambiente de la Junta de Andalucía, Granada

Contreras, S, and Calaforra, J M, 2002 Valoración contingente del patrimonio karstico: el caso del karst en yesos de Sorbas (Almería), in *Karst and Environment* (eds F Carrasco, J J Duran and B Andreo), Fundación Cueva de Nerja, Nerja, 350–68

Cortes, A G, Barettino, D, and Gallego, E, 2000 Inventory and cataloguing of Spain's geological heritage, an historical review and proposals for the future, in *Geological Heritage: Its Conservation and Management* (eds D Barretino, W A P Wimbledon and E Gallego), Sociedad Geológica de España/Instituto Tecnológico Geominero de España/ProGEO, Madrid, 47–67

Elizaga, E, 1988 Georrecursos Culturales, in *Geología Ambiente* (ed Anon), Instituto Tecnológico Geominero de España, Madrid, 85–100

Elorza, M G (ed), 1994 *Geomorfología de España*, Editorial Rueda, Madrid

Fuertes-Gutiérrez, I, García-Ortiz, E, and Fernández-Martínez, E, 2015 Anthropic threats to geological heritage: characterization and management: a case study in the dinosaur tracksites of La Rioja (Spain), *Geoheritage* 7 (1), 1–19

García-Cortés, A, Villar, J A, Suárez-Valgrande, J P, and González, C I S (eds), 2009 *Spanish Geological Frameworks and GEOSITES: An Approach to Spanish Geological Heritage of International Relevance*, Instituto Geológico y Minero de España, Madrid

Gibbons, W, and Moreno, T (eds), 2002 *The Geology of Spain*, The Geological Society, London

Gutiérrez, F, and Gutiérrez, M (eds), 2014 *Landscapes and Landforms of Spain*, Springer, New York

Hose, T A, 2007 Geotourism in Almería Province, southeast Spain, *Tourism: An Interdisciplinary Journal* 55 (3), 259–76

Instituto Geológico y Minero de España, 2004a *Mapa Geológico de España con la inclusión de Portugal continental y Pirineos Franceses, Escala 1:200,000*, Instituto Geológico y Minero de España, Madrid

— 2004b *Mapa Tectónico de España con la inclusión de Portugal continental y Pirineos Franceses, Escala 1:200,000*, Instituto Geológico y Minero de España, Madrid

López-García, J A, Oyarzun, R, Andrés, S L, and Martínez, J I M, 2011 Scientific, educational, and environmental considerations regarding mine sites and geoheritage: a perspective from SE Spain, *Geoheritage* 3, 267–75

Mather, A E, Martin, J M, Harvey, A M, and Braga, J C (eds), 2001 *A Field Guide to the Neogene Sedimentary Basins of the Almería Province, SE Spain*, Blackwell, Oxford

Parques Nacionales, 2012 *Guía de Visita El Parque Nacional de Sierra Nevada*, Parques Nacionales

Romero, J D, 2003 Forward, in *The Gypsum Karst of Sorbas* (ed J M Calaforra), Consejería de Medio Ambiente, Junta de Andalucía, Calle Mayor, 7

Vallejo, M, and Duran, J J, 1999 Serrezuela de Carratraca (Málaga, Southern Spain): A small spot with a diverse geological heritage, in *Towards the Balanced Management of the Geological Heritage in the New Millennium* (eds D Barettino, M Vallejo and E Gallego), Sociedad Geológica de España/Instituto Tecnológico Geominero de España/ProGEO, Madrid, 374–7

Vera, J A (ed), 2004 *Geología de España*, Sociedad Geológica de España and Instituto Geológico y Minero de España, Madrid

Villalobos, M (ed), 2003 *Geología del entorno árido almeriense: guía didáctica de campo*, Tecnología de la Naturaleza, Granada

— 2007 *Geodiversity and Geological Heritage of Andalusia: Geological Excursions through Andalusia, an Educational Field Guide*, Tecnología de la Naturaleza, Granada

Villalobos, M, and Guirado, J, 1999 Tourist promotion and economic use of the geological patrimony on the protected natural spaces of the sub-desertic environment in Almería (Spain), in *Towards the Balanced Management of the Geological Heritage in the New Millennium* (eds D Barettino, M Vallejo and E Gallego), Sociedad Geológica de España/Instituto Tecnológico Geominero de España/ProGEO, Madrid, 425–9

Wimbledon, W A, 1996a National site selection, a stop on the road to a European geosites list, *Geologica Balcania* 26 (1), 5–27

— 1996b GEOSITES: a new conservation initiative, *Episodes* 19, 87–8

— 1998 A European geosites inventory: GEOSITE – an International Union of Geological Sciences initiative to conserve our geological heritage, in *Comunicaciones de la IV Reunión Internacional de Patrimonio Geológico* (eds D Barettino, J J Duran and L Lopez), Sociedad Geológica de España, Madrid, 15–18

Zouros, N, 2006 The European Geopark Network: geological heritage protection and local development – a tool for geotourism development in Europe, in *4th European Geoparks Meeting – Proceedings Volume* (eds C Fassoulas, Z Skoula and D Pattakos), 15–24

15

Geoheritage Case Study:
Geotourism and Geoparks in Scotland

JOHN E GORDON

'Scotland small? Our infinite, our multiform Scotland, *small?*'

Hugh MacDiarmid, *Direadh I*, 1974

Written in the context of Scotland's landscapes, nature and people, the words of Hugh MacDiarmid (1892–1978), poet, polemicist and influential 20th century Scottish writer, apply particularly well to Scotland's geodiversity. Within a land area of 80,060km², including over 900 islands, a coastline extending to some 18,000km, and an offshore area of 88,448km² inside territorial waters extending out to 12 nautical miles (19.3km) (Baxter *et al* 2011), Scotland contains a remarkably rich geodiversity. Rocks from all the periods of the geological column are present (Trewin 2001), from Archean Lewisian gneisses (c. 2,900 million years old) to Quaternary glacial deposits. This geodiversity reflects global plate tectonics and the changing positions of Scotland's terranes in relation to active and passive plate margins during different periods of supercontinent formation and break-up, continental drift from high southern latitudes in the Neoproterozoic (1,000 to 541 million years ago) to present temperate latitudes, and the effects of long-term changes in climate and sea-level, including Quaternary glaciations (Gordon 2010). Large-scale crustal deformation, mountain building, igneous activity (plutonic and volcanic), and erosion and deposition in a range of sedimentary environments, from deep oceans to shallow seas and from tropical swamp to hot desert and cold environments, have all shaped the present landscape. They are represented in the rock record, together with a wide range of geomorphological features associated with glaciation and postglacial coastal, fluvial, periglacial and mass movement processes (Friend 2012; Gillen 2013; McKirdy *et al* 2007; Trewin 2002). In essence, the present landscape of Scotland is a geodiverse and multi-faceted palimpsest, a witness to past and present Earth processes and a product of their cumulative effects over geological time. Scotland's rocks and landforms therefore reveal remarkable stories about the evolution of Planet Earth, with chapters on oceans opening and closing, mountains forming and eroding, tropical seas, volcanoes, deserts and ice ages. These stories have provided the basis for interpretation and promotion of geoheritage in Scotland and for the development of geopark-based geotourism (Gordon 2012a; Gordon and Kirkbride 2009; Gordon *et al* 2004; McKeever *et al* 2006; McKirdy *et al* 2001; Threadgould and McKirdy 1999).

Geodiversity has also been a powerful and varied influence on human activities since Mesolithic people first arrived in Scotland some 10,000 years ago. For example, it has affected people's patterns of movement, settlement and use of land for food production and recreation, supported industrial development, and inspired landscape appreciation, art, architecture, literature and poetry (Gordon 2012b). From the construction of Neolithic stone monuments to the

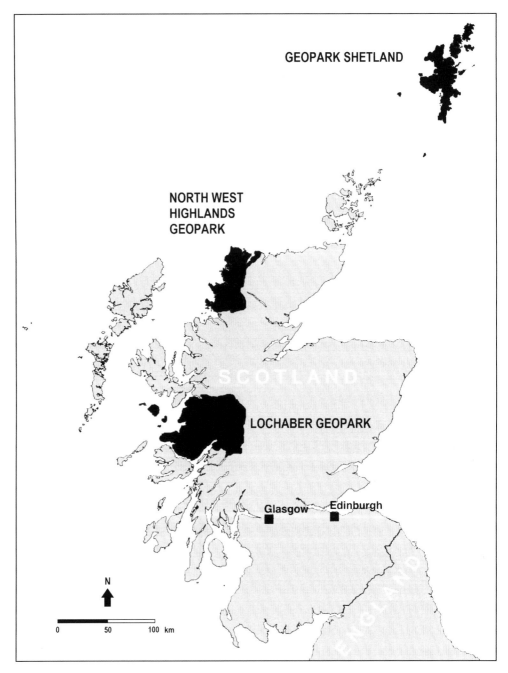

FIGURE 15.1 LOCATION MAP OF SCOTLAND'S THREE GEOPARKS

modern designation of geosites for conservation purposes, people have placed different cultural values on Scotland's geoheritage – those elements of geodiversity considered to have significant intrinsic, scientific, educational, cultural, aesthetic and ecological qualities (Crofts and Gordon 2014). Today, geodiversity is also valued for its wide range of ecosystem services (Gordon and Barron 2013; Gordon *et al* 2012; Gray *et al* 2013). These include the provision of assets for geotourism, which helps to promote geoheritage and the need for its conservation, as well as delivering socio-economic benefits for local communities. This is reflected in the growth of the Global Geoparks Network under the umbrella of UNESCO (McKeever *et al* 2010), the aspirations of which are embedded at a national level in Scotland's Geodiversity Charter (Scottish Geodiversity Forum 2013).

This chapter assesses the contribution and progress of geoparks in Scotland (Figure 15.1) in the context of geotourism and the changing values placed on the physical landscape, from the beginnings of Romantic (geo)tourism in the mid- to late-18th century to the present day, when the concept of geoheritage has become more clearly articulated and formalised.

APPRECIATING SCOTLAND'S GEOHERITAGE: CHANGING SCIENTIFIC AND CULTURAL VALUES

Smout (1993) noted two main progressions in the development of environmental awareness in Scotland: first, from a Romantic view of nature in the 18th and 19th centuries as an aesthetic experience, to a scientific view of nature in the 20th century as a focus for study and conservation in protected areas; and second, from the latter to a recognition of the need for sustainable use of the environment that combines both the Romantic and the scientific viewpoints, and embodied now in the ecosystem approach. Appreciation of geoheritage in Scotland has essentially followed a similar pathway. In the 18th and 19th centuries, geoheritage was valued both for aesthetic and scientific reasons, while in the second half of the 20th century, scientific and educational values were strongly emphasised. Today, alongside the latter, there is better awareness of the wider aesthetic, cultural, historical, economic and ecological values of geoheritage (Gordon and Barron 2012, 2013; Scottish Geodiversity Forum 2013), in line with international trends (Coratza and Panizza 2009; Dowling and Newsome 2010; Gray 2013; IUCN 2012), and this broader approach has been a pillar of geopark development in Scotland.

Scientific values

Many places in Scotland are of great national and international importance to geoscience for their rocks, fossils or landforms. This is recognised in the Geological Conservation Review (GCR), the systematic assessment of key geosites in Great Britain for scientific research and education (Ellis 2011). As noted by Gordon and Barron (2011), these are highly valued for:

- the length of the preserved geological record, plate tectonic history and diversity of palaeo-geographies, palaeoenvironments and geological processes, and the extent to which these phenomena are exposed at surface and therefore accessible for observation and study;
- their role in the history of geology and in the understanding of geological processes in the past (eg volcanism, crustal deformation, glaciation), and their modern counterparts;
- the rich and diverse fossil record that spans different palaeoenvironments and critical moments in the evolution of life.

Evidence from Scotland's geodiversity has played a vital part in the development of modern geoscience, since the conceptual insights of James Hutton in the late 18th century about the great length of geological time, the rock cycle and the continuity of past and present processes shaping the Earth (Hutton 1785, 1788; Playfair 1802). Hutton's ideas provided a foundation for the seminal work of Charles Lyell (1797–1875) and the development of the principle of uniformitarianism (Lyell 1830–33). Among many highlights, Scotland's geodiversity and geoscientists have helped to progress geoscience understanding through fundamental discoveries about geological time and geological processes, including thrust belts and thrust tectonics, cauldron subsidence, metamorphic zonation, Palaeogene volcanism, Quaternary glaciation, and the evolution of the plant and animal kingdoms (Gordon and Barron 2011). Many of the key localities now attract geologists from around the world, both for scientific study and as 'dedicated geotourists' (Hose 2010). Most, but not all, of the GCR sites have statutory protection as Sites of Special Scientific Interest (SSSIs). When appropriately interpreted, many of them, together with non-statutory Local Geodiversity Sites – selected for a broader range of criteria, including their educational, aesthetic and historical significance – can also contribute to raising public awareness and appreciation of geodiversity and its geoheritage values, through enhancing the experience of the more general visitor.

Cultural values: from the 'Romantic sublime' to modern geoconservation

Perceptions of the cultural values of geodiversity and geoheritage have changed over time, and are linked with parallel developments in geoscience, landscape aesthetics and nature conservation. The origins of geotourism in Scotland lie in the Romantic movement of the 18th century and the representation of the landscape in contemporary travel journals, literature and art (Hose 2010; Gordon 2012a; Gordon and Baker 2015). The aesthetics of the sublime and picturesque (Burke 1759; Gilpin 1792) completely changed visitors' perceptions of the Highlands from a bleak and sterile landscape that encouraged emotions of horror and repulsion to one that provoked feelings of either awe and admiration (Burke's 'delightful terror'), particularly in the presence of geological 'wonders', or appreciation of the compositional qualities of the 'natural' landscapes (Andrews 1987; Grenier 2005). Localities such as Staffa (Figure 15.2A), the Falls of Clyde, the Falls of Foyers, the Falls of Braan, Loch Lomond, The Trossachs and Loch Coruisk on the Island of Skye became essential Romantic tourist destinations, recommended in the journals of early travellers and numerous contemporary guidebooks (eg Gilpin 1789; Murray 1805; Pennant 1776; Wordsworth 1874). Many of these localities acquired strong literary associations that greatly enhanced their appeal – for example, through James Macpherson's Ossianic poetry (Macpherson 1765) and later the hugely popular poetry and novels of Sir Walter Scott. Pictorial representations by J M W Turner that accompanied Scott's publications, and those by many other painters in contemporary works of art and picture books, also heightened the sense of wonder and curiosity about the physical landscapes and natural wonders from which writers and artists drew inspiration (MacLeod 2012). Thus, in the popular imagination, geology was appreciated through the physical landscape as interpreted by writers and artists (Gordon 2012a; Hose 2010).

The great expansion of popular tourism in the 19th century, with the advent of organised tours facilitated by steamships and the development of the railway network, was accompanied by a more utilitarian approach to the landscape that declined into Victorian sentimentalism and the superficiality of tartan imagery. At the same time, the wealthier classes, prompted by the acquisition of the Balmoral Estate on Deeside by Prince Albert and Queen Victoria in 1852,

Figure 15.2 Examples of geotourism destinations and on-site interpretation in Scotland
A. Picture postcard by Valentine & Sons of (geo)tourists landing from a steamship at Fingal's Cave on Staffa in 1898. Following a report by Sir Joseph Banks in 1772, Fingal's Cave quickly became an essential Romantic (geo)tourism destination that retains its appeal to the present day.
B. Volcanic geo-art and interpretation board at Stennes, Eshaness, on 'Shetland's Volcano Trail', Geopark Shetland. The installations explain the origins of the Eshaness peninsula, formed from the remains of a large volcano that was active during the Devonian between 400 and 350mya. The bay at Stennes provided a sheltered location for a fishing station that operated up until the end of the 19th century. (Photo copyright: Paul Harvey.)
C. 'An Earth Moving View': interpretation board at Loch Glencoul, on the North West Highlands 'Rock Route', North West Highlands Geopark. The board interprets the Glencoul Thrust, one of the classic structures of the Moine Thrust Zone. A layer of Cambrian rocks on the far side of the loch lies between the Lewisian basement and the overlying Lewisian rocks of the Glencoul Thrust Sheet. (Photo copyright:John Gordon.)
D. Ice Age interpretation board at the Parallel Roads of Glen Roy, Lochaber Geopark. The Parallel Roads, displayed on the far hillside, are the shorelines of a glacier-dammed lake formed during the Loch Lomond (Younger Dryas) Stade (c.12.9–11.7ka). They played a key part in convincing Louis Agassiz during his visit in 1840 that glaciers had formerly existed in Scotland. (Photo copyright: John Gordon.)

purchased Highland estates and hunting lodges in increasing numbers and exploited opportunities for exclusive leisure pursuits of shooting and fishing, as celebrated for example in the art of Sir Edwin Henry Landseer (1802–73). These trends involved stereotyped images and overlooked the human element in the landscape and the deep connections of local people with their environment (Hunter 1995), revealed for example in the Gaelic nature poetry of Alasdair Mac Mhaighstir Alasdair (Alexander MacDonald) (c.1695–1770) and Donnchadh Bàn Mac an t-Saoir (Duncan Ban MacIntyre) (1724–1812). While there were exceptions – for example, in the seascape paintings of William McTaggart (1835–1910) and later in the fiction of Lewis Grassic Gibbon (1901–35), both of whom recognised and responded to the presence of people in the landscape – the Victorian values were most effectively contested in the Scottish renaissance of the first half of the 20th century. This was inspired by Patrick Geddes (1854–1932) and expressed, for example, through the paintings of the Scottish Colourists and those of William Gillies (1898–1973), William Johnstone (1897–1981) and James McIntosh Patrick (1907–98), and particularly in the works of Hugh MacDiarmid (1892–1978) and in the poetry of Somhairle MacGill-Eain (Sorley Maclean) (1911–96), Norman McCaig (1910–96) and others who used nature and the physical landscape as a source of metaphors to reflect on the state of Scotland, the people who lived – or had lived – in the landscape, and the wider human condition (Gordon 2012b).

Over a similar time period, significant changes occurred in approaches to outreach, interpretation and public awareness of geology. The starting point was the recognition by James Hutton that geology 'may afford the human mind both information and entertainment' (Hutton 1785, 30). This broader appeal was developed in the 19th century, notably through the publications of Charles Lyell, Hugh Miller (1802–56) and Archibald Geikie (1835–1924), which stimulated popular interest among the reading public in an increasingly scientific discipline, through infusing a sense of wonder that linked geology and cultural heritage (Gordon 2012a; 2012b); while Geikie's *Scenery of Scotland* (2 and 3 edns 1887, 1901) included the first detailed modern geotourism itineraries in Scotland.

The 19th century, too, saw the first developments in geosite conservation at Salisbury Crags in Edinburgh (1845) and Fossil Grove in Glasgow (1887), and later at Agassiz Rock (1908) in Edinburgh (Thomas and Warren 2008), as well as a scheme to conserve vulnerable erratic boulders (Milne Home 1872). Systematic site conservation underpinned by legislation began in the late 1940s, with the rise of the nature conservation movement in Great Britain and the passing of the National Parks and Access to the Countryside Act (1949) (Prosser 2013). This and subsequent legislation has enabled the designation of SSSIs for fauna, flora and geological and physiographical features. Sites are selected purely for their scientific value and, in the case of geosites since the late 1970s, through the GCR process.

The post-1940s' scientific emphasis was also reflected in a predominantly didactic approach to raising awareness and outreach, based on 'reading the landscape' (Gordon *et al* 2004) and exemplified in the highly successful series of *Landscape Fashioned by Geology* booklets produced by Scottish Natural Heritage (SNH) and the British Geological Survey (BGS) (http://www.snh.gov.uk/about-scotlands-nature/rocks-soils-and-landforms/geodiversity-in-your-area). However, for a more general audience without a background or interest in geology or geomorphology, there has often been an over-emphasis on presentation of information at the expense of appropriate interpretation (Gordon and Kirkbride 2009). Today, with more awareness of best practice in interpretation, the interests of these visitors are better addressed through complementary experiential approaches that engage more with people and draw on cultural and aesthetic values (Gordon

2012a; 2012b). There has also been a shift towards recognising the wider benefits of geocon-servation and its role in sustainable development, as well as its scientific values, and placing geoheritage in the context of the ecosystem services paradigm and the Scottish Government's sustainability agenda (Gordon and Barron 2012). Recognition of such multifunctional values is explicit in Scotland's Geodiversity Charter, which sets out a broader vision and strategic frame-work for geodiversity in Scotland (Scottish Geodiversity Forum 2013), and in the development of Scotland's geoparks, which aim to integrate the promotion of geoheritage and economic development through tourism, while at the same time delivering geoconservation.

GEOHERITAGE, GEOTOURISM AND SCOTLAND'S GEOPARKS

In 2014, more than 15.5 million overnight tourism trips were taken in Scotland, for which visitor expenditure totalled £4.8 billion (VisitScotland 2015). The prime reason for visiting is the scenery and landscape, which motivate more than half of visitors to holiday in Scotland (VisitScotland 2013). Promotion of geoheritage has great potential to enhance visitors' enjoyment and appre-ciation of Scotland and to add value to their experiences (Scottish Geodiversity Forum 2014). Geoheritage can attract people to stay longer in an area, go to areas that they might not otherwise visit, and encourage them to return to explore further. Awareness of geoheritage also leads to better appreciation of other heritage aspects, since it underpins biodiversity, land use, settlement, culture and history. Geoparks are therefore particularly well placed to contribute to tourism through targeted marketing to promote Scotland to those seeking to experience natural landscapes. Such initiatives also accord with the European Union's growth strategy, *Europe 2020*, which recognises the importance of enhancing the competitiveness of the European tourism sector (European Commission 2010a), and with proposals for a new framework for coordinated tourism actions at an EU level to increase the capacity for sustainable growth of European tourism (European Commission 2010b). At a national level, they also align with the Scottish Government's Economic Strategy, which identifies sustainable tourism as a key sector with high growth potential (Scottish Government 2011), with 'nature, heritage and activities' identified as one of the special asset groups in the industry-led National Tourism Strategy (Scottish Tourism Alliance 2012).

Scotland has three geoparks – the North West Highlands, Shetland and Lochaber – all in areas with outstanding geoheritage and spectacular mountain and coastal landscapes (Figure 15.1; Figure 15.2; Table 15.1). The North West Highlands Geopark and Geopark Shetland are members of the European Geoparks Network (EGN), part of the wider Global Geoparks Network (GGN). Membership of the EGN is reassessed every four years, with both Shetland and North West Highlands successfully revalidated in 2013 and 2015, respectively. Lochaber Geopark, a former member, was obliged to default from the EGN in 2011 because of insufficient funding to meet minimum EGN membership requirements, but still functions as a national geopark and plans to apply for re-admission. All three geoparks are geographically located where they should be able to make a significant contribution to sustainable rural development, one of the prime drivers of the geopark movement (Eder and Patzak 2004; McKeever *et al* 2010). They have support from their respective local authorities but operate independently, working in partnership with local and national bodies. Since their inception, they have received funding from a range of local, national and European sources, mainly in the public sector, but essentially on a short-term basis, and they have lacked long-term, sustainable funding. In the face of impending financial difficul-ties and the prospect of North West Highlands and Shetland also having to withdraw from the

TABLE 15.1 SCOTLAND'S GEOPARKS: GEOHERITAGE HIGHLIGHTS AND MAIN ACTIVITIES

Geopark	Year admitted to the EGN	Area (km²)	Geoheritage highlights	Main activities and visitor facilities	Business model
North West Highlands (http://www. nwhgeopark. com/)	2004	2,000	Some of the oldest rocks in the world (3,000 Ma Archean Lewisian gneisses); historically significant classic area for thrust belt tectonics (Moine Thrust Zone); earliest evidence of complex-celled life in Europe (1,200 Ma stromatolites in Torridonian rocks); unique landscape of Torridonian sandstone 'inselbergs' resting on ice-scoured Lewisian basement; glaciated landscapes; beaches and coastal landforms; limestone caves with associated fossil remains including reindeer, polar bear and lynx; strong cultural links through influences of geodiversity on land use, literature, poetry and art.	Geopark visitor centre and central hub with exhibition and cafe at Unapool; Scottish Natural Heritage visitor centre at Knockan Crag National Nature Reserve; North West Highlands Rock Route interpretation; ranger-led guided walks and geo-tours; schools engagement and activities; hosted the European Geopark Network Conference (2007); roadshows; family days; celebration of the centenary of the publication of the classic Geological Survey Memoir on *The Geological Structure of the North-West Highlands of Scotland* (Peach *et al* 1907); distance learning course in Earth sciences delivered through day schools, including Geopark fieldwork; Geodiversity Audit and related Local Geodiversity Action Plan completed; future plans to establish a Geopark quality brand and trail development supported by apps.	North West Highlands Geopark is a social enterprise, limited by guarantee, owned and operated by the 5 Geopark communities; currently appointed Information Manager, Business Manager and Marketing Manager.

Geopark	Year admitted to the EGN	Area (km²)	Geoheritage highlights	Main activities and visitor facilities	Business model
Lochaber (http://www. lochabergeopark. org.uk/)	2007*	4,648	Historically significant classic area for cauldron subsidence (Glencoe, Ben Nevis); Jurassic fossils and historical links with Hugh Miller (Eigg); Palaeogene volcanic activity, palaeoenvironments and fossils (Ardnamurchan, Rum, Eigg, Muck, Canna); all three divisions of the Moine Supergroup of metamorphic rocks are defined in Lochaber, as is the Sgurr Beag Thrust; representation of the Dalradian Supergroup of metamorphic rocks; mining history (Strontian); Great Glen Fault; eroded remnants of Caledonian Mountain chain; classic landforms of glaciation (Parallel Roads of Glen Roy, Glencoe, Rum), including historical significance of the Glen Roy area in the development of the glacial theory by Louis Agassiz and historical link of the Parallel Roads of Glen Roy with Charles Darwin.	Geopark hub at Roy Bridge at the entrance to Glen Roy, with information, display, Darwin-themed café and shop; office in Fort William; 22 interpretation panels, with 'Rock Routes' leaflet to locate them; interpretation leaflets for 8 Geotrails; booklet on *The Birth of Ben Nevis and the Story behind Lochaber's Landscapes*; free video on *The Story of Ben Nevis* available from website; rock safaris, some in support of Wild Lochaber Festival; talks and guided walks for primary and secondary schools; training delivered to rangers from The Highland Council, the National Trust for Scotland, the Forestry Commission and others; support for the Lochaber 'University of the Third Age' Geology Group; interpretation support and special events for local institutions, including the West Highland Museum, the National Trust for Scotland Glen Coe Visitor Centre, Ardnamurchan Lighthouse Museum and the Ariundle Centre; future plans to produce a walker's guide and map to the geology of the Ben Nevis area in partnership with the British Geological Survey and the Nevis Landscape Partnership, supported by the Heritage Lottery Fund.	Lochaber Geopark Association is a community based company limited by guarantee with charitable status; currently appointed Project Manager, Assistant and Business Analyst.

(CONT.)

Table 15.1 (cont.).

Geopark	Year admitted to the EGN	Area (km²)	Geoheritage highlights	Main activities and visitor facilities	Business model
Shetland (http://www.shetlandamenity.org/geopark-shetland)	2009	1,468	Exceptional geodiversity, with rocks of every era from the Precambrian to the Carboniferous brought together in a relatively narrow zone by faulting; the most complete section through ancient ocean crust in the British Isles (the Shetland Ophiolite); the best section through the flank of a volcano in the British Isles (Northmavine Igneous Complex); glaciated landscapes; modern high-energy coastal processes and spectacular coastal landforms (cliffs, storm-beach deposits, tombolo).	Geology themed events; resources and programme of activities for schools; imaginative exhibits, installations (geological walls) and interpretive panels at key geological locations, museums and heritage centres; self-guided trails; lead role in organising the annual Shetland Nature Festival; endorsement scheme for tourism businesses; led development of a tourism app for iPhone and Android with Geopark partners from France, The Netherlands and England; support for lifelong learning through workshops, courses and field trips for schools and adult learning night classes; lead member of the Shetland Environmental Education Partnership – a multi-agency group developing activities, resources, and in-service training to benefit all Shetland schools; Geodiversity Audit and Action Plan completed with 9 geosites designated as Local Geodiversity Sites; currently part of a Shetland Environment Partnership formed to contribute to the Community Plan as part of the Single Outcome Agreement; working on 'Northern Georoutes' project and development of Shetland Sculpture Trail inspired by Shetland's geology and landscape in partnership with Shetland Arts Development Agency.	Geopark Shetland operates under the management of the Shetland Amenity Trust (as a charitable trust); currently appointed Geology Project Officer with access to support for marketing, finance, admin, interpretation, management and business planning through the Trust.

* withdrew 2011

EGN, the Scottish Government provided a total of £377,000 to the three geoparks from 2013 to 2015 to enable the development of a more sustainable approach, recognising the community benefits of geoparks, their importance for the nation and their contribution in helping to deliver the government's strategic objectives (see Gordon and Barron 2012). The funding requires the Scottish geoparks to work together and develop long-term business plans.

The aims of Scotland's geoparks are similar and align with the UNESCO guidelines and criteria (UNESCO 2014). They include conservation of geodiversity and geoheritage; raising awareness of geoheritage and its links with natural and cultural heritage and landscape, both locally and among visitors from the UK and overseas, through provision of appropriate inter-pretation; encouraging educational initiatives, training opportunities and research activities that relate to local geoheritage; promoting sustainable economic and social development linked to geotourism, and the wise use of geoheritage resources in partnership with local communities and businesses and with other geoparks; ensuring long-term financial stability and resourcing for the Scottish geoparks; and providing measurable social and economic benefits to local communities. All three geoparks have committed to Scotland's Geodiversity Charter (Scottish Geodiversity Forum 2013) and to work with the Scottish Geodiversity Forum to help deliver the Charter's objectives.

Geopark activities have focused principally on providing geological interpretative material and facilities for visitors, working with schools and local community projects, delivering life-long learning and promoting their regions (Figure 15.2; Table 15.1) (Barton 2014). Much of the success of these activities reflects the energy and commitment of the few geopark staff able to be appointed on the limited grant-aid budgets available, and supported by dedicated local geology enthusiasts and volunteers. One particular highlight was the hosting by North West Highlands Geopark of the European Geoparks Conference in 2007, which injected an estimated £400,000 into the local economy. While there is no breakdown for individual geoparks, it is estimated that, over the UK as a whole, the seven UK members of the EGN (Shetland, North West Highlands, North Pennines AONB, English Riviera, Fforest Fawr, GeoMôn and Marble Arch Caves) are worth a total of £18.84 million to the UK economy annually (UK National Commission for UNESCO 2013).

There has been more limited progress on targeted geoconservation activities and the system-atic designation of Local Geodiversity Sites, although Shetland Geopark and the North West Highlands Geopark have each produced a geodiversity audit and action plan. There has also been mixed progress in integrating geodiversity into local policies and business and marketing activities. The most successful approach has been in Shetland, where the geopark is adminis-tered through the Shetland Amenity Trust, enabling integration of geointerpretation with that of nature, archaeology and local culture. Future development of the geopark there is based on the Geopark Shetland Strategy and Action Plan, the Geopark Shetland Marketing Strategy and the Geopark Shetland Geodiversity Audit. These are linked closely with the Shetland Islands Council Corporate Plan, the Shetland Structure Plan and the Shetland Cultural Strategy. Nevertheless, despite such integration, Geopark Shetland has still struggled to secure sustainable, long-term core funding. As yet, the North West Highlands Geopark has been unable to embed its geodi-versity action plan into Highland Council policy-making, despite the Council's signing up to Scotland's Geodiversity Charter.

The recent funding from the Scottish Government has enabled the geoparks to initiate a more strategic approach that is essential to secure their future. They also need to set in place a stronger interface with the business community, to attract both local and inward private sector investment as well as public sector funding. Consequently, all three Scottish geoparks are currently working on individual sustainable business plans. North West Highlands Geopark has appointed an information manager, business manager and marketing manager to deliver against a Development and Action Plan and a Local Geodiversity Action Plan, and Lochaber Geopark has appointed a business analyst. Also, as part of a move towards greater cooperation, the Scottish Geoparks Partnership was formed in 2013 to share information and ideas, pursue development opportunities and strengthen Scotland's position within the wider geopark movement.

Significant opportunities exist at an international level to attract funding through collaborative initiatives, particularly for tourism-related projects. For example, Geopark Shetland was the lead partner in 'Heritage Interpretation Using New Technologies' – a LEADER-funded transnational project with North Pennines AONB and Geopark (England), Chablais (France) and Geopark de Hondsrug (the Netherlands). This project, which included development of a map-based tourism app for Geopark Shetland, was completed in 2013 (www.hintproject.eu). Use of new technologies such as iGeology (http://www.bgs.ac.uk/igeology/) and interactive games (eg Rock Operator – https://itunes.apple.com/gb/app/rock-operator/id468043892?mt=8) should also enhance the visitor experience and engagement with young people. Since 2012, Geopark Shetland has been working on 'Northern Georoutes', funded by Nordic Atlantic Cooperation (NORA), in collaboration with Magma Geopark (Norway), Stonehammer Geopark (Canada) and Katla Geopark (Iceland). The project aims to promote the North Atlantic Region as a niche tourist destination and to develop a distinct brand, holiday booking system and travel packages with local tourism providers (http://northerngeoroutes.com) (Gentilini and Barton 2014). Work is currently under way to develop a funding bid to the European Regional Development Fund for a Northern Periphery project, 'Drifting Apart'. If successful, this project will promote and conserve geoheritage in Canada, Greenland, the Faroe Islands, Iceland, Norway, Ireland, Scotland, Finland and Russia, and will develop new opportunities for geotourism through common geoheritage connections and shared experience of the many links to natural, cultural and social heritage. The Scottish geoparks may also benefit from opportunities linked to the extension of the International Appalachian Trail to Europe (see http://iat-sia.com/index.php?page=scotland).

SCOTLAND'S GEOPARKS: AN ASSESSMENT OF PROGRESS

In bringing together geodiversity, biodiversity, landscape and cultural heritage as part of a community-led approach, geoparks aim to foster popular appeal, enjoyment, research, education and conservation of geoheritage through creative activities that also promote sustainable tourism development, particularly in rural areas (McKeever *et al* 2010; UNESCO 2014). They must also have an effective management system and implementation plan in place, as well as sustainable financial support. As part of geopark development, local communities should become more aware of the multiple scientific, educational, aesthetic, cultural and economic values and benefits of their geoheritage, and should be actively involved in its conservation.

It is highly appropriate that geoparks have been set up in Scotland, given the remarkable geodiversity and stories in the rocks, the long tradition of geoheritage-based tourism and recreation, the established record of successful geoheritage interpretation by bodies such as Scottish Natural Heritage (SNH), the British Geological Survey (BGS) and local geoconservation groups, building on the foundations of Charles Lyell, Hugh Miller and Archibald Geikie, and the strong links between geodiversity, landscape and cultural heritage. The Scottish geoparks represent a modern response to appreciating the physical landscape, integrating the values of geoconservation with elements of the Romantic and sublime, and involving local communities in geotourism provision and geoheritage promotion. They are particularly well placed to realise Hutton's aspiration for people to marvel at the sense of order in the geological shaping of the landscape, and also to be inspired by the aesthetic appeal of its natural wonders. The challenge, and one that is largely being met, is to do so through application of best practice in interpretation, addressing a broad spectrum of audiences – from dedicated geotourists to casual visitors – and seeking to enhance their experiences by providing, as appropriate, both information and interactive engagement on an emotional and imaginative level that enables the discovery of a refreshed sense of wonder about the landscape and its natural features, as well as the lives of the personalities involved in past geological discoveries (Gordon and Baker 2015; Pralong 2006; Stewart and Nield 2013).

Scotland clearly has the geoheritage assets and voluntary sector commitment to support geopark and geotourism developments. Much has been achieved, despite the resource constraints, particularly in providing interpretation for visitors and educational activities, but to a lesser extent in explicitly delivering enhanced geoconservation and sustainable business development. The success of Geopark Shetland in integrating interpretation with other components of the natural heritage, and in achieving wider international partnerships and European funding to develop common themes and regional geoheritage identities and connections, provides valuable pointers to the direction of future activities. The value of membership of the UNESCO GGN is also considerable and has helped to raise political awareness. Nevertheless, the three geoparks have struggled to secure long-term core funding, which is particularly challenging in the current economic climate. As a consequence, they have been less successful so far in building sustainable business and marketing models and in integrating geoheritage conservation and geotourism into local, and indeed national, government policies and those of associated agencies. Geopark Shetland, operating under the Shetland Amenity Trust, has had better access to business expertise and partnerships, although it too has faced economic pressures. Geoparks also depend on community engagement, but the level of involvement has been variable across the three geoparks.

Several factors have hindered downstream business development and long-term sustainable funding. They include low populations in the geoparks, relatively small numbers of volunteers (mainly geologists), limited acquisition of business expertise and a lack of marketing, particularly at a national level through VisitScotland, Scotland's national tourism organisation. Landscape is a prime tourist attraction, but there has been little recognition throughout the industry of the major underpinnings of geoheritage and the value that it can add to the quality of the visitor experience. Unfortunately, this is part of a wider picture of low awareness and understanding of geoheritage values among politicians, policy makers and the business sector generally, and is also reflected in the dominance of biodiversity in nature conservation (Crofts 2014). In addition, there is the ambivalent position of Earth Science teaching in schools, limited engagement by the academic community (eg in terms of research on the economic and social values of geodiversity and geoparks, as part of the UK National Ecosystem Assessment of 2011 and its follow-on phase), and a lack of hard data to evaluate the contribution of geotourism and geoparks in Scotland, not only in terms of economic benefits, but also in terms of social, educational, aesthetic and recreational values.

Nevertheless, there are positive signs in the recent progress by the three geoparks to develop a more strategic, business-orientated approach and greater collaboration through the Scottish

Geoparks Partnership. On a wider front, the Scottish Geodiversity Forum is working on a number of initiatives to raise awareness of the multiple values of geoheritage, including the provision of guidance for the tourism industry, and Earth Science activities in schools are being strongly promoted by the Earth Science Education Unit (http://www.earthscienceeducation. com), the Scottish Earth Science Education Forum (http://www.sesef.co.uk) and the Geobus from St Andrews University (http://www.geobus.org.uk). These are particularly important at a time when the Scottish Qualifications Authority is planning to discontinue the suite of geology qualifications, a move that is vigorously opposed by Earth Science Education Scotland (http:// earthscienceeducationscotland.wordpress.com). Following a geoparks outreach event at Our Dynamic Earth visitor attraction in Edinburgh in 2014, universities are beginning to show interest in the development of joined-up research, integrating geology, archaeology, business and tourism in the Scottish geoparks. VisitScotland is also collaborating with the Scottish Geodiversity Forum to add geosites to its website as attractions to visit, including those in the geoparks (http:// www.visitscotland.com/about/nature-geography/geoparks). In addition, the Scottish geoparks are assisting with the outreach programme of Our Dynamic Earth, which is helping to enhance their visibility in the more populated Central Belt.

Conclusion

A revitalised geoparks network in Scotland has an essential part to play in delivering outcomes – not only for local communities, businesses and geoconservation, but also at a national level for Scotland's Geodiversity Charter. The recent funding from the Scottish Government provides an important window of opportunity for the geoparks, locally and in partnership through the Scottish Geoparks Forum, to develop a much stronger business model and greater community involvement. However, without the integration of geoheritage in broader social, economic and nature conservation policies for rural areas, and without its recognition in national and local policy support measures and by the business sector, the geoparks in Scotland may continue to struggle to achieve sustainable funding.

Acknowledgments

Thanks to Robina Barton (Geopark Shetland), Pete Harrison and Laura Hamlet (North West Highlands Geopark), and Jim Blair and colleagues (Lochaber Geopark) for information and helpful comments. However, the author remains responsible for the opinions expressed.

References

Andrews, M, 1987 *The Search for the Picturesque: Landscape, Aesthetics and Tourism in Britain, 1760–1800*, Scholar Press, Aldershot

Barton, R, 2014 Geopark bodies battle for survival, *Earth Heritage* 41, 28–30

Baxter, J M, Boyd, I L, Cox, M, Donald, A E, Malcolm, S J, Miles, H, Miller, B, and Moffat, C F (eds), 2011 *Scotland's Marine Atlas: Information for the National Marine Plan*, Marine Scotland, Edinburgh

Burke, E, 1759 *A Philosophical Enquiry into the Origin of our Ideas of the Sublime and Beautiful* (2 edn), R & J Dodsley, London

Coratza, P, and Panizza, M (eds), 2009 Geomorphology and cultural heritage, *Memorie Descrittive della Carta Geologica d'Italia*, 87

Crofts, R, 2014 Promoting geodiversity: learning lessons from biodiversity, *Proceedings of the Geologists' Association* 125, 263–6

Crofts, R, and Gordon, J E, 2014 Geoconservation in protected areas, *Parks* 20 (2), 61–76

Dowling, R K, and Newsome, D, 2010 *Global Geotourism Perspectives*, Goodfellow, Oxford

Eder, W, and Patzak, M, 2004 Geoparks – geological attractions: a tool for public education, recreation and sustainable economic development, *Episodes* 27, 162–4

Ellis, N, 2011 The Geological Conservation Review (GCR) in Great Britain – rationale and methods, *Proceedings of the Geologists' Association* 122, 353–62

European Commission, 2010a *Europe 2020: A strategy for smart, sustainable and inclusive growth*, European Commission COM/2010/2020 final [online], available from: http://eur-lex.europa.eu/LexUriServ/LexUriServ.do?uri=COM:2010:2020:FIN:EN:PDF [26 August 2014]

— 2010b *Europe: the world's No 1 tourist destination – a new political framework for tourism in Europe*, European Commission COM/2010/0352 final [online], available from: http://eur-lex.europa.eu/legal-content/EN/ALL/?uri=CELEX:52010DC0352 [26 August 2014]

Friend, P, 2012 *Scotland*, Collins, London

Geikie, A, 1887 *The Scenery of Scotland Viewed in Connection with its Physical Geology* (3 edn 1901), Macmillan & Co, London

Gentilini, S, and Barton, R, 2014 Northern georoutes: geoparks develop North Atlantic tourism brand, *European Geoparks Magazine* 11, 15

Gillen, C, 2013 *Geology and Landscapes of Scotland*, Dunedin Academic Press, Edinburgh

Gilpin, W, 1789 *Observations, Relative Chiefly to Picturesque Beauty, Made in the Year 1776, on Several Parts of Great Britain; Particularly the High-Lands of Scotland*, 2 vols, R Blamire, London

— 1792 *Three Essays: on Picturesque Beauty; on Picturesque Travel; and on Sketching Landscape: to which is added a Poem, on Landscape Painting*, R Blamire, London

Gordon, J E, 2010 The geological foundations and landscape evolution of Scotland, *Scottish Geographical Journal* 126, 41–62

— 2012a Rediscovering a sense of wonder: geoheritage, geotourism and cultural landscape experiences, *Geoheritage* 4, 65–77

— 2012b Engaging with geodiversity: 'stone voices', creativity and ecosystem cultural services in Scotland, *Scottish Geographical Journal* 128, 240–65

Gordon, J E, and Baker, M, 2015 Appreciating geology and the physical landscape in Scotland: from tourism of awe to experiential re-engagement, in *Appreciating Physical Landscapes: Three Hundred Years of Geotourism, Special Publication No 417* (ed T A Hose), The Geological Society, London, doi: 10.1144/SP417.1

Gordon, J E, and Barron, H F, 2011 Scotland's geodiversity: development of the basis for a national framework, *Scottish Natural Heritage Commissioned Report No 417*, Scottish Natural Heritage [online], available from: http://www.snh.org.uk/pdfs/publications/commissioned_reports/417.pdf [26 August 2014]

— 2012 Valuing geodiversity and geoconservation: developing a more strategic ecosystem approach, *Scottish Geographical Journal* 128, 278–97

— 2013 Geodiversity and ecosystem services in Scotland, *Scottish Journal of Geology* 49, 41–58

Gordon, J E, and Kirkbride, V, 2009 Reading the landscape: unveiling Scotland's earth stories, in *Proceedings,*

Vital Spark Conference, Aviemore, October 2007 [online], available from: www.ahi.org.uk/include/pdf/TVSpapers/Gordon_J_and_Kirkbride_V.pdf [26 August 2014]

Gordon, J E, Brazier, V, and MacFadyen, C C J, 2004 Reading the landscapes of Scotland: raising earth heritage awareness and enjoyment, in *Natural and Cultural Landscapes – the Geological Foundation* (ed M Parkes), Royal Irish Academy, Dublin, 227–34

Gordon, J E, Barron, H F, Hansom, J D, and Thomas, M F, 2012 Engaging with geodiversity – why it matters, *Proceedings of the Geologists' Association* 123, 1–6

Gray, M, 2013 *Geodiversity: Valuing and Conserving Abiotic Nature* (2 edn), Wiley-Blackwell, Chichester

Gray, M, Gordon, J E, and Brown, E J, 2013 Geodiversity and the ecosystem approach: the contribution of geoscience in delivering integrated environmental management, *Proceedings of the Geologists' Association* 124, 659–73

Grenier, K H, 2005 *Tourism and Identity in Scotland, 1770–1914: Creating Caledonia*, Ashgate Publishing Ltd, Aldershot

Hose, T A, 2010 Volcanic geotourism in West Coast Scotland, in *Volcano and Geothermal Tourism: Sustainable Geo-resources for Leisure and Recreation* (eds P Erfurt-Cooper and M Cooper), Earthscan, London, 259–71

Hunter, J, 1995 *On the Other Side of Sorrow: Nature and People in the Scottish Highlands*, Mainstream Publishing, Edinburgh and London

Hutton, J, 1785 *Abstract of a Dissertation read in the Royal Society of Edinburgh, upon the Seventh of March, and Fourth of April, MDCCLXXXV, Concerning the System of the Earth, its Duration, and Stability*, Edinburgh

— 1788 Theory of the Earth; or an investigation of the laws observable in the composition, dissolution, and restoration of land upon the globe, *Transactions of the Royal Society of Edinburgh* 1, 209–304

IUCN, 2012 *Resolutions and Recommendations, World Conservation Congress, Jeju, Republic of Korea, 6–15 September 2012, WCC-2012-Res-048-EN Valuing and Conserving Geoheritage within the IUCN Programme 2013–2016*, IUCN, Gland [online], available from: https://portals.iucn.org/library/node/44015 [26 August 2014]

Lyell, C, 1830–33 *Principles of Geology*, John Murray, London

MacDiarmid, H, 1974 *Direadh I, II and III*, Kulgin Duval and Colin H Hamilton, Frenich, Foss

MacLeod, A, 2012 *From an Antique Land: Visual Representations of the Highlands and Islands 1700–1880*, John Donald, Edinburgh

Macpherson, J, 1765 *Works of Ossian, the Son of Fingal, in Two Volumes. Translated from the Gaelic Language by James Macpherson* (3 edn), T Becket and P A Dehondt, London

McKeever, P J, Larwood, J, and McKirdy, A P, 2006 Geotourism in Ireland and Britain, in *Geotourism: Sustainability, Impacts and Management* (eds R K Dowling and D Newsome), Elsevier, Amsterdam, 180–98

McKeever, P J, Zouros, N, and Patzak, M, 2010 The UNESCO Global Geoparks Network, *European Geoparks Magazine* 7, 10–13

McKirdy, A P, Gordon, J E, and Crofts, R, 2007 *Land of Mountain and Flood: The Geology and Landforms of Scotland*, Birlinn, Edinburgh

McKirdy, A P, Threadgould, R, and Finlay, J, 2001 Geotourism: an emerging rural development opportunity, in *Earth Science and the Natural Heritage: Interactions and Integrated Management* (eds J E Gordon and K F Leys), The Stationery Office, Edinburgh, 255–61

Milne Home, D, 1872 Scheme for the conservation of remarkable boulders in Scotland, and for the indication of their position on maps, *Proceedings of the Royal Society of Edinburgh* 7, 475–88

Murray, S, 1805 *A Companion and Useful Guide to the Beauties in the Western Highlands of Scotland, and in the Hebrides. To Which is Added, a Description of Part of the Main Land of Scotland, and of the Isles of Mull, Ulva, Staffa, I-Columbkill, Tirii, Coll, Eigg, Skye, Raza, and Scalpa*, Vol 2 (2 edn), printed by W Bulmer & Co for the author, London

Peach, B N, Horne, J, Gunn, W, Clough, C T, Hinxman, L W, and Teall, J J H, 1907 *The Geological Structure of the North-West Highlands of Scotland*, Memoirs of the Geological Survey of Great Britain, HMSO, Glasgow

Pennant, T, 1776 A *Tour in Scotland and Voyage to the Hebrides, 1772*, Benjamin White, London

Playfair, J, 1802 *Illustrations of the Huttonian Theory of the Earth*, William Creech, Edinburgh

Pralong, J-P, 2006 Geotourism: a new form of tourism utilising natural landscapes and based on imagination and emotion, *Tourism Review* 61, 20–5

Prosser, C D, 2013 Planning for geoconservation in the 1940s: an exploration of the aspirations that shaped the first national geoconservation legislation, *Proceedings of the Geologists' Association* 124, 536–46

Scottish Geodiversity Forum, 2013 *Scotland's Geodiversity Charter* [online], available from: http://scottishgeodiversityforum.org/charter/ [26 August 2014]

— 2014 *Taking Scotland's Geodiversity Charter Forward: Tourism* [online], available from: https://scottishgeodiversityforum.files.wordpress.com/2011/12/implementing-scotlands-geodiversity-charter-tourism-june-2014.pdf [26 August 2014]

Scottish Government, 2011 *The Government Economic Strategy*, Scottish Government, Edinburgh [online], available from: http://www.scotland.gov.uk/publications/2011/09/13091128/8 [26 August 2014]

Scottish Tourism Alliance, 2012 *Tourism Scotland 2020: The Future of our Industry, in our Hands: A Strategy for Leadership and Growth* [online], available from: http://www2.jpscotland.co.uk/pdf/tourismstrategy.pdf [26 August 2014]

Smout, T C, 1993 The Highlands and the roots of green consciousness, 1750–1990, *Scottish Natural Heritage Occasional Paper No 1*, Scottish Natural Heritage, Edinburgh [online], available from: http://www.snh.org.uk/pdfs/publications/corporate/occaspapers/Occ%20Pap%201%20the%20highlands.pdf [26 August 2014]

Stewart, I S, and Nield, T, 2013 Earth stories: context and narrative in the communication of popular geosciences, *Proceedings of the Geologists' Association* 124, 699–712

Thomas, B A, and Warren, L M, 2008 Geological conservation in the nineteenth and early twentieth centuries, in *The History of Geoconservation, Special Publication No 300* (eds C V Burek and C D Prosser), The Geological Society, London, 17–30

Threadgould, R, and McKirdy, A P, 1999 Earth heritage interpretation in Scotland: the role of Scottish Natural Heritage, in *Towards the Balanced Management and Conservation of the Geological Heritage in the New Millennium* (eds D Barettino, M Vallejo and E Gallego), Sociedad Geológica de España, Madrid, 330–4

Trewin, N H, 2001 Scotland's foundations: our geological inheritance, in *Earth Science and the Natural Heritage: Interactions and Integrated Management* (eds J E Gordon and K F Leys), The Stationery Office, Edinburgh, 59–67

— (ed), 2002 *The Geology of Scotland*, The Geological Society, London

UK National Commission for UNESCO, 2013 *Wider value of UNESCO to the UK 2012–2013* [online], available from: http://www.unesco.org.uk/uploads/Wider%20Value%20of%20UNESCO%20to%20UK%202012-13%20full%20report.pdf [26 August 2014]

UK National Ecosystem Assessment, 2011 *The UK National Ecosystem Assessment: Synthesis of the Key Findings*, UNEP-WCMC, Cambridge

UNESCO, 2014 *Guidelines and criteria for national geoparks seeking UNESCO's assistance to join the Global*

Geoparks Network (GGN), UNESCO, Paris [online], available from: http://www.globalgeopark.org/UploadFiles/2012_9_6/Geoparks_Guidelines_Jan2014.pdf [26 August 2014]

VisitScotland, 2013 *Scotland Visitor Survey 2011 and 2012* [online], available from: http://www.visitscotland.org/research_and_statistics/visitor_research/all_markets/scotland_visitor_survey.aspx [26 August 2014]

— 2015 *Scotland: The key facts on tourism in 2014* [online], available from: http://www.visitscotland.org/pdf/2015%200729%20Tourism%20in%20Scotland%202014_Final%20draft.pdf [9 November 2015]

Wordsworth, D, 1874 *Recollections of a Tour Made in Scotland AD1803* (ed J C Shairp), Edmonston and Douglas, Edinburgh

16

Geoheritage Case Study:
Canton Valais, Switzerland

EMMANUEL REYNARD

Canton Valais, in south-western Switzerland, covers 5,224km² of mainly upland and mountainous terrain. In terms of land use, just over half (53.5%) of the Canton is unproductive land, almost a quarter (24%) is given over to forestry and almost a fifth (19%) to agriculture (especially dairy, fruit growing and viticulture). Only 3.5% of the land is given over to urban development and, as part of the Rhône Valley, much of this is in small hamlets and villages; these settlements, along with the mountains, are now an essential aspect of its aesthetic tourism appeal. It had 331,763 inhabitants in 2014. Almost three-fifths (58%) of the Canton's workforce are engaged in the tertiary (including tourism) employment sector, and less than 4% work in the primary sector. Two-thirds of the population are French-speaking and a third German-speaking, presenting challenges in widely promoting the Canton's geoheritage. Embraced by two alpine mountain chains – the Bernese Alps in the north and the Penninic Alps in the south – the Canton is characterised by a strong rain-shadow effect and a character somewhat continental in climate, with long periods of sunshine during summer as well as in the winter months, especially favourable to skiing.

The mountains have been a major draw for mountaineers since the 19th century, and they were significant in the development of climbing and skiing as major adventure sports in Europe (notable ascents included the Weisshorn in 1861, the Dent Blanche in 1862, and the Matterhorn in 1865 by Edward Whymper). Switzerland generally, and the Canton in particular, has exploited its physical landscape to attract tourists since the emergence of the Romantic movement. Its scenery has particularly attracted elite visitors, some with literary and artistic aspirations, from England and Germany since the early 19th century. Today, it continues to attract up-market tourists. It is home to classic and much-depicted pyramidal peaks such as the Weisshorn, the Matterhorn, the Dent Blanche, the Dent d'Hérens and the Monte Rosa.

Valais is drained by the Rhône River, flowing over 164km from the Rhône Glacier, situated in the Aar-Gotthard crystalline massif, to Lake Geneva. The catchment is high (mean altitude 2,130m) and much glaciated (glacier land area coverage is 14.3%), with some of the largest Alpine glaciers (including the Aletsch, Gorner, Fiesch, Corbassière, Oberaletsch, Rhône and Findelen). The Rhône River has more than 200 tributaries, of which a large number have a highly torrential character (such as Illgraben, Losentze and St-Barthélémy) and have built imposing alluvial fans. The Rhône flooded the alluvial plain annually, until it was channelised in the second part of the 19th century (Reynard *et al* 2009a). Such terrain is generally prone to landslides, avalanches and floods; hence, risk assessment and management are essential to the development of safe urban and adventure sport expansion and geo-attractions.

Scientific investigation in the Earth Sciences in the region has particularly focused on glacial studies and on structural geology. In the 1970s and 1980s several researchers tried to reconstruct the history of deglaciation, but unfortunately very few studies provided an absolute chronology. Currently, most of the research is carried out in the field of natural hazard assessment and mitigation, with a particular focus on debris flows, on processes related to permafrost melting, and on glacier-related hazards. After a first period of interest in the 19th century that continued until the 1940s, geoheritage studies were quite scarce until a new period of interest began in the 2000s. This chapter explores the main characteristics of the geoheritage of the Canton, which is the basis of focused geoheritage tourism provision (that is, geotourism).

GEOHERITAGE

Thirty-five (11%) out of the 322 Swiss geosites (Reynard *et al* 2012; Reynard 2012a) are situated in Valais (Figure 16.1; Table 16.1). The majority of them are geomorphological and mineralogical sites. The geomorphological sites refer mainly to glacial heritage (nine sites), with some of the largest glaciers in the Alps (including the Aletsch Glacier, classified as a World Heritage Site, the Gorner and the Rhône Glaciers). The latter is equipped with a glacier hole that can be visited; unfortunately, because of climate warming, the exploitation of this hole is becoming increasingly difficult. Other glacial sites are remnants of moraines, in particular the Tortin moraine that is considered a reference site for the Egesen Lateglacial stage in the Western Alps, and the Monthey moraine, very rich in big erratic boulders, and specifically the famous Pierre des Marmettes (Figure 16.2A), which is considered to be one of the blocks that initiated the nature protection movement in Switzerland (Bachmann 1999; Reynard 2004).

Three rockslides are of national significance. The large postglacial rockslide of Sierre has completely cluttered the Rhône Valley and determined – with the debris supply of the Illgraben torrent – a change in the river channel (the Rhône is here a braided river; that is, it cannot transport all sediments and is divided into several channels). This area (with the proglacial margin near the Rhône Glacier and the Rhône delta in Lake Geneva) is one of the only natural river landscapes of the Swiss Rhône River catchment. The Sierre rockslide is also responsible for the high landscape diversity, with dry and wet areas, which has allowed one of the largest pine forests in the Alps (*Pfynwald*) to develop. This high natural diversity allowed the creation of a Natural Regional Park, recognised by the Swiss Federal State in 2009. A second rockslide was deposited in 1714 on an alpine pasture in Derborence, killing 14 people and more than 100 animals; it also destroyed several buildings. A second event, triggered in 1749, partially covered the first one. The total volume is estimated to be about 50 million m³. This geosite is particularly well known because of the novel *Derborence*, published in 1934 by the Swiss writer Charles-Ferdinand Ramuz (who, incidentally, is depicted on the 200 Swiss franc note), and the movie of the same name, directed by Swiss film-maker Francis Reusser in 1985. In this sense, Derborence is clearly a geocultural site, as some remnants of mining and quarrying activities (at Gondo, Goppenstein and Saillon) survive. The third important rockslide (Randa) was deposited quite recently, in 1991, in the Zermatt valley. It initiated significant landscape changes, but is not really promoted through geotourism; this is despite the fact that the Zermatt valley is actually a perfect area for developing educational programs on natural hazards in the Alps, because of the importance of gravity processes related to geology (rockslides), hydrology (debris flows), snow (avalanches),

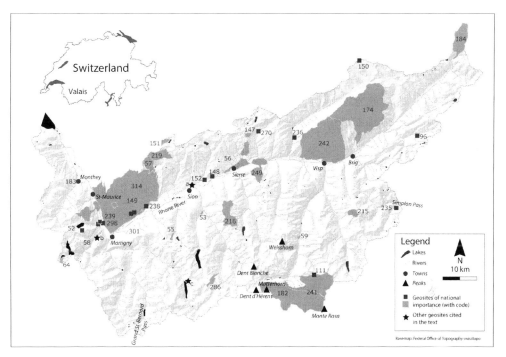

FIGURE 16.1 LOCATION OF THE MAIN GEOSITES IN THE CANTON OF VALAIS
CODES: NUMBERS (FOR CODES OF SWISS GEOSITES) AND LETTERS (OTHER GEOSITES) REFER TO TABLE 16.1

permafrost melting and glaciers (debris flows, ice avalanches). Other geomorphological sites relate to periglacial, fluvial and karstic processes.

Mineralogical sites refer principally to former sites of exploitation of geological resources (quarries and mines). Few of them (such as the Gondo gold mine and Binntal) are equipped for visits. This is particularly the case of Binntal, one of the valleys richest in minerals in the Alps, which became a Regional Nature Park recognised and financed by the Swiss Federal State in 2011. The park has developed several activities aimed at promoting geology. Palaeontological sites are not very common, and are limited to two sites with vertebrate trackways on Triassic deposits (Salanfe and Emosson). The latter was discovered during the dry summer of 1976 and has been protected by the Cantonal State since 1983. New tracks continue to be discovered (Cavin *et al* 2012) and the site is promoted by the Nature Museum of Geneva, which every summer organises a scientific presence on the site, situated at 2,400m (Decrouez and Cavin 2011). Only two structural geosites are in the list of geosites of national importance, especially for their particular importance to the development of the Earth Sciences. The Matterhorn (Figure 16.2B) was the place where the Swiss geologist Emile Argand applied the theory of Alfred Wegener's continental drift in the Alps (Marthaler 2005). The Morcles nappe[1] is one of the most famous Helvetic cover nappes described by the geologists Maurice Lugeon (1914–18) and Héli Badoux (1972);

---

[1] In geology, a cover nappe is a large sheetlike body of rock that has been moved far from its original position (several kilometres).

	Code	Geosite name
Palaeontological sites (2)	64	Vertebrate trackways of Emosson
	52	Vertebrate trackways of Salanfe
Structural sites (2)	182	Matterhorn
	314	Morcles cover nappe
Mineralogical sites (8)	96	Binntal mineralogical quarries
	235	Gold vein of Gondo
	236	Lead and zinc mine of Goppenstein
	238	Slate quarry of Leytron
	239	Granite quarry of Miéville
	298	Conglomerate quarry of Dorénaz
	301	Marble quarry of Saillon
	111	Pillow-lavas of Zermatt
Hydrogeological sites (6)	147	Chamois hole of Varen
	150	Jochloch hole, Jungfraujoch
	149	Karstic network Grand Cor–Poteu, Fully-Saillon
	152	Underground lake of St-Léonard
	148	Crête de Vaas hole, Granges-Sierre
	270	Thermal spring of Leukerbad
Geomorphological sites (15)		
Glacial	174	Aletsch Glacier and Massa valley
	184	Rhône Glacier
	241	Gorner Glacier
	286	Tsijore Nouve Glacier, Evolène
	215	Gruben glacial cirque, Saas Grund
	242	Glacial hanging valleys, Lötschberg
	53	Euseigne pyramids
	55	Egesen stage moraines of Tortin, Nendaz
	183	Monthey moraine (including the *Pierre des Marmettes*)
Periglacial	216	Réchy-Lona periglacial landscape
Rockslides	56	Sierre postglacial rockslide
	57	Derborence historical rockslide
	59	Randa recent rockslide
Fluvial	58	Postglacial valleys between Martigny and St-Maurice (including *Pissevache* waterfall)
	249	Illgraben torrential system
Karstic	219	Glacio-karst of *Tsanfleuron*, Savièse
	151	*Lapi di Bou* karrenfield, Savièse
Other geosites		
Geocultural sites	a	*Clavau* terraced vineyards
	b	*Pierre Bergère* erratic boulder
	c	Giétro glacier

Geotourism interpretative facilities	Geotourism product type and activities
Yes	Brochure, guided tours (summer)
Few	Brochure
Yes	Guided tours; in the offer of Swiss Tourism
Few	Via Geo Alpina itinerary
Yes	Guided tours, Regional Nature Park; in the offer of Swiss Tourism
Yes	Guided tours; in the offer of Swiss Tourism
No	Abandoned mine; no geotourism
No	Abandoned mine; no geotourism
No	-
Few	Educational panel
Few	Book
No	-
No	Only for speleologists
No	Only for speleologists
No	Only for speleologists
Yes	Guided visits
No	Only for speleologists
Yes	Baths
Yes	World Heritage Site
Yes	Glacier hole
Yes	Guided visits, in particular the Glacier Garden
No	On a tourist trail, but no specific geotourist offer
No	-
Few	Numerous tourist offers, but no specific offer in the area of geotourism
Few	Panel
No	-
Yes	Panel, educational trail
Few	Hiking trails but no specific geotourist offer
Yes	Guided tours, Regional Nature Park
Yes	Leaflets; in the offer of Swiss Tourism
No	On a tourist trail, but no specific geotourist offer
Few	Educational trail in the Trient gorges, no specific offer in the other valleys
Yes	Guided tours, Regional Nature Park
Yes	Panels, brochures, no guided tours
No	Only for speleologists
Yes	Guided tours on irrigation channels in the vineyard
Yes	Panels, educational trail
Yes	Museum

Figure 16.2 Examples of geosites in the Canton of Valais

A. The Pierre des Marmettes – a famous erratic boulder, protected since 1908.

B. The Matterhorn – an iconic mountain of Switzerland.

C. The Tsanfleuron glacio-karstic plateau – a geosite of national significance.

D. The Pissevache waterfall – a famous tourist site in the 18th century.

E. The Rhône Glacier – a tourist site visited since the Romantic period.

F. The Euseigne Pyramids – a famous geosite in the Val d'Hérens.

knowledge about the structural geology of the area is included in the Via Geoalpina trail, developed in 2010 within the framework of the International Year of Planet Earth, and sponsored by the Swiss National Geological Survey (www.viageoalpina.eu).

Finally, the Bernese Alps, mainly made of limestone rocks, are rich in sinkholes and underground karstic networks. Situated at high altitudes, some of these sinkholes are glaciated and most of them are only accessible with speleological equipment. The thermal springs of Leukerbad (65 hot springs up to 51°C) have been known since Roman times. Spa tourism developed after 1501, when Cardinal Matthäus Schiner (1465–1522) bought the right to exploit the hot springs; Leukerbad is now one of the most famous thermal and medical resorts in Switzerland. The Tsanfleuron glacio-karstic plateau (Figure 16.2C) is in the Swiss national inventory of geosites for the complex karstic and glacial interactions for which the site became a hotspot of scientific research in karstology and glaciology (Reynard 2008), which induced the University of Lausanne to develop several geotourist products, such as an educational trail, educational panels and a geotourist map (Reynard and Coratza 2016).

Of course, the geoheritage of the Canton Valais is not limited to the sites inscribed in the national inventory. No cantonal inventory has been compiled, although a preliminary study undertaken by Lugon and Reynard (2003) recommended that the Canton should develop an inventory of sites worthy of protection as a basis for territorial planning. Unfortunately, at the moment this inventory does not exist and the Canton Valais is one of the few cantons that have not yet inventoried the geoheritage within their territories (Reynard 2012a). However, several regional inventories have been carried out by the University of Lausanne (eg Maillard and Reynard 2011; Kozlik and Reynard 2013).

GEOHERITAGE AND CULTURE

Because of the role that Valais played in the development of Alpine tourism, the links between culture and geoheritage are quite prominent. In this sense, Valais is a place where the concept of cultural geomorphology – that is, an approach that studies the relationships between cultural goods and geomorphologic aspects (Panizza and Piacente 2003) – can be applied. Indeed, several sites were key tourist places during the first phase of Alpine tourism in the 18th and 19th centuries (Tissot 2000). For example, the Pissevache waterfall (Figure 16.2D) was a tourist 'must-see' when travelling in the Rhône Valley (Reynard et al 2009b; Reynard 2012b), as were the Leukerbad baths on the way to the Bernese Alps through the Gemmi Pass, and also the Rhône Glacier (Figure 16.2E) on the way to the Gotthard Massif through the Furka Pass; all these sites were visited by the German Romantic poet Johan Wolfgang Goethe (1749–1832) during his travels through the Alps (Reichler and Ruffieux 1998; Chiadò Rana 2003). Later, specific mountains were at the core of the development of first-generation resorts focused on summer tourism (such as Zermatt at the foot of the Matterhorn, Champéry near the Dents du Midi, and Finhaut in the Mont Blanc Massif), whereas in other places it was the aesthetically pleasant slopes and expansive panoramas that attracted the first tourists (as was the case in Montana, situated on a large plateau, presenting a combination of small depressions occupied by lakes and hills eroded by the Rhône Glacier).

Several towns are deeply influenced by geological and geomorphological factors. St-Maurice, with its 1,500-year-old abbey, is built on a major glacial crag shaped by the Rhône Glacier when it was advancing to the Lake Geneva area; for this reason, it has been a strategic defensive site for centuries and is sometimes considered the 'door of Valais'. The two castles of Sion, the capital of Valais, are also built on a glacial crag formed by differential erosion; the town itself has expanded on a quite active alluvial fan occupied by humankind since Neolithic times. Sierre developed on the hilly relief formed by a large postglacial rockslide, whereas the cities of Martigny, Visp and Brig were more classically built on alluvial fans of the Rhône's tributaries.

The rural landscapes are also highly linked with geoheritage. The irrigation channels (Nahrath *et al* 2011) – called *Bisses* in the French-speaking part of the Canton and *Suonen* in the German-speaking areas – are dependent on their geomorphological contexts. The most spectacular ones, built in wood and suspended along high cliffs, are situated along U-shaped hanging glacial valleys. The steep slopes of most of the farming terrains (of vineyards and meadows), mainly due to former glacial erosion, frequently necessitated the construction of kilometres of walls and created large surfaces of terraced farmland with high landscape value (Rodewald 2011), as is the case at Clavau vineyard near Sion. Conversely, the flat terrains of the Rhône's alluvial plain and the numerous alluvial fans allowed the development of large areas of farmland, after the rivers were channelled in the 19th century.

One specific place particularly worthy of mention, when exploring the relationships between geoheritage and culture, is the Pierre Bergère erratic boulder, in the village of Salvan. Indeed, the Pierre Bergère and its surroundings form an important geocultural site (Kozlik and Reynard 2013). Not only the block itself, situated at the top of a hill eroded by the Trient Glacier, is culturally significant, but also on the large surfaces of *roches moutonnées* are incised the so-called *cupules* (small petroglyphs thought to date back to prehistoric times) by early humans; it was also used as an experimental site by the future Nobel prize winner Guglielmo Marconi for his first wireless experimentations in 1895 (Reynard 2005a). Marconi used the block as a place for emitting waves, while a young child was moving around in the nearby countryside with flags of different colours to indicate the good or bad reception of the waves at specific spots. Here, the value of the site for Earth Science history joins its value for the history of sciences, a fact that is explained to visitors on two interpretative panels (Kozlik and Reynard 2013).

Valais is also known as one of the key sites for the development of glaciology (Onde 1948); three locations are especially worthy of mention.

1. First of all, in 1818, a glaciological catastrophe – the outburst of the Giétro Glacier – devastated the Val de Bagnes and the city of Martigny. During his stay in the valley to manage the situation, Canton engineer and scientist Ignaz Venetz shared ideas about possible former extensions of glaciers with a local inhabitant, Jean-Pierre Perraudin. Their ideas about former glaciations were then disseminated to the scientific community by Louis Agassiz (1840) and Jean de Charpentier (1841). A small museum presenting the importance of the Giétro catastrophe and Perraudin ideas for the glaciological sciences can be visited in the Val de Bagnes.

2. The Chablais area, near Lake Geneva, is a key site for the history of glaciology. Jean de Charpentier used several erratic boulders and moraines disseminated on the two sides of the Rhône Valley to demonstrate the hypothesis of the glacial transportation of these blocks and, therefore, the glacial theory (de Charpentier 1841), based on three main arguments: the sharp form of blocks, the absence of stratification in moraine sediments, and the absence of size reduction with distance of transportation. Several of these blocks were put under protection after the *Appel aux Suisses pour les engager à conserver les blocs*

erratiques [Call to the Swiss people to engage them to protect the erratic boulders] (Favre and Studer 1867; see Reynard 2004). Unfortunately, it is only during the last decade that recognition of the cultural and tourist value of these blocks has re-emerged (Lugon *et al* 2006).

3. Finally, as already noted, between 1905 and 1908, the Pierre des Marmettes, an erratic block already studied by Jean de Charpentier in the 1830s, was put at the core of a national controversy about the protection of natural heritage versus economic development. This, the largest erratic boulder in Switzerland, situated on private property, was due to be sold to an extraction company to be exploited for building material; after a large mobilisation of Earth scientists and the help of the public administration – and even the population, who could subscribe to parts of the block – it was bought from the private company and then offered to the Swiss Academy of Sciences (Reynard 2005b); the Academy is now preparing a geotourist promotion of the block, for the celebration of its 200-year foundation (Bernard *et al* 2013).

GEOTOURISM

Canton Valais lacks a published account of the history of geotourism. One can consider that the first dedicated 'geotourists' (Hose 2000) were the 18th and 19th century scientists who were travelling through the Canton. This was the case for Horace-Benedict de Saussure (1740–99); in 1779 he made numerous observations on the geology of the western part of the Canton (for example, Grand St-Bernard, Mont Blanc Massif, Rhône Valley). Likewise, in 1843 James D Forbes (1809–68) described the geology and glaciers of most of the valleys of the Mont-Blanc Massif and of the Penninic Alps (including the Val de Bagnes, Val d'Hérens, Zermatt, etc), and Goethe also made several interesting geological observations. These pioneers were followed by generations of scientists and travellers.

A second interesting period is the 1930s, when the Swiss Alpine Postal Coaches published a series of small booklets presenting the geography of the main mountain postal routes. Each booklet contained a text on the natural (geology, geomorphology, climate, flora) and human (settlements, populations, traditions) characteristics of the described valley, and a topographical map, a geological sketch, a panorama showing the main mountains, and a series of lithographs presenting the most interesting sites along the road. Several of these booklets concern the Valais, including the Grand St-Bernard Pass, Val d'Anniviers, Val d'Hérens – in particular, the Euseigne Pyramids (Figure 16.2F), and the Grimsel Pass.

After World War II (1939–45) it seems that interest in geology was limited to the scientific circles, in parallel with a switch of the tourism sector to the promotion of winter season activities. Nowadays, tourism is still dominated by the ski sector – and now also by the snowboard sector – but several initiatives have emerged for developing geotourist products, due to a combination of climate change and the emergence of new interests on the part of the clientele, seeking cultural or ecotourist products (including wine, arts and local traditions), as well as hiking and mountain-biking activities. Most of these initiatives – educational panels and trails, brochures, guided tours – are mainly the work of geoscientists, and the question of the quality of information (simplification versus scientific rigour) is still debated (see, for example, Hose 2005; Martin 2013). The Natural Sciences Society of Valais (La Murithienne), the Hiking Association of Valais (Valrando), the Platform Culture Valais and the Valais Museums have launched an interesting geo-interpretation project. The project's main aim was to assess the quality of cultural and natural educational trails, and then to propose an interactive website (www.sentiers-decouverte. ch), helping the users to prepare their field trip. The ultimate goal is to give a brand label to the achievements that satisfy a set of quality criteria.

Conclusion

Valais is a Canton with some of the most highly diversified geoheritage assets in Switzerland; this is due to the diversity of geological contexts and its situation in the core of the Alps, an area still highly glaciated, as well as demonstrating the dynamics of geomorphological processes. However, this heritage is relatively poorly exploited by the tourism sector, which has very recently rediscovered an interest in geoheritage for the summer season tourist offer. Nevertheless, initiatives to develop geotourism products still remain limited to the geo-scientific community; a concordance between that community and the tourism industry could provide the most sustainable means of implementing future geotourism provision in Switzerland.

References

Agassiz, L, 1840 *Etudes sur les glaciers*, Jent & Gasmann, Neuchâtel

Bachmann, S, 1999 *Zwischen Patriotismus und Wissenschaft: Die schweizerischen Naturschutzpioniere (1900–38)*, Chronos Verlag, Zürich

Badoux, H, 1972 *Tectonique de la nappe de Morcles entre Rhône et Lizerne*, Matériaux pour la Carte géologique de la Suisse, Berne, 143

Bernard, R, Reynard, E, and Jacob, A, 2013 Histoire de la Pierre des Marmettes et rôle de la Murithienne, *Bulletin Murithienne* 130, 13–17

Cavin, L, Avanzini, M, Bernardi, M, Piuz, A, Proz, P A, Meister, C, Boissonnas J, and Meyer, C A, 2012 New vertebrate trackways from the autochthonous cover of the Aiguilles Rouges Massif and re-evaluation of the dinosaur record in the Valais, SW Switzerland, *Swiss Journal of Palaeontology* 131, 317–24

Chiadò Rana, C, 2003 *Goethe en Suisse et dans les Alpes*, Georg, Genève

De Charpentier, J, 1841 *Essai sur les glaciers et sur le terrain erratique du bassin du Rhône*, Ducloux, Lausanne

De Saussure, H B, 1779–96 *Voyages dans les Alpes, précédés d'un Essai sur l'histoire naturelle des environs de Genève*, Barde, Manget et Compagnie, Genève

Decrouez, D, and Cavin, L, 2011 Le géotope du Vieux Emosson (Finhaut, Valais, Suisse), in *Les géosciences au service de la société* (eds E Reynard, L Laigre and N Kramar), Géovisions 37, Lausanne, 91–103

Favre, A, and Studer, B, 1867 Appel aux Suisses pour les engager à conserver les blocs erratiques, *Actes de la Société helvétique des sciences naturelles*, Rheinfelden

Forbes, J D, 1843 *Travels through the Alps of Savoy and other Parts of the Pennine Chain*, Adam and Charles Black, Edinburgh

Hose, T A, 2000 European geotourism – geological interpretation and geoconservation promotion for tourists, in *Geological Heritage: Its Conservation and Management* (eds D Barretino, W A P Wimbledon and E Gallego), Instituto Tecnológico Geominero de España, Madrid, 127–46

— 2005 Writ in stone: a critique of geoconservation panels and publications in Wales and the Welsh borders, in *Stone in Wales* (ed M R Coulson), Cadw, Cardiff, 54–60

Kozlik, L, and Reynard, E, 2013 Inventaire et valorisation de géomorphosites culturels dans les vallées du Trient, de l'Eau Noire et de la Salanfe, in *Managing Geosites in Protected Areas* (eds F Hobléa, N Cayla and E Reynard), Collection Edytem 15, Chambéry, 135–42

Lugeon, M, 1914–18 *Les Hautes Alpes calcaires entre la Lizerne et la Kander*, Matériaux pour la Carte géologique de la Suisse, Berne, 30

Lugon, R, and Reynard, E, 2003 Pour un inventaire des géotopes du canton du Valais, *Bulletin Murithienne* 121, 83–97

Lugon, R, Pralong, J P, and Reynard E, 2006 Patrimoine culturel et géomorphologie: le cas valaisan de quelques blocs erratiques, d'une marmite glaciaire et d'une moraine, *Bulletin Murithienne* 124, 73–87

Maillard, B, and Reynard, E, 2011 Inventaire des géomorphosites des vallées d'Entremont et de Ferret (Valais) et propositions de valorisation, in *La géomorphologie alpine: entre patrimoine et contrainte* (eds C Lambiel, E Reynard and C Scapozza), Géovisions 36, Lausanne, 1–17

Marthaler, M, 2005 *The Alps and our Planet. The African Matterhorn: A Geological Story*, L E P Editions, Lausanne

Martin, S, 2013 *Valoriser le géopatrimoine par la médiation indirecte et la visualisation des objets géomorphologiques*, Géovisions 41, Lausanne

Nahrath, S, Papilloud, J H, and Reynard, E, eds, 2011 *Les bisses. Economie, société, patrimoine*, Société d'histoire du Valais romand, Sion

Onde, H, 1948 Observations glaciologiques en Suisse et en Savoie, il y a un siècle, *Revue de géographie alpine* 36 (3), 399–409

Panizza, M, and Piacente, S, 2003 *Geomorfologia culturale*, Pitagora, Bologna

Reichler, C, and Ruffieux, C, 1998 *Le voyage en Suisse. Anthologie des voyageurs français et européens de la Renaissance au XXe siècle*, Laffont, Paris

Reynard, E, 2004 Protecting stones: conservation of erratic blocks in Switzerland, in *Dimension Stone 2004: New Perspectives for a Traditional Building Material* (ed R Prikryl), Balkema, Leiden, 3–7

— 2005a Géomorphosites et paysages, *Géomorphologie: relief, processus, environnement* 3, 181–8

— 2005b Geomorphological sites, public policies and property rights: conceptualization and examples from Switzerland, *Il Quaternario* 18 (1), 321–30

— 2008 Le lapiaz de Tsanfleuron. Un paysage glacio-karstique à protéger et à valoriser, in *Karsts de montagne. Géomorphologie, patrimoine et ressources* (eds F Hobléa, E Reynard and J J Delannoy), Collection Edytem, Cahiers de Géographie 7, Chambéry, 157–68

— 2012a Geoheritage protection and promotion in Switzerland, *European Geologist* 34, 44–7

— 2012b Le patrimoine géomorphologique de la vallée du Trient et environs à travers l'œil des voyageurs du XVIIIe et XIXe siècle, in *Patrimoines des vallées du Trient et de l'Eau Noire* (ed S Benedetti), Association Vallis Triensis, Finhaut, 45–9

Reynard, E, Berger, J P, Constandache, M, Felber, M, Grangier, L, Häuselmann, P, Jeannin, P Y, and Martin, S, 2012 Révision de l'inventaire des géotopes suisses: rapport final, *Groupe de travail pour les géotopes en Suisse*, Lausanne

Reynard, E, and Coratza, P, 2016 The importance of mountain geomorphosites for environmental education, *Acta geographica Slovenica* 56 (2), 291–303

Reynard, E, Evéquoz-Dayen, M, and Dubuis, P (eds), 2009a *Le Rhône: dynamique, histoire et société*, Cahiers de Vallesia 21, Sion

Reynard, E, Regolini-Bissig, G, Kozlik, L, Benedetti, S, 2009b Assessment and promotion of cultural geomorphosites in the Trient Valley (Switzerland), *Memorie Descrittive della Carta Geologica d'Italia* 87, 181–9

Rodewald, R, 2011 *Vous êtes déporté au-dessus du vide. Les paysages en terrasse du Valais: origine, évolution, perception*, Rotten Verlag, Visp

Tissot, L, 2000 *Naissance d'une industrie touristique. Les Anglais et la Suisse au XIXe siècle*, Payot, Lausanne

17

Geoheritage Case Study:
The Danube Region in Serbia

Djordjije A Vasiljević, Slobodan B Marković
and Nemanja Tomić

The Republic of Serbia is for the most part situated within the Balkan Peninsula and the Pannonian (Carpathian) basin. The country's position on a 'geologically young' area of the Balkan Peninsula has resulted in rocks and superficial deposits representing various climatic, environmental, geodynamic and palaeogeographical events; this geoheritage is recognised as being important at the global scale (Jojić-Glavonjić *et al* 2010). The rich geodiversity of Serbia reveals a dynamic and turbulent geological history: magmatism, sedimentation, the formation of mountain ranges, tectonic depressions, and climate changes (Pantić *et al* 1998). In northern Serbia the most significant geological event was the formation and disappearance of the ancient Tethys and Paratethys marine basins, which were drained by the Danube river system via the south-west Carpathian ranges into the Black Sea. This giant river was the main force in creating the contemporary topography and the landscapes of the surrounding areas.

The Danube is the second longest river in Europe, with an overall length of 2,880km. More than 80 million people populate its catchment area and it connects ten countries, making it the most 'international' river in the world (Weller 2009). This characteristic has provided it with an immeasurable role in the merging of different European nations, cultures, traditions, customs and, above all, regions (Vasiljević *et al* 2014b). Moreover, the diversity of the areas is reflected not only in cultural but also in natural terms. The Danube passes through a wide range of natural and semi-natural surroundings and landscapes, ranging from mountain ranges to vast plains. Accordingly, there are many significant and unique landforms and habitats that have eventually been recognised by the relevant national conservation institutions; they are now widely appreciated as protected areas, such as national parks, nature parks and nature reserves. Serbia's Ministry of Environment and Spatial Planning (2010) has succinctly summarised the nature conservation legislation and achievements of the country; it noted that some 68 sites protected for their geoheritage covered some 76.57km² (less than 0.1% of its land area) of the country. From the Black Forest to the Black Sea, the Danube passes through areas of very rich geodiversity. This chapter provides insights into the geoheritage of this part of the Middle Danube area, and particularly into geotourism in Serbia.

THE DANUBE IN SERBIA AND ITS SURROUNDINGS

The Danube flows through Serbia for some 588km (Figure 17.1) and mostly delineates its borders with Croatia (137km) and Romania (230km). It enters Serbia from the north as it cuts across

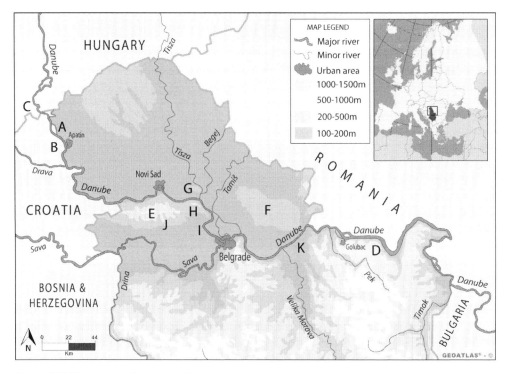

FIGURE 17.1 MAP OF THE DANUBE IN SERBIA, SHOWING THE DISTRIBUTION OF MAJOR GEOHERITAGE SITES

the Serbian-Hungarian border 1,433km from its mouth (Gavrilović and Dukić 2002). Here the Danube is typically a vast and slow Pannonian plain river, 400m to 1,200m wide, maintaining this character until Golubac, where it enters the Đerdap Gorge. On the plain, the river often spreads away from its bed and creates many wetlands and flooded areas on the surrounding lowlands, which have been designated as Special Nature Reserves according to the Law on Nature Conservation (*Official Gazette of the Republic of Serbia*, No 36/2009, 88/2010, Article 29).These are areas whose nature has changed little, or not at all; they are of particular importance due to certain characteristics and natural values, and any resident people live harmoniously with nature. However, areas of unchanged natural features with representative natural ecosystems are protected as purely Nature Reserves. The river's Special Nature Reserves include the Upper Danube Reserve on the Serbian side (A in Figure 17.1 and A in Figure 17.2), Kopački Rit on the Croatian side (B in Figure 17.1), and the Danube-Drava National Park on the Hungarian side (C in Figure 17.1). Further on, it reaches the Fruška Gora Mountain (E in Figure 17.1 and B in Figure 17.2) and changes its course eastwards, cutting into the northern foothills, while its left bank retains in the Bačka district the flat scenery of the plain. The Danube continues eastwards until it reaches the eastern edge of the Fruška Gora Mountain, where it collects the waters of its tributary, the Tisza (or Tisa) River (G in Figure 17.1), and changes its course southwards towards Belgrade. In spite of changing its course, the surrounding topography remains the same, with plain topography on the left bank (now in the Banat district) and with a steep-sided right bank. Throughout its whole course through Serbia, the right bank is characterised by considerable loess

Figure 17.2 Some of the most important geoheritage locations of the Serbian Danube region
A. Special Nature Reserve 'Upper Danube' – note the typical landscape of flooded areas
which characterises the left bank of the Danube up to Belgrade.
B. National Park 'Fruška Gora' – note the wooded slopes intersected by streams, which
create the typical scenery of National Park Fruška Gora.
C. Đerdap Gorge or 'Iron Gate' – note the limestone hills at the narrow point of the river.
D. Deliblato Sands – note the aeolian relief, which is characteristic of desert areas, and the
low dunes in the background.
E. 'Big Gully' at Titel Loess Plateau – note the steep cliffs of loess, with that on the middle
left providing an important research section.
F. Drmno site with excavated mammoth finds – note the *IN SITU* fossils on pillars of loess as
they await their lifting and removal to the on-site museum.

accumulation and sedimentation; this is why there are numerous islands, plateaus, gullies, steep cliffs, and even pseudo-karst landforms on that bank.

In Belgrade, the Danube meets the Sava River and again changes its course eastwards, where it approaches the south-west foothills of the Carpathian range. It slowly cuts through the Carpathians, finally becoming a border river between Serbia and Romania. This is the area where the river has cut the magnificent gorge known worldwide as the 'Iron Gate', and where it is transformed into a typical mountain river. In 1972, after the erection of the Đerdap hydroelectric power plant, a great artificial lake was formed; consequently, in this sector the river's flow was drastically slowed down. In this area the Danube has continuously changed its course, creating several gorges and canyons. This is the area of the Đerdap National Park (D in Figure 17.1), the largest protected area in Serbia. Afterwards, the Danube again gradually turns eastwards when it leaves Serbia and becomes the border between Romania and Bulgaria, in which latter country it eventually discharges into the Black Sea.

The characteristics of the Danube and its geological and pedological (soils) background in Serbia, created and modified by its flow, formed the perfect preconditions for the development of many areas and sites with unique and significant natural resources. The relevant national conservation institutions recognised this potential, and the region now has two (of the five within Serbia) National Parks; Fruška Gora (B in Figure 17.2) was the first to be established, while Đerdap (the second to be established) is the largest in Serbia (C in Figure 17.2). Additionally, in the area adjacent to the Danube there are many other forms of legally protected natural areas, such as Nature Parks, Special Nature Reserves, Natural Monuments and Landscapes of Outstanding Features (all form part of the national categorisation of protected areas designated by the Institute for Nature Conservation of Serbia).

THE DANUBE'S GEOHERITAGE IN SERBIA

The very turbulent geological past of this small region, coupled with the influence of the Danube, has created a rich and complex geodiversity. There is no formal or specifically Danubian geoheritage list for Serbia. However, there are proposals to make an inventory of a wider region (Vojvodina and even Serbia), which is some indication that its geodiversity is worthy of much more attention. The complexity of the various terrains (in terms of their geology and geomorphology) has made the classification of the region's geosites and geodiversity, and the identification of priorities for conservation, a daunting task. Hence, in this chapter we have selected and described specific examples to demonstrate the range of geoheritage sites; these include the Iron Gate (Đerdap Gorge), Fruška Gora Mountain, Deliblato Sands (F in Figure 17.1), and other less well known geosites.

Đerdap Gorge

The landscape character of the territory of Đerdap is naturally dominated and affected by the course of the Danube as it dissects the Carpathian Mountains and connects the Pannonian and Pontian basins. The territory's morphological features indicate that the gorge was created by the Danube as it successively gnawed away at the cliffs of the Carpathian Mountains, creating steep and occasionally vertical cliff faces towering up to 800m above the river (C in Figure 17.2). The territory's most prominent geomorphological feature is the Đerdap Gorge itself, also known as the 'Iron Gate', on the Serbian-Romanian border. The gorge as a whole, at some 100km, is the longest and largest breakthrough gorge in Europe (Dragićević *et al* 2013). It is a composite feature of four narrow points on the river; these are the gorges (Golubac Gorge, Gospodin Vir with the canyon of the Boljetina River, the Kazan Gorge and the Sip Gorge), together with three points where the river opens out into valleys. The whole gorge complex is split into two main parts. The Upper Gorges are the Gospodin Vir Gorge and the Donji Milanovac Valley. The Lower Gorges are the Veliki Kazan, Mali Kazan and Sip Gorges with the Orsova Valley (Božić and Tomić 2015).

The canyon-like forms of the various gorges of the Iron Gates are dominated by limestone massifs with typical karst features (such as sinkholes and the karst grooves that form on steep slopes). The siliceous rocks of the valley-like floors are sharply separated from the limestones of the true gorges. In some parts of the gorge (as at Pecka Bara), metamorphic rocks are sometimes exposed at the surface, but they are generally overlain by a layer of limestone. Besides the various limestones (variously described as massive, red, sandy, marly and silicified), the other rocks include granite (at Brnjica), greenschists (at Dobra), mica-schists (at Boljetin, Tekija and Sip), conglomerates and sandstones (at Pesača and elsewhere), clayey schists and amphibolites (at Boljetin), gabbro (at Donji Milanovac), and sands, clays, sandy clays and pebble formations (at Ključ). The most geologically diverse and complex area is between Bosman and the Boljetinska River (of the Upper Gorges), where there is a sequence of narrow belts of greenschists, conglomerates and sandstones, mica-schists, sandstone, massive and marly limestones, marls, red sandstones and argillaceous (clayey) schists.

The Kazan Gorge (or 'The Cauldron') is the narrowest part (at 150–170m wide) of the gorges, where the Danube is 20–53m deep and has cut through gabbro, Jurassic limestone and crystalline rocks that rise up to 30m above the river. The Sip Gorge is where the Danube cuts through the Miroč Plateau, which is built up almost vertically from shales, Jurassic-Cretaceous limestones and sandstones, together with Upper Miocene conglomerates. At the actual location of the Iron Gate, sharp rocks protrude out of the water, producing the rapids that have always been a great obstacle to the safe passage of boats. In addition to its valuable bio- and geodiversity, Đerdap presents a unique area of archaeological and historical heritage. This dates from the Mesolithic and Neolithic Age (Lepenski Vir archaeological site), through to the Roman period (a memorial plaque, 'Tabula Traiana', commemorates the completion of Trajan's military road, Trajan's bridge and the Diana military fortress) and on into medieval times, with the Golubac and Kladovo (Fetislam) fortresses.

Fruška Gora

The Fruška Gora Mountain is located on the southern rim of the Pannonian basin, along the right bank of the Danube on the south (E in Figure 17.1). It was proclaimed a National Park in 1960 and is the oldest one in Serbia. Although there are only a few peaks higher than 500m (the highest peak is Crveni Čot at 539m), it represents an upstanding dominant geomorphological complex in the mostly flat and rather visually monotonous landscape (B in Figure 17.2) of the Vojvodina region. In addition to its geomorphological significance, this mountain represents the largest formation of geological and pedological diversity in the Pannonian area of Serbia. Furthermore, this relatively small region reflects the very complicated geological evolution of a unique tectonic, lithological and stratigraphical mosaic (Petrović *et al* 2013; Vujičić *et al* 2011).

The present-day appearance of Fruška Gora is a reflection of its long geological history and more recent geodynamic processes. The mountain used to be an island in the Pannonian Sea, and today it is an isolated mountain on the south-east edge of the Pannonian plain. In its core there are Palaeozoic (more than 300 million years old) and Mesozoic sedimentary rocks (from the time of the dinosaurs, from 65–270 million years ago), Mesozoic and Tertiary volcanic rocks and a variety of metamorphic rocks. On its slopes and foothills, Neogene sediments from the former Pannonian Sea can be found, as well as various generic types of Quaternary formations (Petrović *et al* 2013), sediments formed during the Ice Age.

The entire area displays a rich geodiversity, with numerous fossil sites and geological formations that are important for understanding the geological structure and the historical and geological evolution of the Pannonian and Danube regions. The largest number of fossil sites is found in sediments of Neogene (2.5–23 mya) age. During the Ice Age, Fruška Gora was located in the periglacial zone, in which strong winds from the north and north-east deposited huge amounts of fine dust created by the movement of glaciers. These loess deposits cover the fossil remains of numerous Ice Age animals, such as mammoths, rhinos, deer and bison. This area also possesses deposits of various minerals, from semi-precious and ornamental stones, different construction materials (clay, building stone and cement marl), coal and geothermal water.

Deliblato Sands

The Deliblato Sands is an isolated complex of sand masses of aeolian (carried and deposited by wind) origin, often referred to as the 'European Sahara', situated between the Danube and the west slopes of the Carpathians (Hrnjak *et al* 2013). This unique geomorphological feature has an elliptical shape, approximately 35km long and 15km wide, which extends along a south-east to north-west axis (F in Figure 17.1); it represents the single largest accumulation of sand with distinct and conserved elements of aeolian relief characteristic of desert areas (D in Figure 17.2) in Serbia. Aeolian sands in this area were deposited during the late Pleistocene, but sand deposition was continuous during the Holocene as well. During this period, the dominant east wind sculpted and shaped the distinct dune relief which rises between 70m and 200m above sea level. As well as the sandy areas, there are also two attractive loess formation areas: the Dumača plateau and the Zagajica hills (Lukić *et al* 2013).

The main components of the relief are sand dunes and inter-dune depressions, lower valleys, sand slacks, bays, vents and grooves. Dunes and inter-dune depressions occur not only in the sand, but also in the sandy loess throughout the entire area of the Deliblato Sands. The average relative height of the dunes ranges from 10m to 20m, while the length of individual dunes is about 100m. Dunes are representatives of relief in higher parts of the Deliblato Sands, while aeolian relief (formed by wind dynamics) can be found in the lower zone, and it is represented by barhanas (smaller crescent-shaped dunes, where the wind-facing side is gentle and elongated while the opposite side is short and steep), vents and grooves.

Loess geoheritage with Kostolac open mine

Loess consists of windblown dust, deposited over extensive areas, that today covers 10% of the Earth's land surfaces in the mid latitudes (Pecsi 1990; Heller and Evans 1995; Smalley *et al* 2011). The thickest and most important loess sequences are found on plains (such as the Pampean Plain in Argentina and the Russian Plain), plateaus (such as the Chinese loess plateau) and along river basins (such as the Danube basin, the middle Rhine basin, the Mississippi basin, and the middle Yellow River basin), as these are typical geographical zones for its deposition. Loess deposits have gained great importance in the geoscientific community for the reconstruction of past climates; they are important *in situ* palaeoclimatic and palaeoenvironmental archives.

As a significant Quaternary (from 2.6 mya to the present) palaeoclimatic and palaeoenvironmental sediment archive in Europe, the loess-palaeosol sequences (Ice Age deposits reflecting
Pleistocene climatic and environmental dynamics) in the Vojvodina region represent an important element of Serbia's geodiversity (Marković *et al* 2008, 2011; Vasiljević *et al* 2011a, 2014a).
Loess sections along the Danube in Serbia include the most complete and thickest sequences
in Europe. The main Middle and Late Pleistocene loess sites in this part of the Danube basin
include those at Stari Slankamen, Ruma, Batajnica, Titel loess plateau and Nosak (Marković
et al 2011). The Stari Slankamen loess-palaeosol section is located on the north-eastern part of
the Srem Loess Plateau (Vojvodina region, North Serbia), on the western bank of the Danube,
opposite the Tisa confluence (H in Figure 17.1). The roughly 40m high cliff comprises loess
intercalated with nine major palaeo-pedocomplexes (fossil soils), and can be considered one of
the most important Quaternary sections in the Carpathian (Pannonian) basin (Marković *et al*
2011). It contains valuable palaeoclimatic and palaeoenvironmental records of the Middle and
Late Pleistocene, represented in loess and soil particles (quartz, pollen carbon) as well as fossils
(Vasiljević *et al* 2011a, 2011b, 2014a).

The Titel loess plateau is an isolated loess island located in the south of Bačka region, in
the confluence area of the Tisa and Danube rivers (G in Figure 17.1). Thick loess deposits of
between 35m and 55m are intercalated by five main pedo-complexes that were deposited through
the last five glacial/interglacial cycles. Steep loess cliffs expose several profiles ('Big Gully' near
Mošorin village, Feudvar section, Dukatar section, Rogulić gully, Titel old brickyard section and
the Kalvarija section) that are important for understanding climatic and environmental change
in the region during the Middle and Late Pleistocene (approximately 650,000 years ago). This
isolated loess island and unique geomorphological phenomenon, with a rich diversity of loess
landforms (eg loess 'caves', pyramids, gullies and cliffs), dominates its lower surroundings (E in
Figure 17.2), which makes it aesthetically and visually attractive (Lukić *et al* 2009; Vasiljević *et al*
2011b). Additionally, sections such as 'Feudvar' and 'Kalvarija' are of great archaeological value
(Bokhorst *et al* 2009; Marković *et al* 2005; Vasiljević *et al* 2011b). The Batajnica loess section (I in
Figure 17.1) has been recognised as one of the most complete Middle and Late Pleistocene records
in this region. A more than 40m-thick loess-palaeosol succession represents an environmental
transition from semi-humid subtropical environments to temperate forest, and finally towards
landscapes with typical steppe soils. The multi-layered record provides an important link between
the classical Central European loess sites and the Central Asian and Chinese loess provinces that
together provide a detailed history of the past 620,000 years (Marković *et al* 2009).

The loess-palaeosol sequences around Ruma town (35km south of Novi Sad; 63km north-
west of Belgrade) were formed during the later part of the Middle (0.13–0.78 mya) and Late
(0.12–0.012 mya) Pleistocene. Several 1km-long open profiles are exposed in an excavation at a
local brick factory on the left bank of the Jelence River, in the central part of the southern slopes
of the Fruška Gora Mountains (J in Figure 17.1). Six loess units and five palaeosols are preserved
in the 20m-thick exposure of Middle and Late Pleistocene sediments, where the remains of
the large Pleistocene mammals *Ursus deningeri* (Deninger's bear) and *Mammuthus primigenius*
(mammoth) were excavated (Marković *et al* 2004). The profile represents one of the best potential sections in the region for reconstructing detailed palaeoclimates and palaeoenvironments of
the past 350,000 years (Marković *et al* 2006).

The Nosak loess-palaeosol sequence is part of a major Quaternary (last 2.6 mya) site at
Drmno, near Kostolac in north-eastern Serbia (K in Figure 17.1), which contains several

important archaeological and palaeontological remains and geological features. The most spectacular discoveries to date have been the Kostolac steppe mammoth (*Mammuthus trogontherii*) skeleton from Middle Pleistocene fluvial deposits, uncovered in 2009 (Lister *et al* 2012), and the later finding of a rich palaeontological layer, including further mammoth fossils from the latest Middle Pleistocene loess-palaeosol succession, discovered in 2012 (Marković *et al* 2014). The entire profile is 25.4m in thickness and situated within the second largest Serbian lignite opencast mine (F in Figure 17.2). Fortunately, due to lengthy sustainable cooperation and mutual understanding between the management of the Drmno mine and thermal power plant, and the team of archaeologists of the adjacent Roman site of Viminacium, these recently discovered fossils were excavated, relocated, preserved and exhibited in the first-ever palaeontological park in Serbia (Tomić *et al* 2015). The Viminacium Archaeological and Mammoth (palaeontological) Park is located there, as a combined heritage centre and tourist attraction.

Hydrological heritage of the Serbian Danube area

Hydrological heritage sites are an integral part of both nature conservation and geoheritage, as water phenomena are an important aspect of the natural environment and their representative forms also belong to Earth's geodiversity (Simić *et al* 2012). According to Gavrilović *et al* (2008), the classification of the hydrological heritage includes several different hydrological phenomena, such as springs, rivers, lakes, waterfalls, ponds, marshes, swamps, wetlands, stagnant tributaries, peat bogs and hydrographical points.

The Danube, the most important and dominant hydrological feature of Central and Eastern Europe, has created many occurrences of this kind that were eventually recognised and placed under protection. The first and the most impressive is the large marshland complex of the Upper Danube (A in Figure 17.2), protected by the Institute for Nature Conservation of Serbia as a Special Nature Reserve. This area represents one of the last big flood areas on the European continent, as it also extends throughout the neighbouring republics of Hungary and Croatia. Due to its biodiversity, the area has been declared a Ramsar site, an important bird area (IBA) and an important plant area (IPA). In 2012 it was also nominated for UNESCO's Transboundary Biosphere Reserve 'Mura-Drava-Danube' (a cluster of 13 protected areas along the three rivers), with joint sustainable management of the shared river. Similar but smaller areas extend further on the left bank, opposite the Fruška Gora Mountain, such as the Special Nature Reserves Karađorđevo, Bagremara and Kovilj-Petrovaradin Marshes, and the Nature Parks Tikvara and Begečka Jama. All these flooded areas have a distinctive variety of geomorphological forms of river erosion, such as river islands, backwaters, ponds, river beams and fluvial lakes of different ages. Consequently, they have a mosaic of wetland habitats; rare and endemic wildlife and plant species are also present. This combination of biodiversity and geodiversity gives the whole region a spectacular aesthetic appeal and value. Further downstream, towards Belgrade, the left bank still remains rich in similar hydrological phenomena within the 'Forland of Left Danube Bank', which is in the process of gaining the status of Protected Landscape, as designated by the Institute for Nature Conservation of Serbia. This area is rich in river islands, of which the most representative is the Great War Island – a Protected Landscape located at the confluence of the Sava and Danube rivers, its pristine nature contrasting with the urban zone of Belgrade.

In addition to all these classified sites and individual areas that firmly deserve geoheritage status, there are many geosites (such as the Petrovaradin rock, the Profile of the Marine Neogene Sandbank in Kalemegdan, etc) that could be included within the Danube geoheritage inventory of Serbia, including *ex situ* collections within the relevant institutions (for example, those in the Nature Conservation Institute of Serbia, the Provincial Nature Conservation Institute and the Belgrade Natural History Museum).

Geoheritage and geoconservation in Serbia

In spite of its exceptional geodiversity, the concept of geoheritage is relatively new in Serbia. However, nature conservation in Serbia, including geodiversity, has very early roots. In 1349, the Regulation No 123 of Tsar Stephan Dušan's Code forbade Saxon miners from felling trees unless they also provided new plantings. In the early 15th century, the First Code of Mines (issued in 1412 by the Serbian Despot Stefan Lazarević in Novo Brdo – then the economic centre of Serbia and the largest mine in the Balkans) regulated issues concerning the ownership and use of mineral resources. More than five centuries passed after these bylaws before the first practical attempt to conserve elements of geodiversity in 1924; this was when the Natural History Museum of Serbia (in Serbian: Muzej srpske zemlje) suggested that the Zlot Cave should gain legal protection as a unique locality due to its fossil remains of *Ursus speleus* (cave bear) and other Quaternary mammals. Unfortunately, this was postponed due to the political situation and then the outbreak of World War II.

It was not until 1949 that the newly established Nature Conservation Institute of Serbia (then the Institute for Protection and Scientific Research of the Natural Rarities of Serbia) undertook the first practical geosite conservation measures. The first officially protected geosite was the waterfalls of Mala Ripaljka and Velika Ripaljka on the mountain of Ozren (East Serbia); today this is recognised as both geomorphological and hydrological heritage. In the same year, the Institute protected several more sites, mostly speleological and geomorphological ones, including the aforementioned Zlot Cave and Đavolja Varoš (Devil's Town), a natural monument in southern Serbia. This tendency to protect mostly caves continued for several years, when other elements of geodiversity, such as karst landscape and monuments, geological and pedological (soil) profiles and hydrological phenomena, were included in the inventory and placed under its protection. Most of these sites were already within protected areas; thus, no explicit and specific protection was made in order to improve their condition and enhance their sustainability.

Today, however, there are 80 protected geosites, many of which have been designated in the second part of the 20th century. By the end of the 20th century, the most influential developments were admission to ProGEO and the founding of the National Council for the Geoheritage, which resulted in systematic work on the conservation and promotion of Serbia's geoheritage; nearly 650 geosites were subsequently proposed for protection. Unfortunately, true geotourism provision is still rare in Serbia, with the focus mainly on speleological tourism – perhaps unsurprisingly, given its past prominence in recognition and protection. Besides numerous caves arranged for general visits (Lazareva, Ceremošnja, Rajkova, etc), the only geotourism destination, based on its geomorphological attractiveness and rarity, is Devil's Town in south-east Serbia. This area is known for its more than 200 earth pillars (2–15m in height), topped with large andesite (a type of silica-rich volcanic lava, named after the Andes mountains, mineralogically related to granite) cap-rocks; these geological forms helped promote it as one of the nominees for the 'New Seven Wonders of Nature' campaign in 2010.

REALITY AND PERSPECTIVES OF DANUBE GEOTOURISM IN SERBIA

Undoubtedly, Serbia has many remarkable geosites that have both great scientific importance and the potential for geotourism development. Like many developing countries, this immense potential is not adequately utilised in the implementation of geoconservation management or geotourism development. One of the main reasons for this current situation is closely related to the past management of these sites, which made little effort to enhance the experience for existing visitors or to attract more tourists. On-site interpretation is either non-existent or of poor quality, consisting of old and non-professional panels with inappropriate content. Generally, geosites are not even denoted by any signage, and visitor centres do not generally promote geoheritage with either their exhibitions or their printed materials. In addition, the websites of these sites and protected areas (with the exception of Đerdap National Park) do not include any presentations of geoheritage, nor do they propose any possible geo-route.

However, in 2008 an initiative was launched by the Provincial Secretariat for Energy and Mineral Resources to include Fruška Gora Mountain in the Global and European Geopark Networks (which were declared as UNESCO Global Geoparks in late November 2015). Based on previous research, it was determined that Fruška Gora Mountain possesses all of the necessary natural requirements, but it lacks adequate infrastructure; this would need to be constructed before Fruška Gora could be included in the geopark network. Unfortunately, its application was never fully pursued and Fruška Gora still remains only a National Park, without any geopark status. The main reason for this lack of progress can be found in the poor communication between the crucial institutions that manage this area: Vojvodinašume (the provincial forestry management company), the Institute for Nature Conservation, and Fruška Gora National Park. Furthermore, these institutions seldom include relevant academic institutions, or experts from the scientific community, in their management or planning of such projects. This is also one of the main issues related to the management of geosites in other protected areas in Serbia.

Another important issue concerns geoscientific research. Loess profiles currently represent the best studied geoheritage resource in Serbia, and as such they can easily be linked to geotourism and subjected to geoconservation measures. Due to a rapid increase in the extent and volume of Danubian loess research in the past decade, much knowledge and information has been collected about these geosites; hence, they provide very good interpretative possibilities, as they hold the secrets of palaeoclimates and the Ice Ages. Other geoheritage resources, even those with arguably greater geotourism potential, are far less studied and very rarely connected to geotourism and geoconservation. These various shortcomings led to another initiative that was launched in 2008 by the municipality of Inđija; this was to establish the themed museum 'Loessland' at the Čot loess profile, right next to the Danube near Stari Slankamen. The museum could have provided for educational, scientific and tourism functions, whilst containing many exhibitions and interactive contents, a souvenir shop and a café, all with the theme of the Ice Ages, loess and palaeoclimate. However, due to the overly ambitious preliminary project (especially the architectural design of the building) by the Municipality of Inđija, the museum was never constructed as the funding necessary to complete the project was beyond the resources available in the prevailing economic climate.

As a form of special interest tourism (Weiler and Hall 1992), geotourism includes travellers within a small niche tourism market (see Hose 2005). It is still largely based on self-guided tours, which do not usually offer the same quality experience as tours with professional guides.

Another of the problems is that the majority of protected areas, such as National Parks, do not have brochures and maps of geosite attractions to underpin such self-guided tours. Hence, dedicated rather than casual (Hose 2000) geotourists must seek and then examine often quite obscure sources in order to plan their trips.

Conclusion

Although awareness of (and concern about) geoheritage values and geotourism potential is at an inadequate level in Serbia, some initiatives (and specifically research) towards better understanding and promotion of the Danube geoheritage show that, with more efficient collaboration between academia and relevant national institutions, this natural resource could be more widely recognised and appreciated – both within and outside Serbia. Geotourism requires the provision of interpretative and service facilities to enable tourists to acquire knowledge about geosites (Hose 2012); introducing these facilities – especially through measures such as websites, that do not demand major investments in infrastructure – should attract more visitors and provide them with enhanced experiences. Moreover, the geoheritage presented in this study could provide an excellent basis for the creation of a unique Danube geotourism cruise route, linking together a wide array of attractive sites into a unique and complex tourism product. Numerous geological, archaeological and palaeontological sites are faithful witnesses to the long and vivid cultural and geological history of this area. This route could certainly present a good opportunity for creating a new cross-border tourism product, by expanding it into neighbouring countries as an important promotional vehicle to attract more tourists.

Acknowledgment

The work underpinning this chapter is supported by the Ministry of Education, Science and Technological Development, Republic of Serbia (Grant 176020).

References

Bokhorst, M, Beets, C J, Marković, S B, Gerasimenko, N P, Matviishina, Z N, and Frechen, M, 2009 Pedochemical climate proxies in Late Pleistocene Serbian-Ukrainian loess sequences, *Quaternary International* 198 (1–2), 113–23

Božić, S, and Tomić, N 2015 Canyons and gorges as potential geotourism destinations in Serbia: comparative analysis from two perspectives – general tourists and pure geotourists, *Open Geosciences* 7, 531–46

Dragićević, S, Mészáros, M, Djurdjić, S, Pavić, D, Novković, I, and Tošić, R, 2013 Vulnerability of national parks to natural hazards in the Serbian Danube region, *Polish Journal of Environmental Studies* 22 (4), 75–82

Gavrilović, L, and Dukić, D, 2002 *Reke Srbije* [*Rivers of Serbia*], Zavod za udžbenike i nastavna sredstva, Belgrade (in Serbian)

Gavrilović, L, Belij, S, and Simić, S, 2008 Hydrological heritage of Serbia – the preliminary list, *Protection of Nature* 60 (1–2), 387–96 (in Serbian)

Heller, F, and Evans, M E, 1995 Loess magnetism, *Reviews of Geophysics* 33 (2), 211–40

Hose, T A, 2000 Rocks, rudists and writing: an examination of populist geosite literature, in *Proceedings of the Third UKRIGS Conference: Geoconservation in Action* (ed K Addison), Wirksworth, Association of UKRIGS, 39–62

— 2005 Geo-tourism – appreciating the deep time of landscapes, in *Niche Tourism: Contemporary Issues, Trends and Cases* (ed M Novelli), Elsevier, London, 27–37

— 2012 3G's for modern geotourism, *Geoheritage* 4 (1–2), 7–24

Hrnjak, I, Vasiljević, Đ A, Marković, S B, Vujičić, M D, Lukić, T, Gavrilov, M B, Basarin, B, and Kotrla, S, 2013 Application of preliminary Geosite Assessment Model (GAM): the case of the Deliblato Sands, paper presented at *2nd Scientific-Professional Conference Nature Protection of Southern Banat*, 25 October, Vršac

Jojić-Glavonjić, T, Milijašević, D, and Panić, M, 2010 Geoheritage protection of Serbia: present situation and perspectives, *Journal of the Geographical Institute Jovan Cvijić, SASA* 60 (1), 17–30

Lister, A M, Dimitrijević, V, Marković, Z, Knežević, S, and Mol, D, 2012 A skeleton of 'steppe' mammoth (Mammuthus trogontherii Pohlig) from Drmno, near Kostolac, Serbia, *Quaternary International* 276/277, 129–44

Lukić, T, Marković, S B, Stevens, T, Vasiljević, D A, Machalett, B, Milojković, N, Basarin, B, and Obreht, I, 2009 The loess cave near the village of Surduk – an unusual pseudokarst landform in the loess of Vojvodina, Serbia, *Acta Carsologica* 38 (2–3), 227–35

Lukić, T, Hrnjak, I, Marković, S, Vasiljević, D, Vujičić, M, Basarin, B, Gavrilov, M B, Jovanović, M, and Pavić, D, 2013 Zagajica hills as an archive of palaeoclimatic and palaeoecological characteristics and possibilities for geoconservation, *Zaštita Prirode [Nature Conservation]* 63 (1–2), 59–71 (in Serbian)

Marković, S B, Kostić, N, and Oches, E A, 2004 Palaeosols in the Ruma loess section, *Revista Mexicana de Ciencias Geologicas* 21 (1), 79–87

Marković, S B, Jovanović, M, Mijović, D, Bokhorst, M, Vandenberghe, J, Oches, E A, Hambach, U, Zoeller, L, Gaudenyi, T, Kovačev, N, Boganović, Ž, Savić, S, Bojanić, D, and Milojković, N, 2005 Titel loess plateau – geopark, paper presented at *2nd Conference on the Geoheritage of Serbia*, 22–23 June, Belgrade

Marković, S B, Oches, E, Sümegi, P, Jovanović, M, and Gaudenyi, T, 2006 An introduction to the Upper and Middle Pleistocene loess-palaeosol sequences of Ruma section (Vojvodina, Serbia), *Quaternary International* 149, 80–6

Marković, S B, Bokhorst, M, Vandenberghe, J, Oches, E A, Zöller, L, McCoy, W D, Gaudenyi, T, Jovanović, M, Hambach, U, and Machalett, B, 2008 Late Pleistocene loess-paleosol sequences in the Vojvodina region, North Serbia, *Journal of Quaternary Science* 23, 73–84

Marković, S B, Hambach, U, Catto, N, Jovanovic, M, Buggle, B, Machalett, B, Zoeller, L, Glaser, B, and Frechen, M, 2009 The Middle and Late Pleistocene loess sequences at Batajnica, Vojvodina, Serbia, *Quaternary International* 198 (1–2), 255–66

Marković, S B, Hambach, U, Stevens, T, Kukla, G J, Heller, F, McCoy, W D, Oches, E A, Buggle, B, and Zöller, L, 2011 The last million years recorded at the Stari Slankamen loess-palaeosol sequence: revised chronostratigraphy and long-term environmental trends, *Quaternary Science Reviews* 30, 1142–54

Marković, S B, Korać, M, Mrđić, N, Buylaert, J-P, Thiel, C, Stevens, T, McLaren, S, Tomić, N, Petić, N, Jovanović, M, Vasiljević, D A, Sümegi, P, Gavrilov, M, and Obreht, I, 2014 Palaeoenvironment and geoconservation of mammoths from the Nosak loess-palaeosol sequence (Drmno, north-eastern Serbia): initial results and perspectives, *Quaternary International* 334/335, 30–9

Ministry of Environment and Spatial Planning, Republic of Serbia, 2010 *First National Report of the Republic of Serbia to the United Nations Convention on Biological Diversity*, Ministry of Environment and Spatial Planning, Belgrade

Pantić, N, Belij, S, and Mijović, D, 1998 Geo-nasleđe u sistemu prirodnih vrednosti i njegova zaštita u

Srbiji [Geologic heritage in the system of natural values and its protection in Serbia], *Zaštita Prirode* [*Nature Conservation*] 50, 407–13 (in Serbian)

Pecsi, M, 1990 Loess is not just the accumulation of dust, *Quaternary International* 7/8, 1–21

Petrović, M D, Vasiljević, D A, Vujičić, M D, Hose, T A, Marković, S B, and Lukić, T, 2013 Global geopark and candidate-comparative analysis of Papuk Mountain Geopark (Croatia) and Fruška Gora Mountain (Serbia) by using GAM model, *Carpathian Journal of Earth and Environmental Sciences* 8 (1), 105–16

Simić, S, Gavrilović, B, Živković, N, and Gavrilović, L, 2012 Protection of hydrological heritage sites of Serbia – problems and perspectives, *Geographica Pannonica* 16 (3), 84–93

Smalley, I J, Marković, S B, and Svirčev, Z, 2011 Loess is [almost totally formed by] the accumulation of dust, *Quaternary International* 240 (1–2), 4–11

Tomić, N, Marković, S B, Korać, M, Mrđić, N, Hose, T A, Vasiljević, D A, Jovičić, M, and Gavrilov, M B, 2015 Exposing mammoths: from loess research discovery to public palaeontological park, *Quaternary International* 372, 142–50, doi:10.1016/j.quaint.2014.12.026

Vasiljević, D A, Marković, S B, Hose, T A, Smalley, I, Basarin, B, Lazić, L, and Jović, G, 2011a The introduction to geoconservation of loess-palaeosol sequences in the Vojvodina region: significant geoheritage of Serbia, *Quaternary International* 240 (1–2), 108–16

Vasiljević, D A, Marković, S B, Hose, T A, Smalley, I, O'Hara-Dhand, K, Basarin, B, Lukić, T, and Vujičić, M D, 2011b Loess towards (geo) tourism – proposed application on loess in Vojvodina region (North Serbia), *Acta geographica Slovenica* 51 (3), 391–406

Vasiljević, D A, Marković, S B, Hose, T A, Ding, Z, Guo, Z, Liu, X, Smalley, I, Lukić, T, and Vujičić, M D, 2014a Loess–palaeosol sequences in China and Europe: common values and geoconservation issues, *Catena* 117, 108–18

Vasiljević, D A, Marković, S B, and Vujičić, M D, 2014b Appreciating loess landscapes through history: the basis of modern loess geotourism in the Vojvodina region of North Serbia, in *Appreciating Physical Landscapes: Three Hundred Years of Geotourism – Special Publication No 417* (ed T A Hose), Geological Society of London, London (in press) – first published online 3 November 2014, doi:10.1144/SP417.5

Vujičić, M D, Vasiljević, D A, Marković, S B, Hose, T A, Lukić, T, Hadžić, O, and Janićević, S, 2011 Preliminary geosite assessment model (GAM) and its application on Fruška Gora Mountain, potential geotourism destination of Serbia, *Acta geographica Slovenica* 51 (3), 361–76

Weiler, B, and Hall, C M (eds), 1992 *Special Interest Tourism*, Belhaven, London

Weller, P, 2009 The Danube River – the most international river basin, in *Handbook of Catchment Management* (eds R C Ferrier and A Jenkins), Wiley-Blackwell, Oxford, UK, doi:10.1002/9781444307672.ch12

18

Conclusion

Peter Davis and Thomas A Hose

Apart from summing up the volume's contents, this final chapter should also acknowledge past endeavours in exploring and collecting Europe's geoheritage, and in recognising and developing its geotourism provision, as well as pointing the way towards future work. The themed chapters in this volume have sought to establish the basis of modern geoheritage provision (especially Chapters 4 and 6, together with the various case study chapters) and the modern geotourism paradigm (Chapters 1 and 8). The case study chapters provide examples of the application of geoheritage promotion and geotourism provision in conserving and constituency-building for Europe's geoheritage. However, they do perhaps hint at the need for greater knowledge, better understanding, and application of the current literature (to which this volume is intended to make a contribution) on the broad scope of geoheritage material and geotourism provision amongst the various stakeholders. Admittedly, part of the problem is that much of the literature is relatively new, and it is not always widely disseminated beyond the narrow confines of the academic geological and geoconservation communities.

In the UK, only very recently has one Geological Society of London volume for the first time examined travel for the purposes of geological inquiry (Wyse Jackson 2007). However, despite the admirable interest and quality of its generally descriptive papers, their applicability and usefulness to modern geoheritage and geotourism is not made explicit. Other Geological Society volumes have in passing touched upon aspects of, at least dedicated, geotourism (for example, Blundell and Scott 1998; Herries Davies 2007; Lewis and Knell 2009; Oldroyd 2002); these, for example, include synopses of the work and travels of William Buckland (1784–1856), William Morris Davies (1850–1934), Charles Lapworth (1842–1920), Charles Lyell (1797–1875), John MacCulloch (1773–1835), John Playfair (1748–1819) and Adam Sedgwick (1785–1873) – many of whom have been noted in this volume's chapters. A recent Geological Society volume edited by Hose (2016) examines mainly, but not exclusively, Europe's 17th to 20th century travellers, both casual (for leisure and pleasure) and professional (as geologists, mineral and fossil collectors, artists and poets), from a geotourism perspective and develops a historical overview of its development; it very much builds upon his earlier Geological Society paper on the history of UK geotourism (Hose 2008), published within an overview of the history of geoconservation (Burek and Prosser 2008). From Italy, although one paper (Vai and Caldwell 2006) was published by the Geological Society of America, there has been some significant recognition of, and publication on, Europe's early geoheritage (Vai and Cavazza 2003); both volumes contain well illustrated accounts of significant geological travellers – such as Charles Lyell (1797–1875), Luigi Ferdinando Marsigli (1658–1730), Nicholas Steno (1638–86) and Gregory Watt (1777–1804) – and their discoveries and subsequent publications.

Two geology journals whose scope covers both Europe and beyond (the UK's *Proceedings of the Geologists' Association* and Bulgaria's *Geologica Balcania*) have published papers and themed issues on geoheritage and geotourism. Additionally, the geography journal *Acta geographica Slovenica*, published by the Scientific Research Centre of the Slovenian Academy of Arts and Sciences, has published several geoheritage and geotourism papers, and a special part-issue on geotourism in 2011. The *Proceedings* has, since the turn of the 21st century, published a range of papers on the history of geology (for example, Duffin and Davidson 2011 on folklore and geology, and Gallois 2012 on the UK's Norfolk oil shale industry), geodiversity (for example, Crofts 2014; Gray 2008; Gray, Gordon and Brown 2013; Gordon *et al* 2012), geoheritage (for example, Rocha, Brilha and Henriques 2014) and geoconservation (for example, Larwood, Badman and McKeever 2013, in a major themed issue of the journal by Prosser *et al* 2013). *Geologica Balcania*, published by the Bulgarian Academy of Sciences, has had three special issues in 1996 and 1998 relevant to the scope of this volume. The 1996 special issues were specifically on Europe's geological heritage and arose from the first ProGEO regional meeting on south-east Europe's geoheritage, held in Belogradchik, Bulgaria, in 1995. The first issue included papers on the development of geoconservation in south-east Europe (Zagorchev 1996a; Krieg 1996) and a series of papers cataloguing its geosites, together with the Resolution and Declaration that followed the meeting. The second issue included papers on the geological setting of Balkan geoheritage (Zagorchev 1996b) and a major early human occupation cave site in Slovenia (Pohar, Pavlovec and Hlad 1996), indicating the crossover between the purely geological and the archaeological heritages of Europe. The 1998 special issue (Zagorchev and Nakov 1998) was focused on the 'Geological Heritage of Europe', as a memorial volume to ProGEO's first president, George P Black (1929–97), who was also responsible for the direction and development of geoconservation in the UK, as geologist to the Nature Conservancy Council from 1960. Its publication also followed the ProGEO regional meeting at Belogradchik, Bulgaria; its papers included two very short ones on UNESCO geoparks (Patzak and Eder 1998) and geotourism and geoconservation (Larwood 1998), and a much longer paper on effectively communicating geology to the public (Hose 1998). The *Acta geographica Slovenica* part-issue on geotourism included papers on modern geotourism's origins (Hose 2011) and geomorphosite geoconservation (Komac, Zorn and Erhartic 2011).

Three journals that focus specifically on geoheritage and geotourism are regularly published in Europe. *Geoheritage*, the journal sponsored by ProGEO and launched in 2009, has published three special issues relevant to the scope of this volume, on geotourism and geoconservation (Hose 2012a), geomorphosites and geotourism (Reynard, Coratza and Giusti 2011), and protecting and sharing geoheritage (Bentivenga and Geremia 2015). *Geoturystyka*, a Polish journal first published by Krakow's AGH University of Science and Technology in 2004, is well focused on geology-based geotourism. In several issues, its papers discuss the geotourism concept and generally accept (for example, Różycki 2010, 49–50) the essential geological focus of true geotourism. It is also rich in case study material for Britain (for example, Krzysztof and Mazurski 2013, on the UK's Abberley and Malvern Hills), Europe (for example, Bartuś and Kuś 2010, on Poland's Pieniny Geopark project) and further afield (for example, Pawlikowski, Pieprzyk-Klimaszewska and Mikoś 2009, on Egypt's eastern desert). The opening paper (Słomka and Kicińska-Świderska 2004) in its first issue established its essentially geological approach to geotourism. The *GeoJournal of Tourism and Geosites*, a joint Polish and Romanian journal, began publication in 2008. However, its papers cover much more diverse topics than strictly geotourism; for example, Polish cross-border shopping, urban tourism and marketing. Unfortunately, its sole paper defining the scope of geotourism (Lazzari and Aloia 2014) completely misrepresents the work of Hose (especially Hose 2012a), erroneously attributing the introduction and definition of the term to *National Geographic* and suggesting that Hose merely followed their lead. It is evident that dedicated European publications on geoheritage and geotourism date only from the middle of the opening decade of the 21st century, and that they are mainly published in eastern Europe.

THE VALUE OF FIELDWORK AND ITS LEGACY QUESTIONED

Perusal of both the themed and case study chapters shows that the perceived value and relevance of fieldwork to modern geology – and, for that matter, its associated legacy of collections in museums and universities – has become something of an issue during the 21st century. Likewise, perusal of almost any major Earth Science journal shows the increasing trend to publish papers, the research behind which employs analytical methods requiring laboratory and field facilities that even few universities and research institutes can support; a far cry from the heyday of classic field geology (see Chapters 3 and 6), when 'almost anyone could hope to make exciting discoveries … its active pursuit required little more than a reasonable degree of leisure, the means for at least a little travelling, and a taste for the open air and the countryside' (Rudwick 1976, 166). In some ways, that rather encapsulates the appeal of modern geotourism – but then, that too is chiefly an amateur pursuit. The increasing emphasis on the technological collection, analysis and interpretation of geological and geomorphological data has led to the value of the classic geologist's field skills being questioned by some young professional geologists. The matter was explored (in relation to the useful contribution, especially for future employment, of geology undergraduate field mapping in the education of the modern geologist) in a lively manner in 2013 in *Geoscientist*, the magazine of the Geological Society of London. In 2013 it carried two brief opinion pieces for (Harker 2013) and against (Brodie 2013) the usefulness of traditional fieldwork training; most of the in-magazine and online letters favoured, albeit with some updating, retention of traditional geological field training. Fieldwork's value in assessing Europe's geoheritage is at least unquestionable.

LOSING GEOHERITAGE

The shift towards this almost paperless and virtual geology (especially developed and promoted in the UK by the British Geological Survey – see Smith and Howard 2012) reliant upon mobile computing technology, topographical and geological mapping, and database software and GPS systems, has seen the worth and cost of maintaining ready access to (or even retaining) past geological literature and physical collections in universities and museums called into question by some institutional managers. This situation has been considerably worsened by the rationalisation of national geological surveys, university staffs and departments, and museums since the 1990s. The fate of the geological collections held at the University of Newcastle (McLean 1999) is a case in point; after the closure of the university's geology department around 1990, its collections were shunted from one campus building to another before eventually being transferred on permanent loan to the Hancock Museum. This is despite the publication of major UK volumes on the significance of fossils (Bassett 1979; Bassett *et al* 2001) and geology collections (Doughty 1981, but see also Parkes and Wyse Jackson 1998). Geological collections have also disproportionately lost staff, funding and gallery space in most of Europe's refurbished museums, in the move towards supposedly more populist and communicatively interesting subjects, to gain new audiences and tap new funding streams. However, much of the evidence of concern for geology collections is somewhat anecdotal, mainly due to an understandable lack of public comment and publication by curators on such career-sensitive matters, although various reports do indicate the status of geology collections. Likewise, regional library services and university libraries have commonly been stripped of their antiquarian geological and topographic stock, as is evident from perusal of specialist antiquarian book dealers' catalogues (some of whose less expensive stock graces the bookshelves of one of the authors!), in the mistaken belief that citing or using sources more than a decade or so old is now academically inappropriate (Wyse Jackson 1999, 23) – supposedly outstripped and outclassed by new techniques and understandings. Perhaps it is because we are in an age when economic restructuring has made important past locations, structures and skill-sets redundant, and something that just might be regenerated by the heritage and arts industries. Of course, that shift towards arts and heritage promotion invariably leads to, at best, the gentrification – and at worst, the obliteration – of past industrial and mining geoheritage landscapes (Hose 2011); there really is a need to better support geoconservation (as noted in Chapters 7 and 9) and to develop geotourism (Chapters 8 and 18). In many European countries, the emergence of geoparks (Chapters 12, 14 and 15) has been seen as a way to combine geoconservation and geotourism with other heritage and cultural interests.

GEOPARKS AND GEOTOURISM

Should the development of geotourism continue to be limited to the geoscience community, as Reynard (Chapter 16) notes, its broader socio-economic impact is likely to be limited. However, should some sort of agreement be struck, as Reynard (Chapter 16) suggests, between local communities and (partly or wholly) tourism-related businesses, then a truly economically sustainable future geotourism provision would follow. Vasiljević, Marković and Tomić (Chapter 17) propose that integrating geoheritage with archaeological heritage could give the former greater appeal and could also facilitate the development of multi-interest trails for hikers and cyclists – a real form of ecotourism; further, they suggest, such long-distance trails could lead to new transnational geotourism provision as an important promotional vehicle to attract new tourists. The importance of providing new geotourist experiences and updating older facilities, to refresh the public's perception of geoheritage, has been noted by Munt (Chapter 10), who examines the establishment of a dinosaur museum along the lines of those long pioneered in North America, and cites the importance in marketing-speak of a unique selling point; for the Isle of Wight (Chapter 10) it is dinosaurs and sustainable fossil collecting, whilst for Serbia (Chapter 17) it is thick loess deposits with entombed large fossil mammals. The Isle of Wight shows the importance of a varied, including commercial, geotourism offering to meet the changing needs of a diverse range of visitors, and to provide some economic benefit (albeit small, at less than 3% of total tourism receipts) to the local economy; however, such monies can clearly help to cover the cost of geotourism provision and geoconservation measures.

With specific regard to geoparks, it can be noted (Chapter 15) that there have been in Scotland, and indeed elsewhere in Europe, issues over appropriate levels of funding and adequate management arrangements – both of which point to the need for competent long term business plans to be developed prior to a geopark's designation. Specifically for Scotland, but with wider European implications, a geopark network is essential to the delivery of positive outcomes for their host communities and local businesses, as well as promoting geoconservation. But, as Gordon (Chapter 15) notes, unless geoparks and the geoheritage concept are better integrated within broader socio-economic and nature conservation policies for rural areas, they will struggle to find and maintain adequate funding to sustain their growth and future in the long term.

The history of geotourism, as Hose explores (Chapter 8), is relevant to modern geotourism, in terms of the history, development and philosophy of landscape promotion to tourists, when the locality case studies are considered. Geotourism itself initially required the recognition that wild landscapes were worthy places to visit; similar recognition needs to be generated nowadays about urban industrial landscapes; Wrede (Chapter 13) provides some good examples of where this has been achieved in one of Europe's most industrialised (albeit much reduced in recent years) regions. Given that the wild landscapes were promoted by artistic and literary visualisations, the role of cultural geoheritage in the past and present in geotourism's promotion is clear. Accounts of past geotourism describe the impacts of improving physical access and providing specialist guidebooks to encourage tourists to visit previously virtually inaccessible locations; they were not always positive, however, and provide a cautionary tale (when case studies of specific geosites are examined) about why and how modern geotourism might be sustainably managed and the geoheritage protected. Hose (Chapters 8 and 14) and Larwood (Chapter 7) examine the need for, and various approaches to, geoheritage protection and geoconservation measures. The efficacy of widely adopted geoheritage protection measures varies widely across Europe. Hose (Chapter 14) indicates that this can partly be traced to the legacy of past European empires and political blocs.

MUSEUMS AND GEOHERITAGE

The generally minor world of geological and natural science auctions (compared with the art world) has been noted by Hose (Chapter 5), although prices paid for gem quality and spectacular fossil specimens continue to set new records. The fruits of geological inquiry (Chapters 3 and 4) have, in the past, led to lavish and even expensive publications, and to the amassing of large collections for purposes other than those of either modern science or museum visitors' needs. However, it is apparent that there is a developing market for literary geoheritage (Chapter 5), and this would suggest that a similar market for hard-copy and digital reproductions could be created. The study of auctions (Chapter 5) shows the fickle nature of the survival of some museum-based elements of geoheritage and their significant cultural interest. Further, it shows the need for museum staffs to better recognise the replacement costs – let alone the difficulty of actually locating similar specimens – for even quite humble geological collections.

Museums are unquestionably a significant part of geoheritage, but whether appropriate funds will continue to be available to ensure that their collections and exhibitions do not become moribund is very questionable. However, as geology becomes increasingly marginalised in school and university curricula, museums are one of the few places where most people might expect to find geology; it is therefore somewhat unfortunate that there is an increasing trend towards graphics-rich thematic displays, rather than the stratigraphical, palaeontological, petrological and mineralogical displays that formerly helped visitors to identify their geological finds. In some small measure, the collections that these visitors build up are the natural successors to the old 'cabinets of curiosities' that heralded the birth of scientific geology. It is a pity that there are few – and, since the Great War, declining – natural history societies, whilst interest (if not direct engagement) in archaeology is booming. At least the range, quality and availability of geology field guides (Chapter 6) has never been better; their genesis and early development was demonstrably a British one. Many of the locations first described in these in the 19th century have gone, but some classic – perhaps iconic – sites survive to the present day. Study of these guides shows a shift in geoheritage collecting and management emphasis. With their antecedents lying within the mid-18th century antiquarian county accounts that provided topographic and archaeological descriptions, field guides are probably the most enduring and consulted aspect of geoheritage. Indeed, since geological inquiry began as an outdoor observational science (an approach fostered by its pioneer authors), the decline of formal field-teaching should lead to a renaissance of the field guide, and their importance will surely increase. They are, as Hose (Chapter 8) notes, an enduring vehicle in hard-copy and digital formats for introducing landscapes and geosites to geotourists.

Closing thoughts

The volume has given some indication of the origins and richness of Europe's geoheritage, and the importance of geotourism provision in promoting it to a broader audience than the geological community. It should also, for the insightful reader, have indicated possible areas of future study and development; perhaps chief amongst these are the need to plug the historical gaps in the development of the cultural aspects of Europe's geoheritage (especially with regard to the biographies of minor contributors), to develop an up-to-date synthesis of European geoconservation practice, to better document geotourism provision and to ensure that it addresses the needs of actual audiences. Reflecting upon a well-cited paper (Hose 2012b), there should really have been 4G's presented rather than 3G's for geotourism; that is, geodiversity in addition to geoconservation, geo-interpretation and geohistory. It is to be hoped that the mix of fairly scientific, cultural and historical themed chapters and the case studies within this volume will be found interesting, and that they will be of some value in advancing the study of geoheritage and geotourism in Europe and elsewhere in those four fields.

References

Bartuś, T, and Kuś, T, 2010 An exploitation area of Szlachtowa ore as an element of the future Pieniny Geopark, *Geoturystyka* 2 (21), 35–58

Bassett, M G, 1979 *Special Papers in Palaeontology 22: Curation of Palaeontological Collections*, The Palaeontological Society, London

Bassett, M G, King, A H, Larwood, J G, Parkinson, N A, and Deisler, V K, 2001 *A Future for Fossils*, National Museums and Galleries of Wales, Cardiff

Bentivenga, M, and Geremia, F (eds), 2015 Special Issue: 'Geoheritage – protecting and sharing', papers presented at the 7th International Symposium ProGEO on the Conservation of the Geological Heritage, 24–28 September 2012, Bari, Italy, *Geoheritage* 7 (1)

Blundell, D J, and Scott, A C (eds), 1998 *Lyell: The Past is the Key to the Present*, Special Publication No 143, The Geological Society, London

Brodie, M, 2013 Masters of mapping? *Geoscientist* 23 (7), 11

Burek, C V, and Prosser, C D (eds), 2008 *The History of Geoconservation*, Special Publication No 300, The Geological Society, London

Crofts, R, 2014 Promoting geodiversity: learning lessons from biodiversity, *Proceedings of the Geologists' Association* 125 (3), 263–6

Doughty, P S, 1981 *The state and status of geology in United Kingdom museums, Miscellaneous Paper 13*, Geological Society, London

Duffin, C J, and Davidson, J P, 2011 Geology and the dark side, *Proceedings of the Geologists' Association* 122 (1), 16–24

Gallois, R, 2012 The Norfolk oil-shale rush 1916–1921, *Proceedings of the Geologists' Association* 123 (1), 64–73

Gordon, J E, Barron, H F, Hansom, J D, and Thomas, M F, 2012 Engaging with geodiversity – why it matters, *Proceedings of the Geologists' Association* 123 (1), 1–6

Gray, M, 2008 Geodiversity: developing the paradigm, *Proceedings of the Geologists' Association* 119 (3–4), 287–98

Gray M, Gordon, J E, and Brown, E J, 2013 Geodiversity and the ecosystem approach: the contribution of geoscience in delivering integrated environmental management, *Proceedings of the Geologists' Association* 124 (4), 659–73

Harker, S, 2013 Bring back real mapping, *Geoscientist* 23 (10), 9

Herries Davies, G L, 2007 *Whatever is under the Earth: The Geological Society of London 1807 to 2007*, The Geological Society of London, London

Hose, T A, 1998 Mountains of fire from present to past – of effectively communicating the wonders of geology to tourists, *Geologica Balcania* 28 (3–4), 77–85

— 2008 Towards a history of geotourism: definitions, antecedents and the future, in *The History of Geoconservation, Special Publication No 300* (eds C V Burek and C D Prosser), The Geological Society, London, 37–60

— 2011 The English origins of geotourism (as a vehicle for geoconservation) and their relevance to current studies, *Acta geographica Slovenica* 51 (2), 343–60

— (ed), 2012a Special Issue: Geotourism and geoconservation, *Geoheritage* 4 (1–2)

— 2012b 3G's for modern geotourism, *Geoheritage* 4 (1–2), 7–24

— (ed) 2016 *Appreciating Physical Landscapes: Three Hundred Years of Geotourism, Special Publication No 417*, The Geological Society, London

Komac, B, Zorn, M, and Erhartic, B, 2011 Loss of natural heritage from the geomorphological perspective – do geomorphic processes shape or destroy the natural heritage? *Acta geographica Slovenica* 51 (2), 407–17

Krieg, W, 1996 Progress in management for conservation of geotopes in Europe, *Geologica Balcania* 26 (1), 13–14

Krzysztof, R, and Mazurski, K R, 2013 England's Abberley-Malvern Hills Geopark as a tourism project, *Geoturystyka* 1–2 (32–33), 25–36

Larwood, J G, 1998 Geotourism, conservation and society, *Geologica Balcania* 28 (3–4), 97–100

Larwood, J G, Badman, T, and McKeever, P J, 2013 The progress and future of geoconservation at a global level, *Proceedings of the Geologists' Association* 124 (4), 720–30

Lazzari, M, and Aloia, A, 2014 Geoparks, geoheritage and geotourism opportunities and tools in sustainable development of the territory, *GeoJournal of Geotourism and Geosites* (1) 13, 8–9

Lewis, C L E, and Knell, S J (eds), 2009 *The Making of the Geological Society of London, Special Publication No 317*, The Geological Society, London.

McLean, S G, 1999 Rescuing an orphan mineral collection: the case of the University of Newcastle mineral collection at the Hancock Museum, *The Geological Curator* 7 (1), 11–16

Oldroyd, D R (ed), 2002 *Earth, Water, Ice and Fire: Two Hundred Years of Geological Research in the English Lake District – Memoir No 25*, The Geological Society, London

Parkes, M A, and Wyse Jackson, P N, 1998 A survey on the state and status of geological collections in museums and private collections in the Republic of Ireland, *The Geological Curator* 6 (10), 377–88

Patzak, M, and Eder, W, 1998 UNESCO Geopark: A new programme – a new UNESCO label, *Geologica Balcania* 28 (3–4), 33–5

Pawlikowski, M, Pieprzyk-Klimaszewska, K, and Mikoś, T, 2009 Geotourist attractions of the Eastern Desert in Egypt, *Geoturystyka* 1–2 (16–17) 2009, 45–8

Pohar, V, Pavlovec, R, and Hlad, B, 1996 Potocka Zijalka (Slovenia) – upper mountain site of Aurignacian hunters, *Geologica Balcania* 26 (2), 37–62

Prosser, C D, Brown, E J, Larwood, J G, and Bridglan, D R, 2013 Special Issue: Geoconservation for science and society, *Proceedings of the Geologists' Association* 124 (1), 559–730

Reynard, E, Coratza, P, and Giusti, C (eds), 2011 Special Issue: Geomorphosites and geotourism, *Geoheritage* 3 (3)

Rocha, J, Brilha, J, and Henriques, M H, 2014 Assessment of the geological heritage of Cape Mondego Natural Monument (Central Portugal), *Proceedings of the Geologists' Association* 125 (2), 107–13

Różycki, P, 2010 Geotourism and industrial tourism as the modern forms of tourism, *Geoturystyka* 3–4 (22–23), 39–50

Rudwick, M J S, 1976 *The Meaning of Fossils: Episodes in the History of Palaeontology* (2 edn), University of Chicago Press, Chicago

Słomka, T, and Kicińska-Świderska, A, 2004 The basic concepts of geotourism, *Geoturystyka* 1 (1), 5–8

Smith, M, and Howard, A, 2012 The end of the map? *Geoscientist* 22 (2), 19–21

Vai, G B, and Caldwell, W, 2006 *The Origins of Geology in Italy, Special Paper 411*, The Geological Society of America, Boulder

Vai, G B, and Cavazza, W, 2003 *Four Centuries of the Word Geology: Ulisse Aldrovandi 1603 in Bologna*, Minerva Edizioni, Bologna

Wyse Jackson, P N, 1999 Geological museums and their collections: rich sources for historians of geology, *Annals of Science* 56, 417–31

— (ed), 2007 *Four Centuries of Geological Travel: The Search for Knowledge on Foot, Bicycle, Sledge and Camel: Special Publication No 287*, The Geological Society, London

Zagorchev, I, 1996a Information about the first sub-regional meeting, 'Conservation of Geological Heritage in South-East Europe', Bulgaria, 6–12 May 1995, *Geologica Balcania* 26 (1), 3–5

— 1996b Geological heritage of the Balkan Peninsula: geological setting (an overview), *Geologica Balcania* 26 (2), 3–10

Zagorchev, I, and Nakov, R (eds), 1998 Special Issue: 'Geological Heritage of Europe', dedicated to the memory of George P Black, First ProGEO President, *Geologica Balcania* 28 (3–4)

Contributors

Dr John Conway is Director of Research at the Royal Agricultural University, Cirencester, UK. He is a geologist specialising in soil science who has been involved with geoconservation in Wales for many years. John has chaired the South Wales RIGS group, the Association of Welsh RIGS groups and is a director of GeoMon–Anglesey Geopark which he represents on the UNESCO Global Geopark Co-ordination Committee. He specialises in geotourism, having led field holidays for the Snowdonia National Park for over 20 years, and delights in explaining geology and landscape to interested parties.

Dr Kevin Crawford is a senior lecturer in the Department of Geography and Environmental Science, Faculty of Science, Liverpool Hope University. He is currently undertaking research into geoconservation that includes exploring the public understanding of geodiversity and geoconservation, public satisfaction with geological interpretation, and legislative approaches to geoconservation. He has sat on the executive committee of GeoConservationUK and is the secretary for the North West England Geodiversity Partnership. He is also involved with local voluntary RIGS groups that are interested in the conservation and protection of important non-statutory local geological sites.

Peter Davis is Emeritus Professor of Museology at Newcastle University. His research interests include the history of museums; the history of natural history and environmentalism; the interaction between heritage and concepts of place; and ecomuseums. He is the author of several books, including Museums and the Natural Environment (1996), Ecomuseums: A Sense of Place (1999; 2nd edition 2011) and (with Christine Jackson) Sir William Jardine: A Life in Natural History (2001). He has co-edited several volumes in the Heritage Matters Series, including Making Sense of Place (2012), Safeguarding Intangible Cultural Heritage (2012), Displaced Heritage (2014) and Changing Perceptions of Nature (2016).

John E Gordon is an Honorary Professor in the School of Geography and Geosciences at the University of St Andrews, a deputy chair of the IUCN World Commission on Protected Areas Geoheritage Specialist Group and a member of the European Federation of Geologists' panel of experts on geoheritage. He has published on geoheritage and geoconservation, and on his other research interests in mountain geomorphology, glaciers and glaciation.

Dr Thomas A Hose is an honorary research associate in the School of Earth Sciences, University of Bristol. A portfolio career in museums (in geology and natural history curatorial and education posts), secondary schools (mainly teaching biology, geography and geology) and universities (lecturing on arts and heritage management, countryside management, tourism, and social science research methods) led, via an MA in Arts Administration (City University), to his research interests in landscape history and tourism, geoconservation and environmental interpretation, which underpinned his groundbreaking doctoral thesis (University of Birmingham) on geotourism, following his 1990 MA in Museum and Gallery Administration (City University). He was commissioning editor for a special 'geotourism and geoconservation' (2011) issue of *Geoheritage* and serves on the journal's editorial board; he was also editor of the Geological Society of London volume, *Appreciating Physical Landscapes: Three Hundred Years of Geotourism* (2015). He has authored chapters on geotourism for four other Geological Society volumes and another five books, as well as numerous journal and conference papers. He is a Council Member of the British Institute for Geological Conservation and was a long-serving member of his local RIGS group and the Executive Committee of GeoConservationUK (latterly as editor of its national newsletter), and for some years sat on the committee (latterly as its secretary) of the History of Geology Group of the Geological Society.

Dr Jonathan G Larwood is a senior specialist (geology and palaeontology) with Natural England, the UK government's advisory body on the natural environment in England. He has more than 20 years' geoconservation experience, working from an international to a local level. He is currently a volunteer specialist adviser to the National Trust on geodiversity and geoconservation, the archivist for the Geologists' Association, and chair of his local geoconservation group, GeoPeterborough.

Dr Slobodan B Marković is a Full Professor at the Department of Geography, Tourism and Hotel Management, Faculty of Sciences, University of Novi Sad. His research fields are Quaternary geology and stratigraphy, especially loess-palaeosol sequences, but also palaeoclimatology and palaeopedology. He is editor of three established journals – *Quaternary International*, *Acta geographica Slovenica* and the *Central European Journal of Geosciences*. He is president of the Loess Focus Group of the International Union of Quaternary Science, head of the Loess Research Group and Laboratory for Palaeoenvironmental Reconstruction (LAPER), and also manager of national research projects.

Dr Martin Munt is Head of Palaeobiology Collections at the Natural History Museum, London, where he leads the staff managing more than eight million fossil specimens. He is a specialist in Mesozoic and Cenozoic non-marine Mollusca. His principal field localities are Spain, Morocco and southern England. He was curator of geology during the building of Dinosaur Isle Museum on the Isle of Wight, leading on content and display. More recently, he has had a leading role in developing the collections at the Natural History Museum, including the acquisition of the new Stegosaurus specimen on display at the museum.

Dr Emmanuel Reynard is a Full Professor of Physical Geography at the University of Lausanne, and a member of the International Association of Geomorphologists (IAG) executive committee. He has been chairman of the IAG Working Group on Geomorphosites for 12 years (2001–13) and is the president of the Working Group on Geotopes of the Swiss Academy of Sciences. Recently, he joined the Commission on Geomorphology and the Society of the International Geographical Union. His research is focused on mountain geomorphology, geomorphological heritage, landscape geohistorical analyses and water management in mountains. He has carried out several projects in the Canton of Valais (Switzerland).

Dr Nemanja Tomić is a research associate at the Department of Geography, Tourism and Hotel Management, Faculty of Sciences, University of Novi Sad. As a member of the Loess Research Group and Laboratory for Palaeoenvironmental Reconstruction (LAPER) at the Faculty of Sciences, he is involved in research into loess-palaeosol sequences, geoheritage and geotourism on the national level.

Dr Djordjije A Vasiljević is a research associate and assistant professor of Geoecology in the Department of Geography, Tourism and Hotel Management, Faculty of Sciences, University of Novi Sad. He holds a PhD in geosciences and an MSc degree in tourism management. As a member of the Loess Research Group and Laboratory for Palaeoenvironmental Reconstruction (LAPER) at the Faculty of Sciences, his main focus is on geoheritage, geoconservation and geotourism.

Dr Margaret Wood is a geologist and currently managing director of the GeoMôn Geopark. She was formerly monitoring geologist and then North Wales area geologist for the Countryside Council for Wales. Prior to that, she was a research associate to the reader in geochemistry, University of Manchester; in that post she analysed deep-sea rocks and sediments and also Moon rocks.

Dr Volker Wrede was originally a structural and mining geologist. He works with the State Geological Survey of North Rhine Westphalia and is Head of the Natural Resources Department of the Survey; the reclamation of mining areas and preservation of geoheritage are within the scope of this department. Since the 1990s he has been active in the German geoheritage community; he was involved in the definition of the principles of geoheritage in Germany and the organisation of the German National GeoParks Network. Today, he is head of the Ruhrgebiet National GeoPark and until May 2016 was speaker of the working group of German GeoParks. He is also a lecturer at the Halle-Wittenberg University and has published numerous papers on geoheritage issues.

Index

HERITAGE MATTERS